FIBER OPTIC DATA COMMUNICATI
TECHNOLOGICAL TRENDS AND AD

FIBER OPTIC DATA COMMUNICATION: TECHNOLOGICAL TRENDS AND ADVANCES

CASIMER DeCUSATIS
Editor
IBM Corporation
Poughkeepsie, New York

ACADEMIC PRESS
An Elsevier Science Imprint

San Diego   London   Boston
New York   Sydney   Tokyo   Toronto

This book is printed on acid-free paper. ∞

Copyright © 2002, 1998 by Academic Press

All rights reserved.
No part of this publication may be reproduced or transmitted in any form or by any means, electronic or mechanical, including photocopy, recording, or any information storage and retrieval system, without permission in writing from the publisher.

Requests for permission to make copies of any part of the work should be mailed to the following address: Permissions Department, Harcourt, Inc., 6277 Sea Harbor Drive, Orlando, Florida 32887-6777.

The appearance of the code at the bottom of the first page of a chapter in this book indicates the Publisher's consent that copies of the chapter may be made for personal or internal use of specific clients. This consent is given on the condition, however, that the copier pay the stated per copy fee through the Copyright Clearance Center, Inc. (222 Rosewood Drive, Danvers, Massachusetts 01923), for copying beyond that permitted by Sections 107 or 108 of the U.S. Copyright Law. This consent does not extend to other kinds of copying, such as copying for general distribution, for advertising or promotional purposes, for creating new collective works, or for resale. Copy fees for pre-2002 chapters are as shown on the title pages. If no fee code appears on the title page, the copy fee is the same as for current chapters. $35.00

Explicit permission from Academic Press is not required to reproduce a maximum of two figures or tables from an Academic Press chapter in another scientific or research publication provided that the material has not been credited to another source and that full credit to the Academic Press chapter is given.

ACADEMIC PRESS
An Elsevier Science Imprint
525 B Street, Suite 1900, San Diego, CA 92101-4495, USA
http://academicpress.com

ACADEMIC PRESS LIMITED
An Elsevier Science Imprint
Harcourt Place, 32 Jamestown Road, London NW1 7BY, UK
http://academicpress.com

Library of Congress Catalog Card Number: 2001095439
International Standard Book Number: 0-12-207892-6

Printed in China
02 03 04 05 RDC 9 8 7 6 5 4 3 2 1

*To the people who give meaning to my life
and taught me to look for wonder in the world:
my wife, Carolyn, my daughters, Anne and Rebecca, my parents,
my godmother, Isabel, and her mother, Mrs. Crease.* — CD

Contents

Contributors	*xi*
Preface	*xiii*

Part 1 Technology Advances

Chapter 1 History of Fiber Optics — 3
Jeff D. Montgomery

1.1.	Earliest Civilization to the Printing Press	3
1.2.	The Next 500 Years: Printing Press to Year 2000	5
1.3.	Fiber Optic Communication Advancement, 1950–2000	9
1.4.	Communication Storage and Retrieval	17
1.5.	Future of Fiber Optic Communications, 2000–2050	22
	References	31

Chapter 2 Market Analysis and Business Planning — 32
Yann Y. Morvan and Ronald C. Lasky

2.1.	Introduction	32
2.2.	The Need for Applications	32
2.3.	Supporting Technology Infrastructure	33
2.4.	Implementing a Market Survey	34
2.5.	Business Planning	37
2.6.	Summary	41
	Appendix: Market Analysis on a Transmitter Optical Subassembly	42
	Industry Description and Outlook	42
	World Fiber Optics Industry	45
	Target Markets	48
	Competition	58
	Position	60
	Conclusion	61
	References	62

Chapter 3 Small Form Factor Fiber Optic Connectors — 63
John Fox and Casimer DeCusatis

3.1.	Introduction	63
3.2.	MT-RJ Connector	64
3.3.	SC-DC Connector	68
3.4.	VF-45 Connector	71
3.5.	LC Connector	74
3.6.	Other Types of SFF Connectors	77
3.7.	Transceivers	79
3.8.	SFF Comparison	80
	References	87

Chapter 4 Specialty Fiber Optic Cables — 89
Casimer DeCusatis and John Fox

4.1.	Introduction	89
4.2.	Fabrication of Conventional Fiber Cables	89
4.3.	Fiber Transport Services	98
4.4.	Polarization Controlling Fibers	111
4.5.	Dispersion Controlling Fibers	114
4.6.	Photosensitive Fibers	119
4.7.	Plastic Optical Fiber	120
4.8.	Optical Amplifiers	123
4.9.	Futures	125
	References	131

Chapter 5 Optical Wavelength Division Multiplexing for Data Communication Networks — 134
Casimer DeCusatis

5.1.	Introduction and Background	134
5.2.	Wavelength Multiplexing	140
5.3.	Commercial WDM Systems	170
5.4.	Intelligent Optical Internetworking	192
5.5.	Future Directions and Conclusions	207
	References	211

Chapter 6 Optical Backplanes, Board and Chip Interconnects — 216
Rainer Michalzik

6.1.	Introduction	216
6.2.	Frame-to-Frame Interconnections	219
6.3.	Optical Backplanes	230

6.4.	Optical Board Interconnects	241
6.5.	Optical Chip Interconnections	246
6.6.	Conclusion	255
	References	256

Chapter 7 Parallel Computer Architectures Using Fiber Optics 270

David B. Sher and Casimer DeCusatis

7.1.	Introduction	270
7.2.	Historical and Current Processors	274
7.3.	Detailed Architecture Descriptions	283
7.4.	Optically Interconnected Parallel Supercomputers	296
7.5.	Parallel Futures	298
	References	299

Part 2 The Future

Chapter 8 Packaging Assembly Techniques 303

Ronald C. Lasky, Adam Singer, and Prashant Chouta

8.1.	Packaging Assembly — Overview	303
8.2.	Optoelectronic Packaging Overview	315
8.3.	Component Level Optoelectronic Packaging	316
8.4.	Module Level Optoelectronic Packaging	317
8.5.	System Level Optoelectronic Packaging	318
	References	320

Chapter 9 InfiniBand—The Interconnect from Backplane to Fiber 321

Ali Ghiasi

9.1.	Introduction	321
9.2.	Infiniband Link Layer	322
9.3.	Optical Signal and Jitter Methodology	326
9.4.	Optical Specifications	334
9.5.	Optical Receptacle and Connector	345
9.6.	Fiber Optic Cable Plant Specifications	349
	References	351

Chapter 10 New Devices for Optoelectronics: Smart Pixels 352

Barry L. Shoop, Andre H. Sayles, and Daniel M. Litynski

10.1.	Historical Perspective	353
10.2.	Multiple Quantum Well Devices	354
10.3.	Smart Pixel Technology	359

x   Contents

	10.4.	Design Considerations	375
	10.5.	Applications	381
	10.6.	Future Trends and Directions	409
		References	410

Chapter 11 Emerging Technology for Fiber Optic Data Communication   422

Chung-Sheng Li

11.1.	Introduction	422
11.2.	Architecture of All-Optical Network	424
11.3.	Tunable Transmitter	426
11.4.	Tunable Receiver	429
11.5.	Optical Amplifier	433
11.6.	Wavelength Multiplexer/Demultiplexer	436
11.7.	Wavelength Router	437
11.8.	Wavelength Converter	440
11.9.	Summary	443
	References	443

Chapter 12 Manufacturing Challenges   447

Eric Maass

12.1.	Customer Requirements — Trends	447
12.2.	Manufacturing Requirements — Trends	450
12.3.	Manufacturing Alternatives	479

Appendix A	Measurement Conversion Tables	486
Appendix B	Physical Constants	488
Appendix C	Index of Professional Organizations	489
Appendix D	OSI Model	491
Appendix E	Network Standards and Documents	492
Appendix F	Data Network Rates	495
Appendix G	Other Datacom Developments	505

Acronyms	511
Glossary	529
Index	555

Contributors

Numbers in parentheses indicate the pages on which the authors' contributions begin.

Prashant Chouta (303), Cookson Performance Solutions, 25 Forbes Boulevard, Foxborough, Massachusetts 02053

Casimer DeCusatis (63, 89, 134, 270), IBM Corporation, 2455 South Road MS P343, Poughkeepsie, New York 12601

John Fox (63, 89), ComputerCrafts, Inc., 57 Thomas Road, Hawthorne, New Jersey 07507

Ali Ghiasi (321), Broadcom Corporation (formerly SUN Microsystems), 19947 Linden Brook Lane, Cupertino, California 95014

Ronald C. Lasky (32, 303), Consultant, 26 Howe Street, Medway, Massachusetts 02053

Chung-Sheng Li (422), IBM Thomas J. Watson Research Center, 30 Sawmill River Road, Hawthorne, New York 10532

Daniel M. Litynski (352), College of Engineering and Applied Sciences, Western Michigan University, 2022 Kohrman Hall, Kalamazoo, Michigan 49008

Eric Maass (447), Motorola, Incorporated, 2100 Elliot Road, Tempe, Arizona 85284

Rainer Michalzik (216), University of Ulm, Optoelectronics Dept., Albert-Einstein-Allee 45, D-89069 Ulm, Germany

Jeff D. Montgomery (3), ElectroniCast Corporation, 800 South Claremont St., San Mateo, California 94402

Yann Y. Morvan (32), Cookson Electronics, New Haven, Connecticut, 06510

Andre H. Sayles (352), Photonics Research Center and Department of Electrical Engineering and Computer Science, U.S. Military Academy, West Point, New York 10996

David B. Sher (270), Mathematics/Statistics/CMP Dept., Nassau Community College, 1 Education Drive, Garden City, New York 11530

Barry L. Shoop (352), Photonics Research Center and Department of Electrical Engineering and Computer Science, U.S. Military Academy, West Point, New York 10996

Adam Singer (303), Cookson Performance Solutions, 25 Forbes Boulevard, Foxborough, Massachusetts 02053

Preface

"I have traveled the length and breadth of this country and talked with the best people, and I can assure you that data processing is a fad that won't last out the year."

—*Attributed to the chief editor for business books, Prentice Hall, 1957*

"There is nothing more difficult to take in hand, more perilous to conduct, or more uncertain in its success, than to take the lead in the introduction of a new order of things."

—*Machiavelli*

This book arose during the process of revising the second edition of the Handbook of Fiber Optic Data Communication, when it became apparent that one book wasn't enough to contain all of the technology developments in wavelength multiplexing, optically clustered servers, small form factor transceivers and connectors, and other emerging technologies. As a result, we decided to split off the four chapters on futures from the original handbook, combine them with many new chapters, and form this book, which can serve as either a companion to the original book or a stand-alone reference volume.

Many new chapters have also been added to address the rapidly accelerating rate of change that has characterized this field. New component technologies for optical backplanes, parallel coupled computer architectures, and smart pixels are among the topics covered here. Open standards, which to a great extent have created the Internet and the Web (remember TCP/IP?) also continue to evolve, and new standards are emerging to deal with the requirements of the next generation intelligent optical infrastructure; some of these standards, such as Infiniband, are covered in this volume. There are also new chapters on the history of communications technology (with apologies to those who have noted that it remains difficult to determine exactly who invented the first one of anything, and that the history of science is filled with tales of misplaced credit), and predictions of the future, as envisioned by some of the leading commercial technology forecasters. Given the rapid and accelerating rate of change in this field, it is inevitable that some topics of interest will not be covered. As this book goes to press, for example, there is growing interest in microelectromechanical devices (MEMs) and so-called micro-photonics, ultra high data rate transceivers (40 Gbit/s and above), advanced storage area networks and network attached storage, and other areas that are beyond the scope of this text. It is our hope that these and other topics will be incorporated into future editions of this book, just as the original Handbook of Fiber Optic Data Communication has grown through the years.

An undertaking such as this would not be possible without the concerted efforts of many contributing authors and a supportive staff at the publisher, to all of whom I extend my deepest gratitude. The following associate editors contributed to the first edition of the Handbook of Fiber Optic Data Communication: Eric Maass, Darrin Clement, and Ronald Lasky. As always, this book is dedicated to my parents, who first helped me see the wonder in the world; to the memory of my godmother Isabel; and to my wife, Carolyn, and daughters Anne and Rebecca, without whom this work would not have been possible.

Dr. Casimer DeCusatis, Editor
Poughkeepsie, New York

Part 1 | Technology Advances

Chapter 1 | History of Fiber Optics

Jeff D. Montgomery

Chairman/Founder, ElectroniCast Corporation, San Mateo, California 94402

In this review of the history of communication via fiber optics, we examine this relatively recent advancement within the context of communication through history. We also offer projections of where this continuing advancement in communication technology may lead us over the next half century.

1.1. Earliest Civilization to the Printing Press

1.1.1. COMMUNICATION THROUGH THE AGES

All species communicate within their group. The evolution of the human species, however, appears to have been much more rapid and dramatic than the evolution of other species. This human advancement has coincided with an increasingly rapid advancement in communication capability. Is this merely a coincidence, or is there a causal relationship?

The earliest human communication, we assume, was vocal; a capability shared by numerous other species. Archaeological information, however, indicates that, tens of thousands of years ago, humans also began to communicate via stored information in addition to the vocal mode. Cave paintings and cliffside carvings have survived over time, to now, conveying information that at the time was useful. Findings also indicate signal fires existed in those early times, to transmit (via light) information, presumably the sighting of the approach of other humans, the appearance of game animals, or other intelligence. Smoke signal communication emerged along with nautical signal flags, followed by light-beam and flag semaphores.

As civilization advanced (and humans apparently became much more numerous), communication became increasingly complex. Symbols to represent items of interest were conceived and adopted. Techniques were developed to carve these symbols in stone, or to paint them onto media such as walls or sheets made from papyrus reeds — the early communication storage media. The papyrus-enscribed messages were especially significant, in that they were transportable — the early telecommunication ("communication at a distance").

While the development of symbols and media was a major advancement, there were still some major handicaps. Carved messages, in particular, had very low portability. A more general problem was that forming the symbols into the media was a high-level skill that required years of training. Kings and common people could not write (and, in general, could not read). Beyond the limited number of scribes available, and the relatively high cost per message inscribed, was the time required to complete a message; hours to days for a simple scroll; lifetimes for stone carvings. Also, each copy, if wanted, required as much effort and time as the original. These general techniques, however, did not change dramatically over a span of thousands of years. A degree of "shorthand" symbols were developed for commercial messages, and the language became richer through development of more and increasingly refined symbols. Still, it remained a slow form of communication, limited to royalty, wealthy merchants, military leaders, and scholars.

As the need for copies of messages, such as distribution of proclamations, increased, entrepreneurs developed the technique of transferring a symbolic message from the original by applying ink and transferring the message to another surface. Printing! Naturally, as this technique evolved, message originators also evolved to sending out more copies. There also naturally evolved a tendency to create longer, more complex messages. So, although making multiple copies became feasible, crafting the original print master remained the role of a master craftsman and, as messages became longer, more time was required.

Within this period, some messages became long enough to be "books." Creating the print master for a book occupied a crew of engravers for many years. Although communication certainly was advancing, it remained expensive and slow to initiate in transportable, storable form.

1.2. The Next 500 Years: Printing Press to Year 2000

1.2.1. PRINTING PRESS CHANGES THE RULES

The invention of the movable-type printing press by J. Gutenberg, circa 1450, was a major breakthrough. By this time, the language of communication had evolved from pictorial symbols to words formed from a set of characters or other symbols. These were laboriously engraved into printing plates, requiring days to years per plate. With the availability of movable type pieces that could be arranged to construct a clamped-together plate, the time to create a plate was reduced by orders of magnitude; from days to minutes. Of equal importance, the plates could now be constructed by a technician having relatively modest training, instead of by a skilled artisan with years of training and apprenticeship. With the Gutenberg press, the cost of books could be greatly reduced, becoming financially available to a much larger segment of the populace. Over the ensuing 400 years, instruction books became widely available to all students, current news publication flourished, and entertainment books emerged.

1.2.2. THE CASCADE OF INVENTION

With the evolution of the printing press, the worldwide exchange of information between scholars, inventors, and other innovators accelerated. Especially over the most recent two centuries, significant inventions cascaded, often standing on the shoulders of earlier inventions. Some of the key inventions related to the advancement of communication are noted in Fig. 1.1. Signal transmission through space by electromagnetics (Marconi), electrical conductance principles (Maxwell), mechanized digital computing (Babbage), the telephone (Bell) were landmark inventions that set the platforms for the just-completed Magnetic Century. Vacuum tube amplifiers and rectifiers emerged, making radio transmission and reception feasible (and, ultimately, ubiquitous and affordable). Electronic computing evolved, mid-century, from an interesting intellectual concept to become a tool, albeit very expensive, for controlling massive electrical power grids and for tackling otherwise overwhelmingly challenging scientific calculations. (It was visualized that several of these machines, perhaps dozens, might ultimately be useful worldwide; Thomas J. Watson, International Business Machines Chairman, postulated a potential worldwide market for perhaps five of their computing machines.)

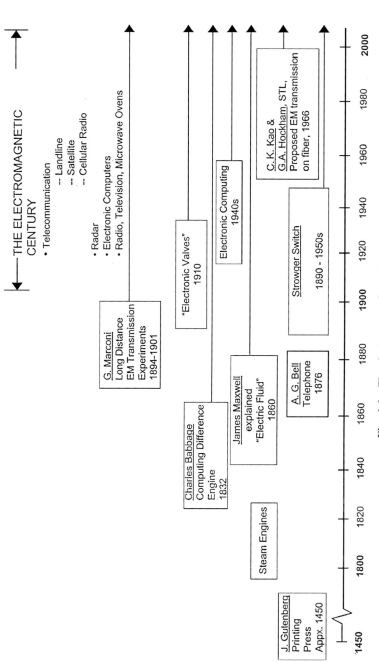

Fig. 1.1 The electromagnetic century.

Over the 1633–1882 span, mechanical computation machines were of continuing interest, with concepts developed by Pascal, Leibnitz, and Schickhard, culminating in the first serious effort to build a mechanical calculator machine (by Charles Babbage, in 1882). The first working electromechanical calculator was built by IBM engineers in 1930 (the IBM Automatic Sequence Controlled Calculator, Mark I), under the direction of Professor Aiken of Harvard University. [1] The first electronic calculator, ENIAC, was built by Eckert and Mauchly, of the University of Pennsylvania, in 1946.

The Strowger switch, invented within Bell Laboratories, illustrates a significant point that keeps recurring in the evolution of communication (and in other fields): When a problem evolves and advances to the point that it threatens the continuing evolution of an important field, inventive minds find a feasible solution. The early wire-line telephone systems required switching, to connect a specific originating telephone to the desired other telephone instrument. This was done by an operator who received verbal instructions from the originator, then plugged a connection cord between the two appropriate receptacles on the switchboard. As the number of subscribers and the number of calls per subscriber steadily increased, it became apparent that within a relatively few years it would no longer be feasible to recruit enough operators to do the switching. Thus, the Strowger switch, doing the same task based on telephone-number-based electrical signals, was developed. This switch occupied a lot less physical space, and did the task faster, at less cost, and with higher 24-hour-per-day dependability and accuracy. The Strowger switch, introduced in the late 1800s, bridged the transition into the Electromagnetic Century.

The advancement of telecommunication technology and facilities was especially dramatic through the first half of the 20th century. Telephone communication advanced from two-wire lines to hundreds of parallel voice grade lines, as illustrated in Fig. 1.2, colliding with another roadblock. The number of open, uninsulated lines routed along city streets and into major office buildings approached the physical space limits. This drove network developers to evolve to "twisted pair" insulated copper wires that greatly reduced the space required for transmission lines. (This was followed by the development of coaxial cables, which could transmit hundreds of voice signals multiplexed onto a single cable.)

This evolved to large cables, "flexible as a sewer pipe," enclosing hundreds of twisted pairs plus several coaxial cables. Most of this cable, installed from about 1930 to date, is still in operation, mainly in metropolitan

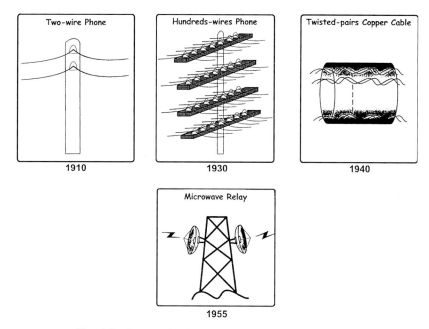

Fig. 1.2 Communication transport 19th-century evolution.

access networks in North America, Europe, and Japan, and still used for long-haul trunk lines in less developed countries.

World War II interrupted the deployment of civilian communication networks, especially through 1940–1945. Paradoxically, however, this global conflict accelerated the technology of microwave technology, deployed initially primarily in radar systems. Rocket vehicle technology also advanced dramatically during this relatively brief interval. With the return to peacetime priorities, point-to-point microwave communication relay products evolved from the radar components base. Led by AT&T, General Electric, and RCA, microwave picked up a large and rapidly increasing share of the long-haul transcontinental telecommunication transport.

Terrestrial microwave relay communication expanded rapidly, but was limited by radio-frequency spectrum space availability, and also by the requirement for line-of-sight transmission. The microwave free space transmission beams also spread, as a function of the antenna dimensions (in wavelengths) and of the distance traversed, limiting networks to links of a few tens of miles. Fortunately, as terrestrial microwave relay became increasingly limited in expansion, especially in the most-developed regions, microwave technology was combined with the continuing rocket launch vehicle development to bring satellite microwave communication to commercial reality.

Satellite microwave communication was a fierce competitor to fiber optic communication, especially through the 1975–1990 era, for long-haul high-volume communication. For transoceanic communication, in particular, before the optical amplifier became commercially available, fiber optic cable was not feasible, and satellite microwave took a major share of this market away from undersea copper telecommunication cable. Satellite microwave also was a strong competitor for terrestrial long haul, overcoming the line-of-sight limitations of terrestrial point-to-point microwave relay. At 2000, satellite microwave remains a major element of long-haul communication, business voice and data transport as well as residential television.

1.3. Fiber Optic Communication Advancement, 1950–2000

Communication technology and facilities advanced rapidly through the first half of the 20th century, evolving from dual open space wires to hundreds of open space wires, then to hundreds of twisted pair wires in cables, then augmented by terrestrial microwave relay, followed by satellite microwave. The expansion of microwave transmission, however, is limited by radio frequency (RF) spectrum availability (although advancing modulation technologies such as CDMA have greatly extended these limits). Copper wire transmission has a severe distance-times-bandwidth limitation. Fiber optic waveguide has become the next-generation transmission media, initially for long-distance, high-data-rate transport. As technologies and production volumes have advanced, with a dramatic fall in cost per gigabit-kilometer of transport, fiber optic networks have now also become the most economical solution for short/medium distance, modest-data-rate transport in new installations, such as residential and business access, displacing copper.

The evolution of fiber-optic-based communication was built upon many different initial concepts that were then advanced through the years by succeeding scientists. The most significant of these were

- Transmission of light through a confining media, with very low loss per length

- A light source, at wavelengths corresponding to low media loss; modulatable at high data rates and having practical lifetime
- An amplifier of the light signal in the media

These three legs of the fiber optic communication stool evolved somewhat in parallel, with impetus in one field coming from advancements in another. These primary advancements were augmented in later years, as the industry evolved to networks of much greater complexity, by advancements in digital computing and in rapidly accessible memory storage.

Serious signal transmission by lightwave was preceded by microwave signal transmission. Thus, it is not surprising that, through the past half century, many of the developers of lightwave communication technology, products, and application have moved over from the microwave field.

1.3.1. LONG ROAD TO LOW-LOSS FIBER

The path to the current low-loss optical glass fiber has had many entrance points. Probably the most significant was the perception by Charles K. Kao, a native Chinese engineer working for Standard Telecommunications Laboratories (STL), UK, that eliminating impurities in glass could yield glass having very low light transmission loss, less than 20 dB per kilometer, building on the concept of total internal reflectance of light in a glass fiber core with a glass cladding of lower index of refraction. Working with microwave engineer George Hockham (1964–65), supporting data was collected, results were published in 1966, and application to long-distance communication over single-mode fiber was proposed. This effort was in the STL optical communications laboratory, headed by Antoni E. Karbowiak, who resigned in 1964 and was succeeded by Charles Kao. (Dr. Kao later retired from STL to return to China, joining the staff of a major university, where he has been a major influence in the current strength of China in the fiber optic communications field.)

The research and initiative by Charles Kao leveraged from earlier work by Elias Snitzer at American Optical Corporation, and Wilbur Hicks (who started and developed the optical fiber components program at AO in 1953). Will Hicks later started his own nearby company, Mosaic Fabrications, which evolved into Galileo. He subsequently started Incom, which was acquired by Polaroid. Snitzer and Hicks demonstrated, in 1961, waveguiding characteristics in glass fiber, total internal reflectance, comparable to theory developed earlier in microwave dielectric waveguides. Elias Snitzer also is credited with being first to propose the principal of the optical fiber amplifier, in 1961 at AO, with first publication in 1964. The concept had to remain on the shelf, however, until an adequate light pump became available.

Will Hicks was an early proponent of Raman amplifiers, and of wavelength-division multiplexers (WDM) based on circular resonant cavities as the wavelength-selective element.

The low-loss glass fiber concept, demonstration, and promotion by Charles Kao was followed by the development of commercially feasible optical fiber production in 1970 by Donald Keck, Robert Maurer, and Peter Schultz at Corning Glass Works. High-purity glass core was achieved by depositing solids from gasses inside a heated quartz tube to form a rod from which fiber was drawn. This was soon followed by low-loss fiber producibility demonstrated at AT&T Bell Laboratories, by John MacChesney and staff.

The development of optical communication fiber also drew from various glass fiber and rod experiments through the first half of the 20th century. Clarence W. Hansell in the United States, and John L. Baird in the UK, developed and patented the use of transparent rods or hollow pipes for image transmission. This was followed by experiments and results reported by Heinrich Lamm (Germany) of image transmission by bundles of glass fiber. This was followed, in 1954, by fiber bundle imaging research reported by Abraham van Heel, Technical University of Delft, Holland, and by Harold H. Hopkins and Narinder Kapany at Imperial College, UK. This in turn was followed, in the late 1950s, by glass-clad fiber bundle imaging reporting by Lawrence Curtiss, University of Michigan.

1.3.2. LIGHT POWER TO THE CORE

For practical communication transport over single-mode fiber, the light signal must be coupled into the fiber core (which is only a few microns diameter), must be modulatable at high data rates (initially, a few megabits per second; will reach 40 gigabits per second, commercially, in 2001), and must have a long lifetime (to ensure high reliability of the network). Early experiments used flash lamps, and the earliest optical communication links (generally in industrial or other specialized applications, rather than telecommunication) used light-emitting diodes (LEDs) with relatively large core multimode fiber. There was broad acceptance of the concept that the semiconductor laser diode was the most promising long-term candidate, but its availability lagged behind the optical fiber. Theodore Maiman, of the Hughes Research Laboratory of Hughes Aircraft (U.S.), built the first laser, based on synthetic ruby, in 1960. Reliable solid state lasers still had many years of development ahead.

Laser concepts go back to quantum mechanics theory, outlined in 1900 by Max Planck and advanced in 1905 by Albert Einstein, introducing the photon concept of light propagation and the ability of electrons to absorb and emit photons. This was followed by Einstein's discovery of stimulated emission, evolving from Niels Bohr's 1913 publication of atomic model theory.

Microwave research during and after World War II set the stage for the next phase of laser development. Charles Townes, in 1951, as head of the Columbia University Radiation Laboratory, pursuing microwave physics research, working with James Gordon and Herbert Zeiger, built a molecular-based microwave oscillator to operate in the submillimeter range (evolving from mechanical-based centimeter-wavelength oscillators). Townes named this MASER (Microwave Amplification by Stimulated Emission of Radiation). (Succeeding laboratories, pursuing government-funded research and development contracts, were heard to refer to this as Money Acquisition Scheme for Extended Research.) Further experimentation by Townes determined the concept could be extended into the lightwave region; for which, "Light" was substituted for "Microwave"; thus, LASER. Results were published in 1958, leading to the 1960 ruby laser announcement by Theodore Maiman.

Early research and development of gas and crystal lasers used flashlamp pumps. Meanwhile, semiconductor device development was proceeding, sparked by the transistor invention in 1948, at AT&T Bell Laboratories, by William Schockley, Walter Brattain, and John Bardeen. Heinrich Welker, of Siemens, Germany, in 1952 suggested that semiconductors based on III-V compounds (from columns III and V of the periodic table) could be useful semiconductors. Gallium arsenide, in particular, appeared promising as a base for a communication semiconductor laser. Work on GaAs lasers progressed on several fronts, leading to operational GaAs lasers demonstrated in 1962 by General Electric, IBM, and Lincoln Laboratory of Massachusetts Institute of Technology.

These early GaAs lasers, however, had very short life; seconds, evolving to hours, due to creation of excessive heat in operation. Cooling attempts were inadequate to solve the problem. The solution was to confine the laser action to a thin active layer, as proposed in 1963 by Herbert Kroemer, University of Colorado. This led to a multilayered crystal modified GaAs structure, doped with aluminum, suggested in 1967 by Morton Panish and Izuo Hayashi at AT&T Bell Laboratories. This led, after years of research and development by several leading laboratories, to operational trials of semiconductor communication lasers by AT&T, Atlanta (1976), and the first commercial laser-driven fiber communication deployment, in Chicago (1977).

A major contribution to the lifetime of communication semiconductor laser diodes was the development of molecular-beam epitaxy (MBE) crystal growth by J. R. Arthur and A. Y. Cho at AT&T Bell Laboratories. This achieved much greater precision of layer thickness, permitting higher operational efficiency, thus less heat and longer life (one million hours).

Early deployment of telecommunication fiber links progressed slowly. The 1977–1978 deployment totaled only about 600 miles. The first significant thrust was the AT&T Northeast Corridor, 611 miles Boston to Washington, DC (later extended to New York City); plans submitted in 1980, deployment completed and service "turned up" in 1984. The Northeast Corridor was planned for 45 Mbps; upgraded to 90 Mbps during deployment.

1.3.3. LIGHT AMPLIFICATION FOR LONG REACH

Early fiber communication systems, powered by laser diode transmitters, achieved relatively short link distances between regenerators (which detected the electronic signals from the light beam modulation, reconstituted the pulses, and re-transmitted). Higher power transmitters were not an early alternative, due to linearity and life problems. The regenerators were (and still are) expensive and a source of system failure. Thus, there was a need for periodic amplification of the light power level, along the trunk line. The concept of the optical fiber amplifier had been proposed earlier by Elias Snitzer at American Optical, and in 1985 the concept of using erbium-doped optical fiber as the pumped amplification medium was discovered by S. B. Poole at the University of Southampton, UK. The short length of erbium-doped fiber functions as an externally pumped fiber laser. This concept was developed into a demonstration laser by David Payne and P. J. Mears at the University of Southampton and by Emmanuel Desurvire at AT&T Bell Laboratories. This was demonstrated by Bell Laboratories in 1991. These efforts led to the further development, rigorous life testing, and ultimate deployment of ultra-reliable optical amplifiers in the AT&T/KDD joint transpacific submarine cable.

The preceding summary highlights only a few key efforts that contributed to the dynamic advancement of fiber optic communication. The *City of Light*, by Jeff Hecht, details many more of the breakthroughs and the recognized researchers. Behind this recognition were thousands of unrecognized researchers, worldwide, who moved this technology to the marketplace. (For a detailed, linear chronology of the development and commercial realization of communication-grade optical fiber, the reader is referred to the excellent, thoroughly researched book, *City of Light*, by Jeff Hecht.[2] The following discussion will highlight some of the key elements of this progression, and place them in context with other, parallel developments. The historical (and continuing) development of optical communication fiber is, indeed, impressive. Its commercial feasibility, however, has also benefited greatly from the serendipitous development of lasers (particularly, laser diodes), which supported optical amplifiers, as well as the development of magnetic data storage, semiconductor integrated circuits, and other technical advancements. In the final analysis, also, all of these technical advancements have been greatly accelerated by the fact that they could be used to enable attractive returns on invested capital.)

The manufacture of glass began thousands of years ago; initially, a precious commodity for decorative purposes, evolving very slowly (until the middle of the last millennia) to commercial use. Techniques for producing glass fibers emerged hundreds of years ago, and they were applied to practical light transmission for various illumination applications by the late 1800s. The concept and principals of using a bundle of glass fibers for image transmission was outlined by Clarence W. Hansell in 1926, laying the basis for a thriving imaging product industry. American Optical (AO), in Massachusetts, was an early leader in this field, with production accelerating from the early 1950s. AO was an early base for Will Hicks, a scientist/entrepreneur who has subsequently boosted fiber optic communication development in several contexts. He was a founder of the first fiber optics company, Mosaic Fabrications, in 1958, but continued to cooperate with AO, particularly in developing single-mode fiber, which subsequently was advanced by Elias Snitzer of AO.

As already discussed, the post-World War II era brought the realization, by AT&T Bell Laboratories, Standard Telecommunication Laboratories (STL), and other major communication firms, that twisted-pair copper cable was approaching its limit as an economically viable long-haul transmission media. Driven by economics, numerous alternatives were explored; cylindrical millimeter microwave waveguide, satellite microwave, . . . and optical fiber. Charles K. Kao and George Hockham, in the STL optical communication program, were early pioneers, particularly through the late 1960s, in advancement of the technical and economic arguments for optical fiber communication.

Moving into the 1970s, the commercial feasibility arguments advanced by Kao/Hockham at STL (later acquired by Northern Telecom), and others, convinced Corning to support the development of commercially producible low-loss optical fiber. A team of Robert Maurer, Donald Keck, Peter Schultz, and Frank Zimar achieved rapid successive breakthroughs in low-loss fiber development, especially through the early 1970s.

Meanwhile, there remained the practical realization that low-loss optical transmission of signals was only an intellectual exercise, unless a light source capable of high-speed modulation could be developed. (The photodetector capability was also evolving, with less drama.) The primary candidate was the laser (first demonstrated by Theodore Maiman, of Hughes Research Laboratories, in 1960). More specifically, semiconductor diode lasers; but, early laser diodes had almost zero lifetime. Numerous parallel laser diode development programs proceeded, with Robert N. Hall's group at General Electric first to demonstrate operation, in 1962 (but with short life, and only by operating in liquid nitrogen temperature). STL demonstrated 1 gigabit per second (Gbps) laser diode modulation in 1972. Bell Labs demonstrated 1000 hours laser diode lifetime in 1973, and Laser Diode Labs (a spinoff from RCA Sarnoff Labs) demonstrated room-temperature operation of a commercial CW laser diode in 1975. In 1976, Bell Labs demonstrated 100,000-hour life of selected laser diodes, at room temperature. Also in 1976, Bell Labs demonstrated 45 megabits per second (Mbps) modulation of laser diodes, coupled with graded-index optical fiber.

Thus, driven by major economic imperatives, the development of optical fiber and laser diodes advanced dramatically through the 1960–1975 span. This laid the base for the dawn of commercial deployment of optical fiber communication networks, starting with the AT&T Northeast Corridor project (Boston–New York–Washington, DC; initially planned as 45 Mbps transmission; deployed as 90 Mbps). (The first independent consultant forecasts of fiber optic communication deployment were published in 1976, led by *Fiber Optic and Laser Communication Forecast,* Jeff D. Montgomery and Helmut F. Wolf, Gnostic Concepts, Inc.)

The last quarter century, 1975–2000, has seen the explosive development of technology and commercial realization of fiber optic communication. The global consumption of fiber optic cable and other components, for example, advanced by about 5 orders of magnitude, from $2.5 million in 1975 to $15.8 billion in 2000. Component development has proceeded through hundreds of laboratories, handing off to hundreds of factories, large and small. Fiber loss continued to drop, laser modulation speeds increased; an old concept, wavelength division multiplexing, found economic justification and catapulted into the marketplace.

Much of this advancement, however, would not have occurred without support from the sidelines: the optical fiber amplifier. The optical fiber amplifier concept was first outlined by Elias Snitzer, in 1961, but for many years it went nowhere, for two reasons: (1) little commercial need was seen; (2) pump laser diodes, an essential component of the amplifier, were not available. The travails of the transmitter laser diode were previously discussed. The pump diode experienced similar difficulties. It needed to operate at much higher peak powers than the transmit diodes of that time; thus, lifetime was an even more severe problem. Also, it needed to operate at a significantly different wavelength than the transmit diodes, so it could benefit less from the earlier diode development.

Early developmental pump diodes had lifetimes of milliseconds; gradually expanded to minutes, then hours, then thousands of hours. The evolution of the optical fiber amplifier, in the context of other related components, is illustrated in Fig. 1.3.

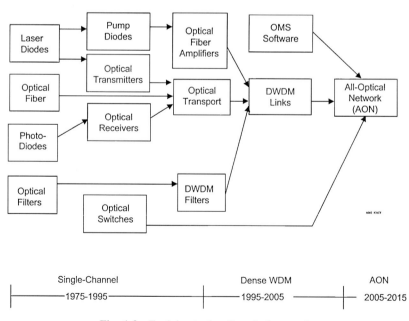

Fig. 1.3 Evolving to the all-optical network.

As with many other breakthroughs, the optical fiber amplifier became a commercial product because of its apparent economic payoff. Commercial realization was retarded by the pump lifetime problem previously mentioned, plus the very high cost of final development and life test/demonstration, plus the expected high cost of production, after demonstration of technical feasibility. This barrier was broken by a partnership of AT&T Submarine Cable Systems and KDD (Japan), in development of a transpacific submarine fiber cable. Calculations indicated that, if the amplifiers met specifications and had sufficient lifetime, they could substantially boost the cable performance/cost ratio. The team funded the design, production, and life test of about 200 amplifiers, at an estimated cost of $40–50 million (an impressive amount at the time). The amplifiers were produced, demonstrated long life, and were deployed.

This demonstration led other network developers, both submarine and terrestrial long haul, to consider optical amplifiers. Although they were initially quite expensive, they could eliminate a substantial share of the also-expensive optical/electrical/optical regenerator nodes in the network, so deployment accelerated. With increased production, costs dropped, opening additional markets.

A point of this is, without the optical fiber amplifier, dense wavelength division multiplexing (DWDM) over interexchange (long-haul) networks would not be feasible. So, although DWDM was not a dominant element in the initial amplifier development, it benefited and opened up a major new market.

With DWDM, plus evolutionary developments (such as 40 Gbps transmission per wavelength), terabit-per-fiber transmission becomes feasible, as shown in Fig. 1.4. With cables now commercially available with over 1000 fibers, this provides a base for petabit-per-cable systems.

1.4. Communication Storage and Retrieval

It is important to recognize that modern communication depends greatly on the storage of messages and other information, as well as on the technology of transmitting this intelligence from one location to another. With the early storage of communication by carving symbols into stone, transport was essentially impossible, the recipient had to travel to the message. Storage by inscription onto papyrus sheets was transportable, but required a lot of time to create and, generally, a lot more time (weeks to months) to deliver the scroll to the recipient. The printing press drastically shortened

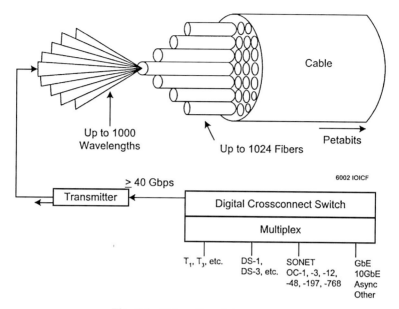

Fig. 1.4 Future fiber data transport.

the printing time, and the parallel evolution of physical transportation shortened delivery time. Still, however, delivery of this stored message typically required hours to weeks.

The telephone was the dramatic answer to shortening delivery time; essentially instantaneous. However, it simultaneously lost the storage capability. It was necessary for the recipient to handwrite the perceived message, typically in abbreviated and possibly erroneous format, or else have no storage at all. So, the telephone was a very useful advance for the transmission of personal viewpoints and general information, but problematic for conveying precise business data.

As electronic computers emerged, integral data storage was essential for their operation. Earlier mechanical computers used exotic mechanisms for rudimentary storage, but this was infeasible for major machines. This requirement drove the early development of the individual-switch magnetic wound core memory and the development of magnetic tape memory.

The early computers (with much less capability than year 2000 electronic pocket calculators) depended on wound core and tape memory, vacuum tube switches and punched card instructions. This was adequate for the perceived demand of a few additional machines per year. Thus, there was relatively little pressure to pursue other than evolutionary advancement of storage (or switching, or input instruction) technologies.

Quite unrelated, AT&T encountered a serious and steadily increasing problem with vacuum tubes. These tubes were key elements in the microwave relay transmit/receive equipment that, leveraging from military-developed microwave technology during World War II, were used for transporting much higher volume of communication than had been feasible over the earlier wire line. Thousands of tubes were in operation in a system, and the average lifetime of the tubes was a few thousand hours, so, statistically, a tube failure in the system could be expected at an average hourly interval but, at the statistical edges, any minute. The solution to this problem was to deploy crews of technicians that continually replaced tubes in the system with new tubes, statistically far in advance of their expected failure time. It could be extrapolated that this, like the earlier switchboard operator roadblock, would become an impossible handicap within a few years. Substantial research was applied to improving tube lifetime, but with limited results.

This dilemma drove AT&T Bell Labs to approach the signal amplification problem from a new base: semiconductor effects. This, within a relatively short period, emerged as the transistor. The transistor, serendipitously, turned out to be much more than the solution to the amplifier tube problem. Early on, it superseded the vacuum tubes in next-generation computers. The long operating lifetime of the transistor, its much smaller size, and its much lower cost in high-volume production, led to developing data storage based on the transistor.

As transmission data rates inside computers and other digital machines move up to gigabits per channel, interconnect links are evolving from copper to optical. Guided wave internal optical interconnect links were widely used in digital crossconnect switches, servers, and other machines in 2000, and terabit free space links are being developed, under US DARPA sponsorship, for military/aerospace ultracompact systems.

Historically, most computer internal interconnect has been from digital signal processors (DSPs) to memory, over copper; the various data streams combined into a single TDM stream by a serializer IC chip, and separated at the other end by a deserializer chip. As DSPs have progressed from 4-bit to 64-bit chips, and data rates per pin have advanced from a few megabits to gigabit level, the cost of serializer/deserializer sets has increased exponentially, and the reach of the copper link falls inversely proportional to data rates. With the cost per gigabit of optical links falling rapidly, optical links will dominate future short reach communication, as well as long haul.

The early transistor, though a major advancement over the vacuum tube for most applications, still in its early years was an individually packaged and socketed device, with multiple transistors connected by discrete wiring, evolving to conductive patterns printed on "printed wiring" boards. A significant supporting technical advancement was the perception, by a team (led by Jack St. Clair Kilby) at Texas Instruments, that it might be useful to process the interconnection between transistors on the parent silicon wafer itself, rather than separating the individual transistor chips out of the wafer, connecting wire leads to the transistor, then connecting these leads to printed wiring board conductors to reach another transistor, etc. This interesting, demonstrated concept attracted little interest or enthusiasm for (in retrospect) a very long time, but it finally burst onto the commercial market as the "integrated circuit." It found an early home in computers (which by this time had progressed beyond the market concept of a few units per year). Since then, the number of interconnected transistors has been doubling about every 18 months. This wasn't very impressive in the early years (few equipment designers could envisage a need for 64 transistors in a single package). The interconnected transistors per chip, however, in 2000 had advanced well beyond the million-device level, and continuing.

As fiber optic communication links advance to production of tens of millions per year, including internal interconnects numbering hundreds per equipment, pressure is increasing for both reduction of physical space per transceiver and reduction of cost per gigabit. In 2000, several major fiber component producers shifted into automated packaging and test of transceivers, optical amplifiers, photonic switches, and other components, achieving volume reduction of 75 to 99 percent. Parallel links also have evolved to production (by Infineon, in Germany); 12 transmitters per module. These trends demand major advancements in package design to meet heat transfer, optical, and electrical isolation requirements in micron-tolerance low-cost packages, as well as advanced assembly/test equipment. These trends are evolving to hybrid optoelectronic integrated circuit (HOEIC) packaging, which in turn will evolve to monolithic optoelectronic ICs over the 2000–2010 decade, supporting hundreds of optical channels per module.

Closing the loop back to communication: about the same time that the transistor phased into widespread application and production, and the integrated circuit (IC) began its market multiplication, AT&T approached another "switchboard operator" roadblock with the Strowger switch. Telephone company central offices had large rooms filled with these electromechanical marvels, clanking away. Projections of switching demand indicated a new solution was needed, and soon. The integrated circuit became the key to telephone signal switching; quite similar technology to computer applications.

Parallel to the application of the IC to telephone signal switching, computers (and other digital machines) found that digital data storage could be accomplished in ICs. The earliest application was as a replacement of the wound core memory, but this soon evolved for use also for archival storage. Also in parallel, however, magnetic disk memory superseded magnetic tape memory, with lower cost per memory bit, much less physical space, and faster store/retrieve. The IC-based memory has become a key element in year 2000 communication equipment, as well as some computer sections. The magnetic disk, however, through aggressive increases in storage efficiency (bits per square inch), access speed, and size options, has maintained a strong commercial position in computers.

1.4.1. TASK NETWORKING AUGMENTED BY FIBER

Along with the continuing advancement of digital signal processor speed, and the trend to harnessing many DSPs to build a mainframe computer, a parallel trend of computer networking has emerged. Networking has taken two forms:

1. Synchronized interconnection of a number of separately located mainframes, for simultaneous processing of different elements of a single problem
2. Digital communication between computers; data transfer, analogous to telephone voice networks

Several interconnected high-end workstation computers can provide computing power matching supercomputers. This has evolved from concept to practice, 1965–2000, in both government laboratories and commercial organizations. Optical fiber is the only transmission media that is practical for this interconnection, due to the increasingly high data rates and long connection distances.

The advancement of task-sharing computer networking owes much to the funding of research and development in this field by the U.S. Advanced Research Project Agency (ARPA; now DARPA), starting in the late 1960s.

This led to the 1970 inauguration of the ARPAnet, precursor to the Internet, interconnecting four U.S. west coast universities. In 1972, the first International Conference on Computer Communication (ICCC) was held in Washington, DC, to discuss progress of these early efforts. The chairman of this conference was Vinton Cerf, who, along with Robert Kahn, would release the standard Internet protocol TCP/IP just four years later. It was also Cerf who proposed linking ARPAnet with the National Science Foundation's CSNet via a TCP/IP gateway in 1980, which some consider to be the birth of the modern Internet.

The National Physics Laboratory in the United Kingdom and the Societe Internationale de Telecommunications Aeronotique in France, in the 1960s, also explored similar concepts.

1.5. Future of Fiber Optic Communications, 2000–2050

Fiber optic networks in 2000 were transmitting several hundred gigabits per fiber; terabit per fiber capability had been demonstrated. The throughput of fiber cables deployed in 2000 could be increased by typically two orders of magnitude by a combination of increased DWDM (more wavelengths per fiber) within currently developed spectral bands plus higher data rate modulation. Beyond this, the available spectral bandwidth will probably be expanded by at least 10X over the 2000–2010 span. Beyond 2010, new fibers can extend the low-loss spectrum by hundreds of nanometers. And, of course, it will always be feasible to deploy additional cables.

Unlike RF/microwave communication, which is limited by available spectrum, and copper cable transmission, which is limited by low data rate capability, high cost per transmitted bit, and the large physical space consumption, future fiber optic communication expansion is relatively unlimited. The key question is: who needs it? Is there a long-term commercial demand for rapidly increasing bandwidth per subscriber, at a rapidly falling price per transmitted bit? To those with a long-term background in integrated circuits and in computers, this is a familiar and long-since-answered question.

To phrase the question differently: will there be new services, enabled by greater bandwidth, at declining cost per gigabit-kilometer, for which subscribers will see economic justification for purchase? Can higher data rate global communication, at little additional cost, enable a business to increase revenues; decrease costs; better negotiate business cycles through greater agility? Will there be bandwidth-dependent residential services offering greater subscriber satisfaction, at little additional cost? Will business and residential interests tend to merge? The history of commercial and residential communication services over the past 50 years, and especially the past 15, supports an affirmative response.

E-business is now emerging, with fits and starts, but the underlying logic seems clear; there are numerous ways it can reduce costs and risks. E-business will require increasingly voluminous instant global transfer of numeric and graphic data. Global subscriber-to-subscriber internets will proliferate, with rapidly falling communication costs per bit.

Personal videoconferencing, on the "back burner" for personal communication as an augmentation of the telephone over the past 40 years, is now entering through the back door; on the personal computer terminal, which is rapidly evolving to an internet data/voice/video communication terminal. This will evolve to higher resolution, color video, using orders of magnitude more bandwidth compared to voice grade lines. It will be affordable, because costs will drop rapidly with increasing volume use, and it will make communication more effective, thus of greater value.

Residential entertainment video is still in an early stage of development. With analog broadcast TV, the TV set designer has been limited to whatever functions she could accomplish within about 4.5 GHz of bandwidth. Very impressive advancements have been made within this restriction; evolution to color TV, and increased definition and/or larger screens. Another restriction has been the vacuum tube display screen; limited in size, although advancements have been achieved. With virtually unlimited bandwidth at very low cost, and with continuing evolution of high-resolution flat panel displays, life-sized three-dimensional "virtual window" displays will become technically and economically feasible. Coupled with entertainment content improvements made feasible by computer graphics/animation, 3D effects, and other innovations, a major future digital TV market appears likely; requiring megabits now; tens of megabits (per receiver) by 2010; gigabits within 50 years.

1.5.1. DYNAMIC, CREDIBLE VIDEO CREATION

Major advancements were made, over the 1990–2000 decade, in the perception of quality of created (versus recorded/reality) video. Much of this was driven by the economic returns from video games; their increased perception of reality added to popularity and profits. In broadcast entertainment and advertising, also, major advances were made in three-dimensional effects, animation, and nuance. This impressive advancement, however, was restricted by the high cost of massive computational power, size, and resolution limitations of video display screens, and the television receiver bandwidth restriction imposed by the requirement to transmit all program information within an imposed narrow segment of the radio frequency (RF) spectrum. Associated with this was the inevitable time required for the learning process, each incremental step building upon the accumulated knowledge. These restrictions are yielding to the continuing advancements of technology on many fronts.

1.5.2. ANIMATION TECHNOLOGY ADVANCING

Within the 2000–2050 span, animators will create libraries of "actors" of virtually unlimited scope, with totally lifelike visual and voice replication of a Marilyn Monroe, John Wayne, or any other then-known or fantasy actor. These animatons will be programmable for seamless performance. They will not be temperamental or temporary, and will not demand huge salaries. They will be instantly, continually available to program creators. Copyright laws will expand to protect the rights of both "real" and created actors.

1.5.3. SURGE OF VIDEO SCREEN DEVELOPMENT

The capability of "flat panel" video screens advanced impressively, 1990–2000, while the cost/performance ratio drastically dropped, driven by the economic returns available from improved portable computers, cellular phones, and other instruments. (This followed the path of semiconductors, which were accelerated into wide usage by Sony's high-volume sales of semiconductor-based portable personal radios.)

1.5.4. FIBER BANDWIDTH SUPPORTS RECEIVER REVOLUTION

The use of a substantial segment of the RF spectrum for entertainment video transmission will be displaced, within the 2000–2050 period, in the more industrially advanced nations, by signal transmission over fiber. (Regions that cannot economically be served by fiber will receive TV broadcasts via satellite microwave and by local, low-power terrestrial microwave transmitters.) This will free television receiver designers from the current restriction of accomplishing all signal conversion and application within a bandwidth of about 4.5 MHz. By 2010, over 100 million homes globally will receive television entertainment over fiber at data rates up to 622 Mbps. By 2050, this will stretch to over one billion homes at up to 10 Gbps. This will enable the design and profitable supply of receivers presenting apparent life-size, three-dimensional television via instant selection from remote storage servers offering tens of thousands of programs.

While television viewers continue to show strong acceptance of fictional/fantasy presentations, there also is rapidly increasing demand for reality-based programs. The evolving technology of video program creation using computer-enabled script and character creation will make feasible lifelike evening programs based on same-day morning news.

1.5.5. MAJOR ADVANCEMENT OF NON-ENTERTAINMENT VIDEO

The 2000–2050 advancements in video content creation, including animation plus 3D screen presentation, enabled by broadband signal fiber transmission to homes and businesses, will support strong growth in video instruction for both formal education and instructional "how to" markets. The rapid (and accelerating) advancement of technology on all fronts means a person cannot complete their formal education and then remain proficient in their chosen field for their entire career through only workplace exposure. Continuing education will be required; fiber-enabled broadband video increasingly will become the most economical medium to deliver this education. Instruction manuals will be displaced by video clips.

1.5.6. DYNAMIC FUTURE OF COMMUNICATION FIBER OPTICS INDUSTRY

The economics-driven growth of global, regional, and local fiber communication networks over the past 25 years has supported, and been enabled by, dynamic expansion of fiber network deployment. The global consumption of fiber optic components in communication networks exploded from only $2.5 million in 1975 to $15.8 billion in 2000. Continued growth to $739 billion in 2025 is forecast.[3] The value of fiber optic communication equipment and trunk lines incorporating these components in 2000 was more than twice the component value. While year-to-year growth rates varied significantly over the past quarter century, the overall trend was a pattern of gradually slowing investment growth rate, while communication transport capacity expanded at a relatively steady exponential rate.

1.5.7. STANDARDS, INTEGRATION, MASS PRODUCTION KEYS TO FIBER UBIQUITY

Transistors, in their early market, cost more than vacuum tubes. If that price relationship had continued, the advantages of semiconductors nevertheless would have supported a growing market. However, over the past 50 years the average price of a transistor has dropped by about six orders of magnitude. This enabled transistors to take market share away from vacuum tubes but, much more importantly, supported the design and economic production of a dazzling range of products that would not have been technically or economically feasible with vacuum tube technology. This includes nearly all of the computer market, portable electronics, most vehicular and space electronics, and much more. Keys to the unprecedented price drop achieved by transistors have included increasing integration (2001 laboratory demonstration: 400 million transistors on a chip) and related miniaturization, plus mass production enabled by substantial standardization and automation of processes, packaging, and testing. The production of fiber optic cable has become relatively mature by 2000, comparable to the copper communication cable of 1975. Other fiber optic components are much less mature; many new components entered the market 2000–2001, and many are yet to be conceived and developed.

1.5.8. COMPONENTS ARE FAR FROM MATURE

Fiber optic active and integratable products in 2000 were at a maturity stage comparable to semiconductor integrated circuits circa 1970 (a few transistors on a chip). Integration of active photonic components is now evolving aggressively in hybrid format, including multiple parallel channels, with production of over one million units projected for 2001. This will expand to over ten million units in 2010. Monolithically integrated optoelectronic circuits are now mainly developmental, but will exceed one million production units by 2010. Wavelength-division multiplexing (WDM) and parallel channel integration will combine to achieve ten-terabit interconnection fiber links, terminated in single small transmit/receive modules, well before 2050.

Fiber optic component production has evolved, 1975–2000, from small quantities assembled by engineers plus semi-skilled labor, to quantities of hundreds per day assembled by highly skilled, tooling-assisted labor (mostly in low labor cost regions). The industry, in 2000, was at a very early stage of high-volume automated assembly, packaging, and testing along the path pioneered circa 1975 by the semiconductor industry. With the evolutionary progress through 2000, the price of optoelectronic transmitters and receivers (in current dollars), measured in megabits per dollar, dropped by about four orders of magnitude. An even larger decrease will occur over the 2000–2050 span. To achieve this, automation of alignment and attachment, especially of optical beam transmission components, to submicron tolerances is required. Standardized packages in several formats, metallic and nonmetallic, will be necessary, along with precision pick-and-place machines. Heat transfer plus both optical and electrical crosstalk challenges must be met.

The historic growth of the economically significant fiber optic component categories 1975–2000, and forecasted growth 2000–2050, are presented in Table 1.1. The total global consumption of fiber optic components in 2050 is forecast to reach about $28 trillion. While fiber optic cable has dominated component consumption value to date, active components will steadily move to dominance over the next half century as many passive functions are integrated into monolithic OEICs and as consumption quantities are dominated by much shorter links, requiring little fiber, compared to year 2000 mix.

1.5.9. ADVANCEMENT WILL BE EVOLUTIONARY

The advancement of fiber optic components over the next 25 and 50 years will be evolutionary, in a pattern resembling the evolution of the semiconductor device industry over the past half century. The underlying trends will be toward

- Higher performance
- Higher integration (more functions per component)
- Miniaturization; function/volume ratio increased by more than four orders of magnitude
- Lower cost per function, by more than four orders of magnitude

1.5.10. OPTICAL PACKET SWITCHES BY 2010

Amazing progress along this trendline occurred over the 1998–2001 span, with two orders of magnitude size reduction along with one order of cost reduction in commercially available transmitters and receivers of

Table 1.1 Global Fiber Optic Component Consumption Growth Rates, by Function

Component type	1975 $Million	1975 %	2000 $Billion	2000 %	2025 $Billion	2025 %	2050 $Billion	2050 %	Average annual growth rate % 1975–2000	Average annual growth rate % 2000–2025	Average annual growth rate % 2025–2050
Fiber Optic Cable	1.78	71	25.20	70	430	40	3750	13	47	12	9
Active Components	0.40	16	7.40	20	490	45	20090	72	48	21	16
Passive Components	0.16	6	2.45	7	124	11	3290	12	47	19	14
Other Components	0.15	6	1.20	3	39	4	665	2	43	15	12
Total Consumption	2.49	100	36.25	100	1083	100	27795	100	47	15	14

fixed performance. Components with further dramatic advancements were demonstrated, in 2001, in numerous industry laboratories and will become commercially available 2002–2003. These include dozens of transmitter/receiver pairs in a single small package; dozens of optical amplifiers on a single small chip; photonic transparent nonblocking matrix switches with 10,000 × 10,000 port capability. The low-loss fiber transport spectrum, supported by available and emerging fiber and other components, will expand by more than an order of magnitude over the 2000–2025 span. Optical packet switching capable of 40 Gbps per channel, terabits per fiber throughput, supported by holographic memory and sub-nanosecond semiconductor optical switches, will be deployed in long-haul network packet switches by 2010.

The anticipated trend of fiber optic communication over the 2025–2050 span becomes more hazy. It is highly probable that the basic function will continue to grow; rapidly in terms of capacity (throughput; gigabits × kilometers), and at a declining but still-positive constant dollar (i.e., after deducting inflation effects) investment rate. Throughput of a fiber cable will continue to increase, while the cost per gigabit-kilometer continues to drop. Expanded services that consume much more bandwidth will continue to emerge. Miniaturization, upward integration, and quantity increases will continue. Standard monolithic optoelectronic integrated circuits (MOEICs) will become the mainstream standard component category, but a major market in application-specific MOEICs also will emerge. Technical breakthroughs that have already been conceived, and perhaps demonstrated, as well as inventions through the 2000–2025 span, will have a major impact, not now identifiable, on 2025–2050 components.

Who, in 1950, could project that transistors would largely supersede vacuum tubes, with many orders of magnitude per-function cost reduction, thus opening the way for major new, useful applications requiring multibillion-dollar annual equipment production? Who, even in 1975, could forecast that millions of transistors could be processed on a single tiny chip selling for a few dollars each, by 2000? (Indeed, there were scholarly treatises demonstrating this would be impossible.) Fiber optic communication will not progress in a competitive vacuum. Although copper cable will decline in relative significance (as have vacuum tubes over the past 50 years), it will still be deployed in some situations. More competitive will be radio-frequency wireless and unguided optical spacebeam communication, strongly supported by semiconductor-based encoding that greatly expands the available channels and usable bandwidth within a served area.

By 2050, satellite-supported wireless will make video and high data transfer rates accessible to miniature portable units having immense memory, anywhere in the world, at a cost that is economically attractive to businesses and to professional people.

1.5.11. COMMUNICATION EVOLUTION WILL CONTINUE

Communication with, between, and among professional individuals in 2050 will extend commonplace technology beyond the 2000 perceived outer limits. At an economically justifiable cost, the professional, wherever he or she travels in the world, can continuously send and receive voice, data, graphic, and video communication. This will include transfer of massive data and text, which can be stored in the personal terminal. Verbal language translation will be automatic. An immense library of public and private information will be instantly accessible. Voice-to-hardcopy translation in any commonly used language can immediately be printed.

All known prospective contacts can be contacted/connected in seconds, and desired but unidentified contacts can quickly be identified and connected. This will be accomplished by the global communication network's combination of fiber optic, satellite, and cellular transport.

If these 2000–2050 forecasts seem aggressive, they should be judged in the context of advancement over the past 500 years, not the past 50. (This is the rule, mentioned earlier, that over half of the progress of any period has occurred during the last 10 percent of the period.) They should be judged alongside the probability that wealthy individuals (of which there will be millions) in 2050 can (and some will) have several hundred children, with or without genetic relationship. The probability that nuclear fusion power (the 20-year future breakthrough forecast since 1960) can provide virtually unlimited low-cost clean electrical energy that can displace fossil fuels in transportation, heating, and industrial processes; and can achieve low-cost water purification and transport, supporting major greening of the planet and alleviation of hunger. The probability of space travel to solar planets. The probability of dramatic advancement in the treatment of common diseases. These and many other possibilities, viewed from 2000, appear more likely than the world of 2000 appeared to the thinkers of 1500.

Communication is not a fad. Modern society considers it a necessity, and will see economic justification for better, more responsive, ubiquitous communication. Fiber optics technologies and related markets will continue growth over the next 50 years, and beyond.

References

1. DeCusatis, C. 1998. *Handbook of Fiber Optic Data Communication.* San Diego: Academic Press.
2. Hecht, J. 1999. *City of Light, The Story of Fiber Optics.* Oxford, UK: Oxford University Press.
3. Montgomery, J. D. 1999. *Fifty Years of Fiber Optics.* Nashua, N.H.: Lightwave/PennWell Corp.

Chapter 2 | Market Analysis and Business Planning

Yann Y. Morvan
Cookson Electronics, New Haven, Connecticut 06510

Ronald C. Lasky
Consultant, Medway, Massachusetts 02053

2.1. Introduction

Performing an effective market survey and business plan may be the most important aspect of any successful technology. However, because most engineers have little experience with this topic, one is seldom performed until late in a program and when done is usually poorly performed. This weakness has been especially true in optoelectronics. There are numerous examples of sound technology that did not get implemented because of the lack of market analysis or a sound business plan. It is hoped that this chapter will help the reader to avoid pitfalls.

2.2. The Need for Applications

When Apple shipped its first MacIntosh in 1984 there were no customers crying for a product with its features. However, it was phenomenally successful. This success argues strongly against the philosophy of providing what the customer wants. Masseurs Job *et al.* anticipated what the customer would want and delivered a complete workable system. In today's high-technology world, this approach is the true winner. The lackluster companies will be content to provide only what the customer wants. On the other end of the spectrum is a technology that is impressive and anticipates a need but is not complete. A good, but perhaps facetious example would be the delivery of a Pentium integrated circuit (IC) in 1985. It is truly a great accomplishment, but there is no infrastructure to support it or make it useful. No circuit boards or buses exist that can support the high clock speeds, and there are no applications that require the fast speeds or supporting software. This example stresses the need to have technology support a complete system that performs a useful function.

Thus, when delivering a technology component there must be a need that it fills in the technology infrastructure for an existing or future application. An optoelectronics example would be a very low-cost fiber distributed data interface (FDDI) transceiver module. The FDDI standard, although its use has not emerged as rapidly as hoped, now has applications and a complete technology infrastructure. Hence, someone producing this transceiver can find customers that have a use for it. In 1992 a major company developed a single-mode (SM) Gb/s transceiver module. It was initially developed for internal use but was eventually targeted for the external market. There was much excitement for it because it seemed to support the information superhighway, but in 1992 no applications existed. Unfortunately, it did not attract much serious interest because the needed infrastructure did not exist.

2.3. Supporting Technology Infrastructure

The issue of supporting technology infrastructure needs some clarification. The previous example of the FDDI transceiver is used. A transceiver such as this one requires much technology to support it. Printed circuit boards (PCBs) that can support Gb/s data rates typically need to handle 20 channels of 50 Mb/s. In 1992 PCBs of this complexity were not common. The transceiver did have a built-in serialize/deserialize (similar to mux/demux) function. Many transceivers do not, hence the availability of low cost ICs to perform this function is crucial. These ICs are often available. Other interface hardware to the system of interest is also needed. At leading-edge data rates it will typically not be available unless there is a driving application. Assuming that the hardware issues are settled, what about software? To be useful, a transceiver usually has to talk to systems with different types of protocols. An example might be IBM microchannel architecture to AppleTalk. This software did not exist.

All these issues arose in late 1992 when a team of electrical engineering students was commissioned to use two of the Gb/s transceiver modules to enable an IBM RISC workstation to communicate with a DEC workstation. This application was extremely simple compared to likely use for this product, in something like a complex switch for the Fibre Channel Standard. The team initially felt that the assignment would be easy. However, after 8 months it could only produce a report on how it would solve the problem if it had more time and money. The following were their conclusions:

1. The data rate is so high that the team had to buffer the data between computers.
2. No software drivers existed that could handle the two different types of systems: IBM vs DEC. It would be a 10-man year effort to develop them for this simple application.
3. The team could find no PCBs or ICs to use to provide the hardware interface support needed to connect the module to the two computers.

It should not be surprising that the developers of this transceiver could not sell it. There were few applications requiring this data rate and no supporting hardware and software that would be needed to implement an entire, sellable function.

The technology that one is supplying must support a complete, sellable function for which there is a current or future application. If the technology predates the necessary hardware and software that are needed to support it, it may fail. One takes a tremendous risk if one produces a technology element for which the supporting elements are undeveloped.

2.4. Implementing a Market Survey

Performing a marketing survey should be one of the first tasks in a technology program. Many technologists take pride in their understanding of the market, but this pride is often misplaced because a thorough knowledge of a technology market is a full-time job.

Marion Harper Jr., a marketing executive, once said: "To manage a business well is to manage its future: and to manage the future is to manage information" [11, p. 103]. A market analysis is the prelude of planning, implementation, and control of a marketing strategy. The launch of any new product must go through this process in order to be successful. Knowing your market is not an option, it is a prerequisite for anyone involved in the product's design. The intent of this section is to offer a basic understanding of the major steps to follow when conducting a market analysis.

2.4.1. FIRST STEP: INDUSTRY DESCRIPTION OUTLOOK

The Standard Industrial Classification (SIC) manual is the first source to consult when one conducts market research. The researcher must give a detailed description of the primary industry in which the product is manufactured: size (present and over time), characteristics and trends, main customers (industrial, individuals, or government?), etc.

2.4.2. SECOND STEP: SEGMENT, TARGET, AND POSITION

Segmenting the market helps to obtain a clear view of one's customer. A segment consists of a group of potential customers (individual or organization) who are similar in the way that they value the product, in their patterns of buying, and in the way they use the product.

In order to be effective, the segmentation must follow Kotler's rule: It must be measurable, accessible, substantial, and actionable [11, p. 252]. In every segment, all the following items should be checked:

Price levels
Purchasing behavior
Helpful sources of information
Best media
Demographic information
Competitive products
Distribution channels

For a clearer view of the target market, it is interesting to establish a segmentation grid that summarizes all the characteristics of each segment (Table 2.1).

Once all the segments of the market have been identified according to different variables (demographic, geographic, psychographic, or behavioral), it is pertinent to select the segment that represents the most suitable opportunities to the company. To do so, the researcher must assess its size and anticipated growth as well as the company objectives and resources. In a target segment, not only strengths are required but also superiority over competition.

Anticipating the market penetration can be achieved after the interpretation of the data collected during the market research. Thus, the company can estimate its market share, the number of its customers, and its geographic coverage. Any trend or expected change should always be kept in mind

Table 2.1 **Segmentation Grid**

| | Usage rate | | |
| | Segment No. | | |
	1	2	3
Products purchased	……	……	……
Product benefits and attributes	……	……	……
Key information sources and influences	……	……	……
Best media	……	……	……
Price limit	……	……	……
Best distribution channels	……	……	……
Demographic information	……	……	……
Segment size (example)	35%	50%	15%
Market size = 100%			

Note. From Harmer, R., *New Product Development Project Workbook.* Boston: Boston University School of Management, 1994, pp. 3–4. Comment: The highest percentage for segment 2 does not mean that it is the most enticing but simply that it represents 50% of the entire market. The biggest segment of a market is not always the best suited to the company's resources.

when making a decision. The environmental analysis shall never be forgotten. Its main parameters are economic forecasts, governmental regulations, demographic and social trends, and technology breakthroughs.

Even if the company greatly focuses on its target segment, it is wise to collect precious information, such as needs, demographics, and significant future trends, on the secondary target segments because they may become attractive in the future.

The last assignment is the position step. This final stage can be achieved with the analysis of one's competition. Identifying competitors and listing their strengths and weaknesses is a good way to refine one's positioning strategy. The company must place its product, in the customer's mind, at a higher value than any other competing products. It should emphasize the appropriate competitive advantages that the product may offer and base its communication on them.

2.4.3. THIRD STAGE: DEFINING THE MARKETING STRATEGY

The third stage consists in defining the marketing strategy based on all the information collected throughout the market analysis (Fig. 2.1). It is clear that the marketing strategy is not a static formula but should always

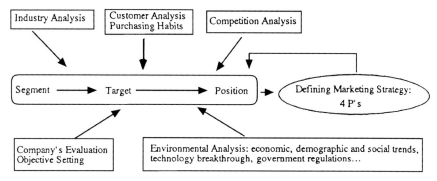

Fig. 2.1 The market analysis diagram. From Morvan, Y., Université Jean Moulin, Lyon, France. Copyright 1996.

be modified according to the market evolution and the life cycle of the product. The "four P's," in marketing terminology, represent the four pillars of the strategy: product, price, promotion, and place (i.e., distribution). A market analysis on the transmitter optical subassembly (TOSA), presented in the Appendix, gives an illustration of most of the points discussed previously.

2.4.4. NOTE ON DATA COLLECTION

A market analysis consists of collecting data that come in two forms: primary and secondary. Primary data are collected for a specific goal through methods such as personal interviews, focus groups, questionnaires, experiments, or observation. On the contrary, secondary data consist of information available at some source without a specific goal. Secondary data (Table 2.2) include the company's internal information, government publications, periodicals and books, commercial data, and international data. Most secondary data can be obtained from public libraries or university libraries specialized in business. Let us point out that, nowadays, the Internet represents a wonderful source of information, often at no charge.

2.5. Business Planning

Jeffry Timmons wrote in his book, *New Venture Creation Entrepreneurship in the 1990's,* "Planning is a way of thinking about the future of a venture; that is, of deciding where a firm needs to go and how fast, how to get there,

Table 2.2 **Sources of Secondary Information**

Industry analysis
    Directories: SIC manual, *U.S. Industrial Outlook* (USDOC), *Standard & Poor's Industry Surveys, Census of Manufacturers* (USDOC), *Industry Norms and Ratios* (Dun & Bradseet), *Market Share Reporter*
    Databases: Nexis, Investext, ABI/Inform, F&S Index: United States and International
    Others: Books, periodicals, Internet

Competition analysis
    Directories: *Ward's Business Directory, Million Dollar Directory, Standard & Poor's Register of Corporations, Moody's Manual, Principal International Business, International Directory of Corporate Affiliations*
    Databases: Nexis, Dialog databases, Dow Jones News/Retrieval, Global Vantage PC Plus, American Business Disk
    Others: Company brochures, books, periodicals, Internet

Environmental analysis
    Government publications: *Statistical Abstract of the U.S., County and City Data Book*
    Periodicals and books: General business and economic magazines and newspapers such as *Business Week or The Wall Street Journal*, specialized newsletter and periodicals

Company's evaluation
    Internal source: Income statement, balance sheets, ratio analysis, prior research reports

Note. From Morvan, Y., Université Jean Moulin, Lyon, France. Copyright 1996.

and what to do along the way to reduce the uncertainty and to manage risk and change."

Business planning aids in developing the marketing strategy to be carried out based on the market analysis described previously and on the resources and objectives of the company. It is commonly included in a document called "business plan." Let us now evaluate the purpose of such a document and study its content.

2.5.1. THE PURPOSE OF A BUSINESS PLAN

Figure 2.2 describes the main advantages of having a business plan. It sets the objectives of the company based on the knowledge of its resources (physical, financial, and human), its market, and its environment. It offers

2. Market Analysis and Business Planning   39

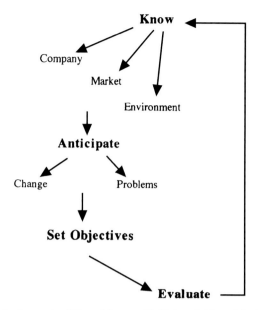

Fig. 2.2 Why a business plan? From Morvan, Y., Université Jean Moulin, Lyon, France. Copyright 1996.

a long-term view for the manager, which helps to anticipate future trends. The business plan is a document that anyone inside the company can access to evaluate performance. It can also be utilized externally by the firm to attract investors or to communicate with outside players such as suppliers.

The purpose of a business plan is to

1. Evaluate capacity and potential
2. Set objectives
3. Evaluate performance
4. Communicate

2.5.2. THE CONTENTS OF A BUSINESS PLAN

A traditional business plan includes the following topics:

Executive summary
Company description
Product description

Market analysis and marketing strategy
Financial analysis

A business plan must be clear, comprehensive, and concise. Always keep in mind that your reader does not necessarily know the nature of your business; therefore, it is useless to overwhelm him or her with too many technical terms.

The following outline can help to understand what a business plan should include. It is only a suggestion and should be regarded as such. Not all business plans are alike, simply because every company is unique:

Business plan outline
1. Executive summary
    a. Purpose of the plan
    b. Market analysis
    c. Company and product descriptions
    d. Financial data
2. Company description
    a. Nature of business
    b. Ownership
    c. History
    d. Location and facilities
3. Market analysis
    a. Industry
    b. Market segmentation and target segments
    c. Competition analysis and positioning
    d. Environmental trends affecting the market
4. Product description
    a. Benefits and competitive advantages
    b. Product life cycle
    c. Copyrights and patents
5. Marketing strategy
    a. Market penetration strategy
    b. Pricing strategy
    c. Distribution strategy
    d. Promotion strategy
6. Management and ownership
    a. Legal structure and organization chart
    b. Brief description of key managers (compensation, skills, etc.)

       c. Board of directors
       d. Shareholders
   7. Financial data
       a. Historical documents
           Income statement
           Balance sheet
           Cash flows
           Key ratios analysis
           Break-even analysis
       b. Prospective documents
           Income statement
           Balance sheet
           Cash flows
           Key ratios analysis
           Break-even analysis
           Capital budgets
   8. Appendices
       a. Resumes of key managers
       b. Pictures of products
       c. Professional references
       d. Market studies
       e. Pertinent published information
       f. Patents
       g. Significant contracts

Writing an effective business plan is not an easy task but you can obtain precious help from college business schools, another business owner, the local chamber of commerce, library/bookstore, or business association. Many software programs are also available to assist you in this work, especially with regard to the financial aspects.

2.6. Summary

The key steps in market analysis are market segmentation, market targeting, and market positioning. The segment, target, and position process is the prerequisite for a successful commercialization. The market analysis is often included in the business plan. Business planning gives an evaluation of the capacity and potential of the company; it also sets objectives, evaluates performance, and represents an internal and external medium of communication.

Appendix: Market Analysis on a Transmitter Optical Subassembly

Introduction

This summary presents information obtained through literature research concerning the market analysis on a TOSA.
The product has the following characteristics:

- Data rates above 1 Gb/s; distance without repeaters >20 km
- Single-mode light 1.3 or 1.55 μm in wavelength; rise and fall times under 0.3 ns; spectral width under 1 nm
- Most likely applications will be asynchronous transfer mode (ATM) application at Gb/s speeds and Fibre Channel Standard applications at Gb/s speeds

The purpose of this work is to assess the overall industry of fiber optic components, differentiate the segments of this market, define the target segment and its trends, and initiate a study of the competition.

Industry Description and Outlook

DESCRIPTION

The optical fiber, characterized by low transmission loss and immunity to electromagnetic interferences, provides the transmission speeds and high-volume data handling that are currently driven by the consumer needs.

The U.S. fiber optics industry comprise cables, optoelectronic components (transmitters, receivers, and fiber amplifiers), connectors, and passive optical devices. Telecommunications, data networks, and cable television are the main users of fiber optic equipment.

The transmission of all types of signal, including voice, and video data, will be predominantly digital beyond the year 2000. In 1993, 50% of the

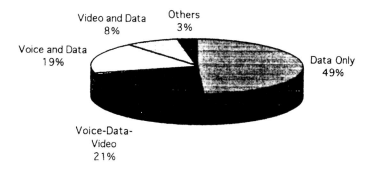

Fig. A.1 Fiber optic networks. Source: Kessler Marketing Intelligence, 1993.

fiber optic networks carried only data, whereas 21% carried voice, data, and video at the same time [1] (Fig. A.1).

INDUSTRY CHARACTERISTICS AND TRENDS

More than 50% of U.S. fiber optic networks have been installed since 1991. The change from copper to fiber indicates that a greater capacity, or bandwidth, is required to deliver the broad array of services proposed for the superhighway. The need for greater bandwidth is evidenced in part by the fact that transmission of data over telephone is increasing by approximately 20% a year [2].

This industry is expected to enjoy continued growth throughout the 1990s. In 1993, the North American consumption of fiber optic components was worth $1.93 billion and is estimated to reach $9.14 billion in 2003. Therefore, the consumption of optoelectronics in North America is expected to grow at an average annual rate of 18.75% between 1993 and 2003 [3] (Table A.1). Although telecommunications still leads demand (Table A.2), cable TV (CATV) and data communications are growing more rapidly (Table A.3), and have become significant markets with different needs because they consume more optoelectronic interfaces than does the telecommunication industry. The fastest fiber optic optoelectronics growth over the next decade should be in enterprise or premises data communications network applications. In this market, fiber optics must compete against the cost of unshielded twisted-pair copper wire, coaxial cable, and electronic terminations and connectors. The total North American consumption of optoelectronics components in enterprise networks should expand by 32% per year between 1993 and 2003. Over the

Table A.1 North American Consumption of Fiber Optic Components

Component type	1993		1998		2003		Average annual growth rate (%/year)	
	$ (Billion)	%	$ (Billion)	%	$ (Billion)	%	1993–1998	1998–2003
Cables	1.19	61.6	2.28	55	4.9	53.7	13.9	16.5
Optoelectronics	0.57	29.5	1.35	32.6	3.18	34.8	18.8	18.7
Connectors	0.12	6.4	0.31	7.3	0.61	6.6	20.8	15.1
Passive optical devices	0.05	2.5	0.21	5.1	0.45	4.9	33.2	16.5
Total production	1.93	100	4.15	100	9.14	100	16.5	17.1

Table A.2 **North American Fiber Optic Installation Apparatus**

Telecommunications	75.0%
Premises data networks	18.0%
Cable TV	7.0%

Note. Distribution of fiber market apparatus installations is shown in percentage. Total market in 1992 was $384 million. Source: Lightwave, 1993 (August), p. 24, from Electronicast Corp.

next 10 years, the most important growth of optoelectronics-electronic demand in enterprise networks should be in premises ATM switches (Table A.4). Light-emitting diodes and laser diode-based transmitter/receiver unit prices are expected to drop over the next decade. The laser diode-based transmitter/receiver share should rebound to 59%, or $1.67 billion, in 2003.

New equipment used for multimedia applications (full-motion video and supercomputer visualization), digital high-definition TV, video conferencing, distributed high-speed supercomputer networks, and other wide-bandwidth applications will enter the market within 5 years and drive an explosive expansion of bandwidth demand. Telecommuting and distance learning are also promising applications.

World Fiber Optics Industry

The U.S. Department of Commerce notes that Asian-Pacific countries, with more than two-thirds of the world population, represent one of the fastest growing markets for fiber optic equipment. Emerging countries in the Caribbean, Latin America, and Eastern Europe will also provide opportunities as these countries upgrade antiquated networks (Brazil, Mexico, etc.). The world's largest underdeveloped telephone market is China. The Chinese government wants to raise the ratio of 10 lines/100 people to 40 lines by 2020 [4].

U.S. producers and consumers of fiber optic equipment appear to be increasingly optimistic about the potential demand for the expanded service offerings possible through fiber optic technology.

Table A.3 North American Consumption of Fiber Optic Optoelectronics (by Application)

Application	1993		1998		2003		Growth rate (%)	
	$ (Million)	%	$ (Million)	%	$ (Million)	%	1993–1998	1998–2003
Telecommunications	789	75	1148	56	3180	48	8	23
Enterprise (premises) data	166	16	627	31	2853	43	30	35
Cable TV	38	4	107	5	260	4	23	19
Military/aerospace	45	4	102	5	212	3	18	16
Specialty applications	18	2	49	3	158	2	22	26
Total consumption	1056	101	2033	100	6663	100	14	27

Source: Electronicast Corp.

Table A.4 **Enterprise Network Consumption of Fiber Optic Optoelectronics (by Network Product Category)**

	1993		1998		2003		Growth rate (%)	
Network product	$ (Million)	%	$ (Million)	%	$ (Million)	%	1993–1998	1998–2003
Premise ATM	2.3	1	191	30	1704	60	142	55
FDDI	36	22	83	13	40	1	18	−14
Fast Ethernet	0	0	96	15	364	13		31
Fiber Channel	0	0	62	10	441	15		48
Conventional data Communications	128	77	194	32	303	11	9	9
Total consumption	166.3	100	626	100	2852	100	30	9

Source: ElectroniCast Corp.

Target Markets

The optical communication market includes applications in telecommunication, computing cable television, and automobiles. Currently amounting to more than $3 billion, this market is supposed to grow to more than $30 billion by 2003, which represents an annual growth of 26% between 1993 and 2003 (Fig. A.2). Over the next 20 years, the technology used in both telecommunications and data communications is going to evolve. (Fig. A.3) presents the optical communications product road map.

The data communications market for fiber optic equipment, which was just $0.8 billion in 1993, is forecast to grow to $2.6 billion in 1998 and to $6.3 billion in 2003 (i.e., annual growth of 23% from 1993 to 2003). The Optoelectronics Industry Development Association (OIDA) divides the optical communication equipment market in four segments:

1. Long-distance telecommunications (trunk and municipal area networks): This is the high-performance segment of optical communications, characterized by repeaterless spans of 5–100 km

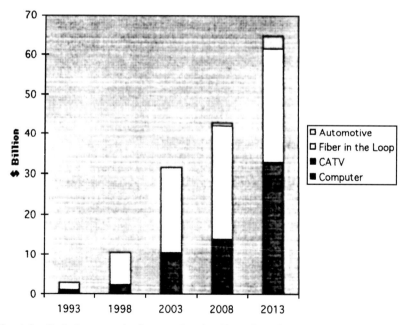

Fig. A.2 Optical communications market size. From Optoelectronics Industry Development Association [5].

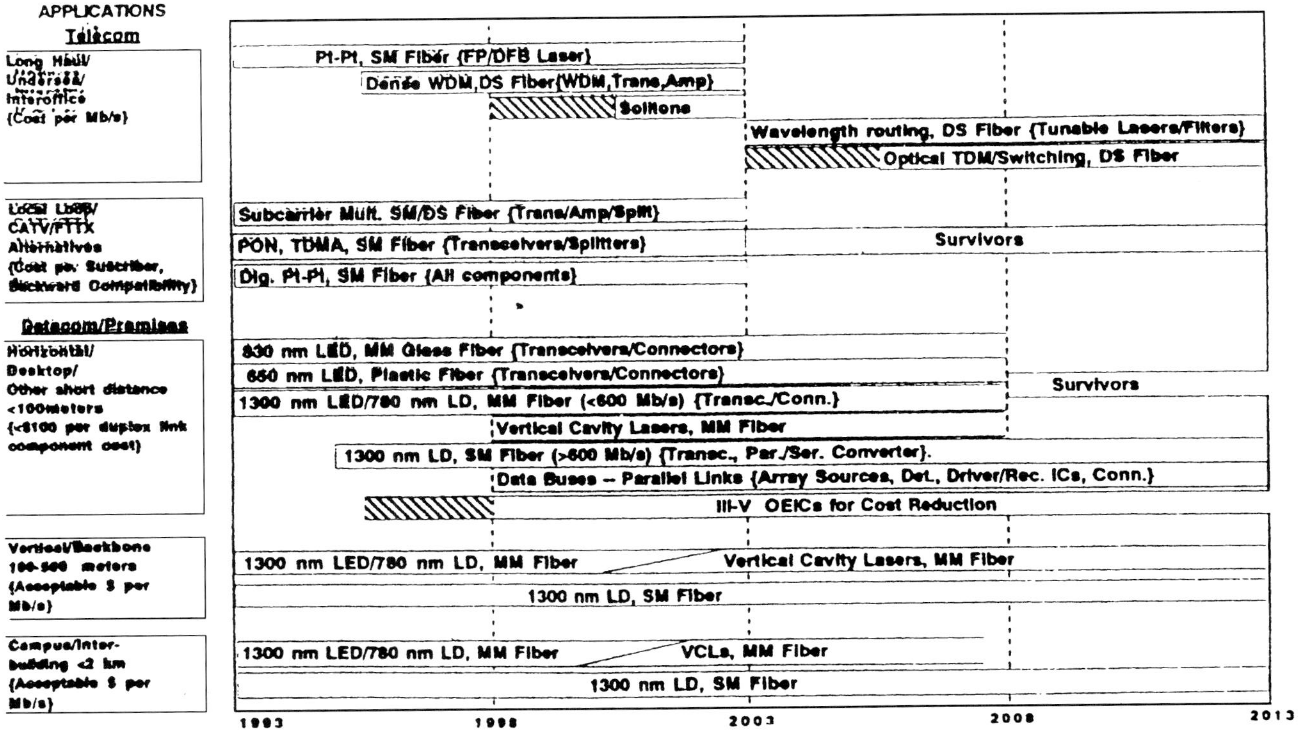

Fig. A.3 Optical communications product road map.

and high speeds (0.6–2.5 Gb/s). A key requirement is high-performance lasers.

2. Shorter distance telecommunications (such as fiber in the loop): This segment is characterized by spans of 1–10 km and speeds of 50–622 Mb/s. This market segment is growing and is highly cost sensitive. U.S. industry lags behind Japan in this segment and needs improvement, primarily in the form of lower cost manufacturing, packaging and alignment technologies, and better epitaxial growth and processing of lasers.

3. High-performance data communications: This segment is characterized by long distances (for data communications) of 300–2000 m or relatively high speeds (200–1000 Mb/s). It is cost sensitive. U.S. industry is competitive, but improvements in the form of cost reductions are needed.

4. Low-cost data communications: This segment is characterized by short distances (< 300 m) or low speed (< 200 Mb/s). This segment is very cost sensitive. Component costs need to be competitive with wire or small relative to installation costs, and currently are in many cases. Individual U.S. suppliers are strong, but users would like more domestic suppliers. The main barrier of this segment is the lack of familiarity with optical technology on the part of installers and users.

Much of the market growth is expected to be in segments 2–4, which are all characterized by higher volumes and higher cost sensitivity than the first segment, which has been the largest segment to date [5].

The segment, high-performance data communications, which we are targeting, is expected to experience considerable growth during the next decade.

HIGH-PERFORMANCE DATA COMMUNICATION SEGMENT

Characteristics

The main drawback that the fiber optic industry is facing is the high cost of optoelectronic interfaces. The cost of optoelectronic transmitters, in particular, can add significantly to the cost of hardware such as hubs, concentrators, and network cards. In 1992, according to Frost & Sullivan, the

2. Market Analysis and Business Planning 51

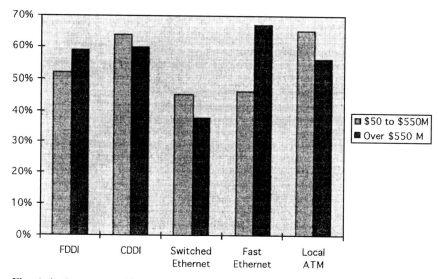

Fig. A.4 Percentage of larger and smaller companies that expect to be evaluating higher speed transmission. Source: Sage Network Research & RFTC, Inc., 1993.

average price of any kind of fiber optic transmitter was $600, whereas the other components cost much less: an eight-fiber cable cost approximately 59¢/ft, an average of $100 for a receiver, an average of $20 for a connector, and an average of $170 for a coupler [6].

Optical communication technologies are being used increasingly for shorter distance communication, where much of the growth is expected and where lower cost components are required. Therefore, the reduction of the transmitter cost is a crucial factor for the growth of the high-performance data communication segment.

A faster local area network (LAN) is of concern to not only large companies but also small companies. According to Fig. A.4, companies earning between $50 and $500 million per year are also considering high-speed LAN solutions, in some cases more aggressively than larger firms.

Fiber and optoelectronic component manufacturers are also expecting a local loop fiber explosion (Table A.5). Deepak N. Swamy, a senior analyst for KMI Corp., predicts that,

By 2002, the U.S. market for fiber optic cable and equipment for loop applications will increase to more than $5 billion. By then, local exchange carriers will have installed 7.9 million local fiber access lines that may each connect hundreds

Table A.5 **The Future of Fiber Applications**

Application	% 1993	% 1999
Long-haul terrestrial	52.3	38.2
Under sea	22.6	7.5
Subscriber loop	14.7	35.3
MAN	4.3	6.8
LAN	6.1	12.2

Note. According to "World Fiber Optic Communication Markets," a report released by Market Intelligence Research Corp., although long-haul fiber will continue to present the majority of total worldwide fiber cable revenues, it is expected to lose market share. Long-haul applications include terrestrial and undersea cabling. The strongest foreseen growth application for fiber optics is the subscriber loop. Its revenues are expected to jump to well over double its present market share by the end of the decade. Other gainers are expected to be LANs and metropolitan area networks (MANs). MANs are already enjoying popularity as high-bandwidth communication links and are expected to continue to do so. Fiber deployment in the LANs is expected to continue increasing to address the growing needs of private network users. Source: Telecommunications, 1993 (September), p. 12.

of customers. Meanwhile, cable television companies will connect more than 40 million subscribers to small-radius nodes [6].

In short, the critical need for this segment is low-cost optoelectronic components able to sustain the increasing transmission of data. The TOSA would fully satisfy this need.

High-speed telecommunications standards are synchronous optical net works (SONET) and synchronous digital hierarchy (SDH). The standard of the data communication segment is Fibre Channel Standard (FCS).

SONET is a North American telecommunication standard for a high-capacity fiber optic transmission system. The transport system is based on the principles of synchronous multiplexing, which offers compatibility with today's data rate and tomorrow's capacity. The traditional speed rates are 51.84, 155.52, and 622 Mb/s, and 2.48832 and 9.95328 Gb/s.

SDH is an international standard similar to SONET, but it uses the terminology transport module to refer to an optical transmission rate.

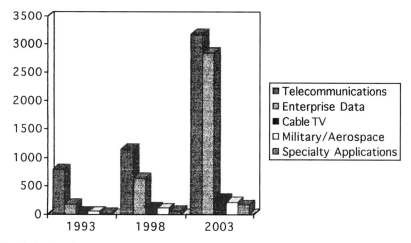

Fig. A.5 North American Consumption of Fiber Optic Optoelectronics. Source: Electronicast Corp., 1993, *Lightwave*.

Fibre Channel is an example of a MAN that is optimized for transfers of large amounts of data between high-performance processors, disk and tape storage systems, and output devices, such as laser printers and graphic terminals. It converts bytes of data into a serial transmission stream at signaling rates of 132.813 and 256.625 Mbauds and 1.0625 Gbauds.

Trends

Although telecommunications remains the main customer of optoelectronic components much of the growth is expected to occur in enterprise data networks (Fig. A.5).

On April 19, 1995, *The Wall Street Journal* stated,

Institutions are rushing to build computer networks to use as competitive tools. . . . Much of the buoyancy in the industry's fortunes also comes from changes in the way corporations use computers. The trend toward hooking computers into vast networks is accelerating, boosting sales of the "server" computers — like Sun's and H.P.'s — that tie networks together as well as the software — such as Novell Inc.'s — that runs on networks. Sun attributed its earning surprise to strong sales in networking and Internet markets.

This trend is confirmed in Table A.6, in which the feeder/local telecom sector is expected to grow by 31.5% every year until 1998.

The following chart shows the forecast revenues for sending high speed data communication over fiber optic cable:

	$(Billion)
1993	0.8
1998	2.6 (Annual growth rate 22.9%)
2003	6.3

Source: Data over fiber, OIDA, 1994, Computer Reseller News, p. 61. Copyright ©1994 by CMP Publications, Inc., 600 Community Drive, Manhasset, NY 11030. Reprinted from Computer Reseller News with permission.

This chart demonstrates that high-speed data communication has a bright future because its revenues are expected to grow at an annual compound rate of 22.9%.

The future of high-speed rate lasers or transmitters is in the short data transfer of the high speed computer interconnects. Distances are on the order of several thousands feet compared with telecommunication's thousands of kilometers. The speed rate determined by the High Speed Committee of the American National Standards Institute (ANSI) will be 500 Mb/s to 1 Gb/s. Video is another promising application for high-speed transmitters, especially CATV.

The use of optical fibers and, therefore, the use of transmitters in subscriber loop systems will continuously increase. Indeed, the optical data

Table A.6 **Forecast Market for Fiber Optics (by Application Sector)**

	Year		Annual growth rate (%)
	1992	**1998**	
Long-haul telecom	1668	1,715	0.5
Feeder/local telecom	1364	7,048	31.5
Multimode	1807	2,053	16.8
Cable TV	198	800	26.2
Other	120	560	29.3
Total	4157	12,176	19.6

Source: KWI Corp., 1993 (December 1), Lightwave, p. 9. Copyright 1993 by Lightwave, PennWell Publishing Co., Nashua, NH, USA.

communications business should experience a tremendous development during the next decade as fiber is extended from the central office further toward business and residential customers and as the use of higher capacity local area networks increases. The different configurations of subscriber loops are the following:

Fiber-to-the-curve (FTTC)
Fiber-to-the-office (FTTO)
Fiber-to-the-home (FTTH)

FTTC and FTTO use metallic cables near subscribers, whereas FTTH brings the optical fiber directly to the home of the subscribers. Residences and small businesses would be the beneficiaries. FTTH is considered to be the final step. However, FTTH is difficult to realize without lowering the cost of optical receivers and transmitters.

Among the optical subscriber loop networks using single-mode fibers, the following different methods of multiplexing are used:

Space-division multiplexing (SDM)
Wavelength-division multiplexing (WDM)
Directional-division multiplexing (DDM)
Time-compression multiplexing (TCM)

SDM is the most expensive because two fibers are needed; one for the upstream transmission and one for the downstream transmission. In WDM, two different wavelengths are performed over a single fiber (usually 1.3 and 1.5 nm) that require two lasers, increasing the cost. In DDM, the laser diode and photodiode are combined by using a coupler. A TCM system is close to DDM, when a diode is in the transmission mode and the other is in the receiving mode ("ping-pong transmission"). The maximum information bit rate in the TCM system is 1.5 Mb/s.

SURVEY

In order to present a clear evaluation of the high-speed data communication segment, a survey has been designed (Fig. A.6). Thanks to the results, we will be able to acquire precious information on the following:

SURVEY

Hello,

Our company is interested in developing an optical transmitting subassembly for use in high-speed fiber datacommunication. We would like to develop a product with the following characteristics:

- Data rates > 1 gigabit/sec, distances without repeaters >20 km.

- Single-mode light 1.3 or 1.55 microns in wavelength, rise and fall times < 0.3 nanoseconds, spectral width < 1 nanometer.

- Most likely applications will be:
    Asynchronous Transfer Mode (ATM) applications at Gb/s speeds
    Fiber Channel standard applications at Gb/s speeds

Would you please take the time to respond to the following questions? Thank you.

I) How would you describe your organization?
a) Telecom Industry
b) Enterprise Datacom Industry
c) Cable TV
d) College/University
e) Government
f) Other_____

II) How many 1 Gbit/s optical subassemblies does your organization currently purchase annually?
a) 0
b) 1-99
c) 100-1000
d) 1000-10,000
e) >10,000

III) How many Gbit optical subassemblies does your organization foresee purchasing in 1996?
a) 0
b) 1-99
c) 100-1000
d) 1000-10,000
e) >10,000

IV) How much would you expect to pay for an optical subassembly as described above?
a) >$5000
b) $3000-$4999
c) $1500-$2999
d) $500-$1499
e) <$500

V) Do you believe that low price implies low quality?
a) Yes
b) No

Fig. A.6 Survey.

VI) Which characteristic of current commercial optical subassemblies are you most dissatisfied with?
a) Speed
b) Price
c) Other_____

VII) Which of the following applications does your organization presently use?
a) FDDI
b) ATM
c) Fiber Channel
d) ESCON
e) Hybrid (explain) _____
f) Other_____

VIII) Which type of fiber conection would you prefer?
a) FC
b) ST
c) SC
d) ESCON
e) Other_____

IX) What wavelength will your organization be using in 1996?
a) 1.55μ
b) 1.3μ
c) $700\text{-}900\text{n}\mu$

X) When do you envision such a transceiver would be desireable for your network?
a) Immediately
b) 1-2 years
c) 3-5 years
d) over 5 years
e) never

XI) From which of the following sources do you receive information on optical transceivers?
a) Professional Journals
b) Personal references
c) Trade Shows
d) Trade Magazines
    d1) Laser Focus World
    d2) Lightwave
    d3) Lasers and Optronics
    d4) Photonics Spectra
    d5) Fiberoptic Product news
    d6) Other_____
e) Other _____

XII) How would you describe your position within your organization?
a) Product Development
b) Purchasing
c) Sales
d) Management
e) Research
f) Other_____

Fig. A.6 *Continued*

Size of the target market
Critical needs
Price level
Weight of each application (FDDI, ATM, Fiber Channel, ESCON, etc.)
Media to use to reach this segment
Measure of the desirability of the TOSA
Forecasted sales
Potential customers

Competition

COMPANIES CONTACTED

Approximately 25 companies in the United States were identified as manufacturers of digital transmitters in the *Thomas Register* and the buyer's guide 1995 from *Laser Focus World* magazine. Receiving catalogs and price lists of their products will enable a clear identification of the competition by product line. The following is a list of the companies contacted:

Name of Company	*Phone No.*
Advanced Fiber Optic Technologies (CA)	818-357-0159
Analog Modules Inc. (FL)	407-339-4355
AT&T Microelectronics (PA)	610-712-5133
Broadband Communications Products Inc. (FL)	407-728-0487
Fiber Optic Center (MA)	800-473-4237
Fiber Options Inc. (NY)	516-567-8320
Force Inc. (VA)	703-382-0462
Hewlett-Packard Co. (DE)	800-545-4306
Laser Diode Inc. (NJ)	908-549-9001
Lasertron Inc. (MA)	617-272-6462
Litton Poly-Scientific (VA)	703-953-4751
Math Associates (NY)	516-226-8950
MRV Technologies (CA)	818-773-9044
NEC Electronics Inc. (CA)	415-960-6000
Optical Communication Products Inc. (CA)	818-701-0164
SI Tech (IL)	708-232-8640

Name of Company	Phone No.
United Technologies Photonics (CT)	203-769-3000
Ross Engineering (CA)	800-654-3205
Philips Broadband Networks (NY)	800-448-5171
Telecommunication Techniques Corp. (MD)	800-638-2049
Telco Systems (MA)	617-551-0300
LNR Communication, Inc. (NY)	516-273-7111
Fiber Com, Inc. (VA)	800-537-6801
Optek Technology Inc. (TX)	214-323-2301

NEW PRODUCTS

In this very high-technology market, it is important to be aware of the new projects undertaken by other companies that could create either direct or indirect competition.

Scientists at Bell-Northern Research (BNR), the R&D arm of Northern Telecom, have developed a low-cost optoelectronic device. Known in scientific circles as a Mach–Zehnder (MZ) optical modulator, made of III–V semiconductor, it has a speed rate of 10 Gb/s. In a BNR test, scientists used the semiconductor MZ prototype to successfully maintain 10 Gb/s transmission over conventional glass fiber at distances exceeding 100 km with minimal distortion or signal loss. The BNR model is less than 1% of the size of existing units. Its semiconductor design requires less power and is more resistant to temperature and vibration than other units. BNR's MZ also has the potential to be economically mass produced [7].

At Supercomputing '94, on November 15, the IBM Research Division demonstrated its new Rainbow 2 all-optical network. The network prototype can support 32 nodes, each running at 1 Gb/s. IBM said that,

> They are developing low-cost high-speed networks to answer increasing demands in scientific, medical, university and other research environments for networks with one Gbps per node throughout rates.... Multimedia applications, in particular, will take advantage of this speed in areas such as full-motion video and supercomputer visualization.... University and other research environments already need local area networks and metropolitan area networks that will provide supercomputer support of high-performance workstations with live-motion color graphics and supercomputer visualizations. Medical imaging is another important applications [8].

Position

The TOSA uses vertical cavity surface-emitting laser (VCSEL) technology, which has become, according to the OIDA, a noticeable technology in data communication:

The Fibre Channel Systems Initiative (FCSI) and the Fibre Channel Association (FCA) announced on March 14, 1995 a major leap in possible data communication speeds with the ANSI Committee adoption of Fibre Channel standards for 2 and 4 Gbps data rates. This effectively quadruples the previous Fibre Channel ANSI standard of up to 1 Gbps and provides the standards for the fastest data communication speed possible to date. This major advancement in data communication speed is made possible through Fibre Channel's enabling implementation of Vertical Cavity Surface Emitting Laser (VCSEL) technology, a new technology that provides a practical, cost-effective means of using lasers to transmit data at ultra-high speeds. The ultra-high data transfer speed is ideal for such applications as motion picture and video production where large amounts of audio and other digitized information is manipulated [9].

VCSELs offer several important advantages over conventional edge-emitting visible diodes, including surface-normal output, ease of fabrication into two-dimensional arrays, a circular beam with little angular divergence, wafer-level testing and additional control over the lasing wavelength. Visible VCSELs may be an enabling technology for many advanced applications such as plastic-fiber-based communications for LANs, 2-D visible arrays for displays (including laser-projection), printing applications and optical memory [10].

According to an article published in *Photonics Spectra* in February 1995, VCSELs appear to be the "key technology for future data communications." It was first introduced on the market in 1992 by Photonics Research Inc. and Bandgap Technology Corp., both of which later became a single company called Vixel Corp.

VCSELs offer many advantages:

Easy to manufacture
Very low cost
High performance
Multichannel lightwave transmitters
Compact

Low-divergence circular beams
High wall-plug efficiency
Low drive current of a few milliamps
Low drive voltage of a few volts

The primary applications of VCSELs are the following:

Fiberoptic communication
Optical storage
Laser printing
Laser scanning
Optical sensing

Applications of VCSELs in data communications include:

Massive parallel processing
Interconnections of workstations and high-performance PCs and video file servers
Premises switching
Optical LANs
Inter- and intracabinet switching and Fiber Channel switching

VCSELs lower the cost and increase the performance of optoelectronic devices. Their wavelength ranges from 630 to 1050 nm and is expected to increase from 480 to 1300 nm. The primary function of VCSELs is the transport of information at high speed over single or multiple channels in which the use of this technology will expand. They fulfill the needs created by the high-performance data communication segment.

Conclusion

This analysis shows that the fiber optic industry will experience large growth during the next decade. The North American consumption of fiber optic components is expected to increase at an annual rate of 20% in the period leading up to 2003. The global market provides a similar perspective with countries such as China, Brazil, Mexico, or the Asian Pacific countries. Although telecommunications remains the main customer of this industry, the most rapid growth will occur in data communications. By segmenting the overall market, we realize that the high-performance data communication segment shows a critical need: a reduction in the cost of optoelectronic interfaces. Indeed, the current trend is toward the local networks that require lower cost in order to be profitable. This need could be satisfied by the low-cost TOSA.

The market analysis is certainly not finished. The results of the survey will enable us to measure the desirability of this product and give us valuable information on this particular market segment. However, personal interviews and focus groups will need to be conducted. Personal interviews will include a discussion with professionals in the field or with potential customers, whereas a focus group will give many reactions to the presentation of the concept. A study of the competition and an evaluation of the regulatory restrictions are other aspects of the market analysis that should be completed.

References

1. *Lightwave.* January 1994. p. 15. Nashua, N.H.: Lightwave/PennWell Corp.
2. *Standard & Poor's.* October 1994. p. T43. Reprinted by permission of Standard & Poor's, a division of The McGraw-Hill Companies.
3. *Lightwave.* December 1993. p. 42. Nashua, NH: Lightwave/PennWell Corp.
4. *U.S. Industrial Outlook.* 1994.
5. Optoelectronics Industry Development Association. 1994. *Optoelectronic technology road map.*
6. *Photonics Spectra.* 1994.
7. *Canada Newswire.* 1994.
8. *Business Wire.* 1994.
9. *OIDA.* Edge Publishing. 1995. No. 347.
10. *Photonics Spectra.* 1993.
11. Kotler, P., and Armstrong, G. *Principles of Marketing,* 6th ed. Englewood Cliffs, N.J.: Prentice Hall.

Chapter 3 | Small Form Factor Fiber Optic Connectors

John Fox
ComputerCrafts, Inc., Hawthorne, New Jersey 07507

Casimer DeCusatis
IBM Corporation, Poughkeepsie, New York 12601

In this chapter, we will describe the similarities and differences between four of the industry's Small Form Factor connectors. Although there are certainly more than four connectors that have been developed to meet the criteria of a smaller form optical interface, the MT-RJ, SC-DC, VF-45, and LC connectors are the focus of this chapter based on their popularity and general acceptance across the industry. The connectors' physical characteristics, coupling techniques, and optical performances will be discussed in detail in order to provide a comprehensive comparison of these predominant connector types.

3.1. Introduction

The natural progression of technology will inevitably drive new product designs to be faster, smaller, and less expensive. This is also the case in the world of fiber optic interconnects, and a new generation of connectors, collectively known as Small Form Factor connectors, have now been developed with these goals in mind. Although the MT-RJ, SC-DC, VF-45, and LC connectors have all significantly reduced the size and cost of optical interfaces, in comparison to the standard SC duplex connector, each design approach is unique and therefore there are significant physical and functional differences that can make the connectors application dependent.

Table 3.1 **EIA/TIA Proposed SFF Connector Standards**

Connector type	FOCIS document number	FOCIS author*
MT-RJ	12	Amp
SC-DC[a]	11	Siecor
LC	10	Lucent
SG (VF-45[b])	7	3M
Fiber Jack	6	Panduit

Note. The authoring company listed does not necessarily support or manufacture only one connector type, nor does this table include all of the supporters or manufacturers of each connector type.

[a] SC-DC is a trademark of Siecor.
[b] VF-45 is a trademark of 3M.

Various types of next-generation SFF optical interfaces have been proposed to the Electronics Industry Association/Telecommunications Industry Association (EIA/TIA), for inclusion in developing standards such as the Commercial Cabling Standard TIA-568-B. While no single connector has been selected for inclusion in this standard, it requires that connectors be defined by a reference document called a Fiber Optic Connector Intermatability Standard (FOCIS), which defines the connector geometry so that the same connector build by different manufacturers will be mechanically compatible. The EIA/TIA also requires connectors to meet some minimal performance levels, independent of connector design; test methodologies are defined by the EIA/TIA Fiber Optic Test Procedures (FOTPs). Other industry specifications are also relevant to these connectors; for example, Bellcore spec. GR-326-CORE defines fiber protrusion from a ferrule to ensure physical contact and prevent back reflections in the connector. The relevant FOCIS documents defined for connectors that have currently been proposed to the TIA are given in Table 3.1.

3.2. MT-RJ Connector

The MT-RJ connector utilizes the same rectangular plastic ferrule technology as the MTP array-style connector first developed by NTT, with a single ferrule body housing two fibers at a 750-um pitch (Fig. 3.1). These ferrules are available in both single-mode and multimode tolerances, with the lower-cost multimode version typically comprised of a glass-filled thermoplastic and the critically tolerance single-mode version comprised of a glass-filled thermoset material. Unlike the thermoplastic multimode ferrules, which can

3. Small Form Factor Fiber Optic Connectors

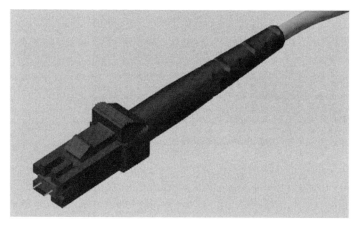

Fig. 3.1 Standard MT-RJ male connector.

be manufactured using the standard injection mold process, the thermoset single-mode ferrules must be transfer molded, which is generally a slower but more accurate process.

By design the alignment of two MT-RJ ferrules is achieved by mating a pair of metal guide pins with a corresponding pair of holes in the receptacle (Fig. 3.2). This feature makes the MT-RJ the only Small Form Factor connector with a distinct male and female connector. As a general rule, wall outlets, transceivers, and internal patch panel connectors will retain the guide pins (thus making their gender male) and the interconnecting jumpers will have no pins (female). In the event that two jumper assemblies require mating mid-span a special cable assembly with one male end and one female end must be used. However, some unique designs do allow

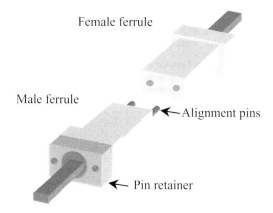

Fig. 3.2 MT-RJ ferrule alignment method.

Fig. 3.3 Two fiber ribbon construction.

for the insertion and extraction of guide pins in the field, affording the user the ability to change the connector's gender as required.

Latching of the MT-RJ connector is modeled after the copper RJ-45 connector, whereby a single latch arm positioned at the top of the connector housing is positively latched into the coupler or transceiver window. Although this latch design is similar in all the MT-RJ connector designs, individual latch pull strengths may vary depending on the connector material, arm deflection, and the relief angles built into the mating receptacles. For this reason it is recommended to evaluate connector pull strengths as a complete interface, depending on the specific manufacturer's connector, coupler, or transceiver design, as the coupling performances may vary.

MT-RJ connectors are typically assembled on 2.8-mm round jacketed cable housing two optical fibers in one of three internal configurations. The first construction style consists of the two optical fibers encapsulated within a ribbon at a 750-um pitch (Fig. 3.3). This approach is unique to the MT-RJ connector and designed specifically to match the fiber spacing to the pitch of the ferrule for ease of fiber insertion. Although this construction style may be ideal for a MT-RJ termination, it can cause some difficulty when manufacturing a hybrid assembly, and availability may also be an issue based on its uniqueness. A second design which is more universal, utilizes a single 900-um buffer to house two 250-um fibers (Fig. 3.4). This construction is more conducive to hybrid cable manufacturing but the fibers

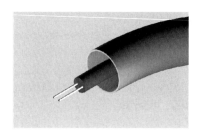

Fig. 3.4 Dual 250 micron construction.

3. Small Form Factor Fiber Optic Connectors 67

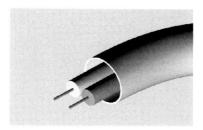

Fig. 3.5 Dual 900 micron construction.

will naturally maintain a 250-um pitch, thus making fiber insertion rather difficult. The third design is considered a standard construction and is used across the industry (Fig. 3.5). In this configuration each individual fiber is buffered with a PVC coating. The coating thickness is typically 900 um, but as in the previous case this does cause a mismatch of the fiber to ferrule pitch. To compensate for this, some connector designs incorporate a fiber transition boot, which gradually reduces the fiber pitch to 750 um, while others simply use a non-standard buffer coating of 750 um.

In general the assembly and polish of the MT-RJ factory-style connector is considerably more difficult than the other small form factor connectors. Typical MT-RJ designs have a minimum of eight individual components that must be assembled after the ferrule has been polished, allowing for a number of handling concerns (Fig. 3.6). As with the case of most MT-style

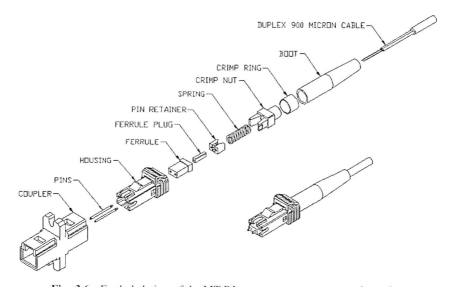

Fig. 3.6 Exploded view of the MT-RJ connector components and coupler.

ferrules, the perpendicularity or flatness of the ferrule endface with reference to the ferrule's inner shoulder is critical and this cannot be accomplished if the connector is pre-assembled. Another unique requirement of the MT-RJ polish involves fiber protrusion. Although the ferrule endface is considered to be flat, depending on the polishing equipment, fixtures, and even contamination some angularity may occur. Therefore it is recommended that the fibers themselves protrude 1.2 to 3.0 ums from the ferrule surface in order to guarantee fiber-to-fiber contact.

For the reasons previously mentioned, the factory-style MT-RJ connector is not a good candidate for field assembly and polishing, so a number of similar yet unique Field Installable Connectors have been developed. All of these field solutions utilize a pre-polished ferrule assembly, which is mated with two cleaved fibers and aligned by v-grooves, with the entire interface filled with index matching gel to compensate for any possible air gaps. One of the major differences between the various solutions is in the mechanism used to open and close the spring clip that maintains constant pressure on the two sandwiched halves of the channeled interface. In one design a cam, which is integrated into the connector body, is used to separate the halves while in other designs a separate hand tool is required. The other general difference between the field solutions is the application design — a number of the connectors are designed to be just that — a field connector with a distinct gender that can be terminated onto distribution style cable — while others are designed to be male receptacles only that must be wall or cabinet mounted. Because of the inherent difference between the MT-RJ fields solutions one solution may be better suited to a given application than another.

3.3. SC-DC Connector

The SC-DC (dual contact) or SC-QC (Quattro contact), developed by Siecor, has a connector body design resembling a SC simplex connector with a round thermoset molded ferrule that is capable of handling either two or four optical fibers (Fig. 3.7). The connector is designed to support both single-mode and multimode cabling applications. However, the SC-DC is one of the few connectors that is used exclusively for cable interconnecting and therefore has no transceiver support.

As in the case of the MT-RJ connector the SC-DC fibers are on a 750-um pitch while the SC-QC fibers are on a 250-um pitch. The same ferrule and

3. Small Form Factor Fiber Optic Connectors

 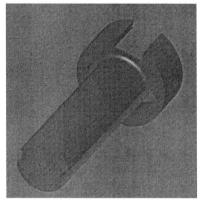

Fig. 3.7 Siecor SC-DC/SC-QC connector and ferrule assembly.

housings are used in both connectors so the only feature that distinguishes between the two is simply the number of fibers used. The four ferrule holes of approximately 126-um diameter are placed on 250-um centers along the ferrule center line to form a SC-QC connector. By using only the two outermost ferrule holes and leaving the inner two empty the SC-DC is created.

Although the ferrule composition is very similar to that used across the MT technologies, the geometry and alignment methods are very different. The SC-DC ferrule has a standard round shape with a 2.5-mm diameter, but unlike its ceramic counterparts there are two semicircular grooves with a 350-um radius positioned along each side of the ferrule at a 180° interval. This feature provides the ferrule-to-ferrule alignment when mated with the corresponding ribs of the coupler. The design of this feature within the couplers is different for a single-mode and multimode connection. The multimode coupler uses a one-piece all-composite alignment insert with molded ribs, while the single-mode coupler incorporates two precision alignment pins captured inside the sleeve (Fig. 3.8). By design the ribs of the coupler and grooves in the ferrule are on the same 2.6-mm pitch as the pins of a male MT-RJ connector to allow for possible hybrid mating of the two connector types.

The latching mechanism of the SC-DC connector is the same push-pull style used on the industry standard SC simplex connector and the outer housing dimensions are identical. Because the two connectors physically appear the same but are not functionally interchangeable, the housing alignment key of the SC-DC is offset to help distinguish between the two

Fig. 3.8 Siecor SC-DC/SC-QC multimode and single-mode coupling devices.

connector types and prevent any confusion in the field. By following the same basic footprint as the SC connector this new SFF optical connector already has a familiar look and feel with proven plug reliability.

SC-DC connectors and cable assemblies are solely manufactured and sold by Corning, so the variability of design and cable material is not an issue as is the case with other SFF connectors. The connectors are typically assembled onto 2.8-mm round jacketed fiber incorporating a single-ribbon fiber populated with either two or four fibers, which assists in the alignment of the fiber to the ferrule holes. The internal ferrule geometry can accommodate the standard single-fiber cables and this may be an option for the fabrication of hybrid cable assemblies.

As previously stated the size and shape of the SC-DC ferrule is similar to the standard ceramic and therefore can be polished in much the same manner and still produce endface geometries akin to the MT-RJ. A flat endface polish is the desired result, much like with the MT-style ferrules, however the perpendicularity is referenced to the sides of the ferrule rather than to an internal feature, so the SC-DC connector can be pre-assembled and polished as a complete connector. The ability to pre-assemble a connector significantly reduces complexity of manufacturing and typically results in better yields.

The SC-DC connector is available in a field installable version, the SC-DC UniCam, which utilizes pre-polished fiber stubs much like other SFF solutions. Alignment of the fiber stubs to the in-field cleaved fiber is achieved through a gel-filled mechanical splice element. The splice element is opened and closed with a mechanical cam and is retained within a connector housing. Termination of the standard SC-DC connector can also be accomplished in the field by using conventional equipment and methods much like the field termination of the SC-style connector.

3.4. VF-45 Connector

The VF-45 connector, developed by 3M, is perhaps the most innovative SFF connector design, in that it eliminates the need for precision ferrules and sleeves altogether. The overall look and feel of the "Plug-to-socket" design closely resembles the standard telephony RJ-45 "Connector-to-jack" system whereby the cable assembly mates directly to a terminated socket, reducing the need for couplers (Fig. 3.9). Although this concept has been around for years in the copper industry, the creation of a bare fiber optical interface, using alignment grooves and no index matching gels, requires some revolutionary techniques.

The VF-45 connector incorporates two 125-um optical fibers, suspended in free space on a 4.5-mm pitch protected by a RJ-45 style housing with a retractable front door designed to protect the fibers. The connector design supports both single-mode and multimode tolerances by relying on the inherent precision of the optical fibers within the two injected molded v-grooves of either the transceiver or a VF-45 socket. The design of the interconnect allows the natural spring forces of the optical fibers to align the fibers within the v-grooves as well as ensuring physical fiber-to-fiber contact.

Because of the uniqueness of this interconnect the geometry of the endface polish of both the plug and receptacle fibers has been modified to provide optimum performance. The VF-45 optical connection relies on the spring force created by the bowing of the optical fibers to provide a physical contact force of approximately 0.1 N and this force coupled with an 8-degree angle polished endfaces produces the optimum connection and return loss results. The tips of the plug fibers are also beveled at 35 degrees,

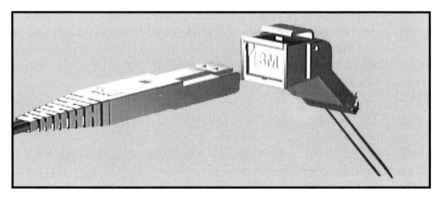

Fig. 3.9 3M VF-45 "Connector-to jack" system.

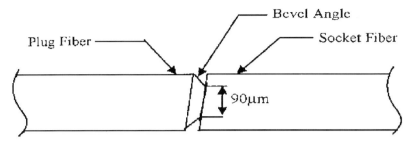

Fig. 3.10 VF-45 plug fiber to socket fiber interface.

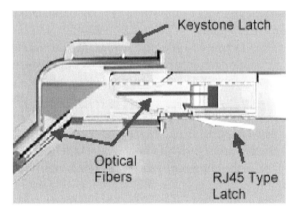

Fig. 3.11 Insertion of the VF-45 plug.

allowing a 90-um contact area and providing a relief for the fiber to slide into the v-grooves with no damage to the core region (Fig. 3.10). This chamfer is not required on the receptacle fibers since they remain stationary while the plug fibers may have to endure multiple insertions (Fig. 3.11). As previously mentioned, the contact force created at the optical interface directly influences the optical performance of the plug-socket connection. This downward compressive force is generated when the two fibers of the plug engage with the resident fibers of the socket and cause a slight "bow" in the plug fibers (Fig. 3.12). Because of the constant stress on these fibers, long-term reliability on standard optical fibers became a concern and therefore a specialized high-strength optical fiber was developed for this application, called GGP (glass-glass-polymer) fiber. GGP fiber consists of 100-um glass fiber with a polymeric coating applied to bring the outer diameter to 125 um. By reducing the outer diameter of the glass the tensile

3. Small Form Factor Fiber Optic Connectors 73

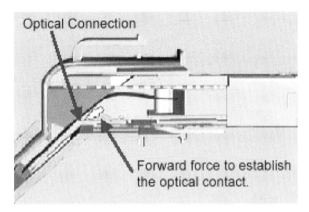

Fig. 3.12 VF-45 optical connection.

stress on the fiber is minimized and the additional coating also provides protection against abrasions from the v-grooves and reduces the chance of damage to the glass during the mechanical stripping process used in the termination process.

The factory termination of the VF-45 jumper plugs is considerably different than the conventional ferrule-based connectors. The process of threading a 125-um fiber into a precision ferrule hole filled with epoxy is now eliminated and replaced with a mechanical fiber holder that grips the fibers in place. The fibers are then cleaved and polished to the endface geometry previously described and the cable strain relief slid into place. The fibers and holder are then placed into a protective shroud and the front door cover installed. The relative simplicity of this manufacturing process makes the VF-45 connector one of the best candidates for a fully automated production line.

The socket of the VF-45 was specifically designed for termination in the field with minimal effort and training. After preparing the fibers for termination by removing outer buffer material, they are inserted into a mechanical fiber holder that retains the fiber by gripping them inside a deformable aluminum crimp. The fibers are then cleaved and hand-polished to an 8-degree angle with a slight radius generated by the durometer of the polishing pad. The fiber holder with the polished fibers is guided into the socket v-grooves and the housing plate snapped into place (Fig. 3.13). Although this method of field termination does vary from the other pre-polished SFF connectors, the total termination time and the complexity of the process is very similar.

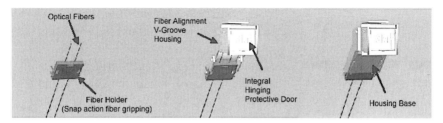

Fig. 3.13 Field termination of the 3M VF-45 socket.

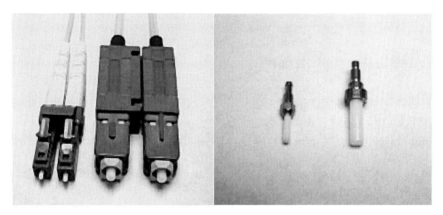

Fig. 3.14 LC duplex connector and ferrule compared with a SC duplex connector and ferrule.

3.5. LC Connector

The LC connector developed by Lucent Technologies is a more evolutionary approach to achieving the goals of a SFF connector. The LC connector utilizes the traditional components of a SC duplex connector having independent ceramic ferrules and housings with the overall size scaled down by one half (Fig. 3.14). The LC family of connectors includes a stand-alone simplex design; a "behind the wall" (BTW) connector; and the duplex connector available in both single-mode and multimode tolerances, all designed using the RJ-style latch.

The outward appearance and physical size of the LC connector varies slightly depending on the application and vendor preference. Although all the connectors in the LC family have similar latch styles modeled after the copper RJ latch, the simplex version of the connector has a slightly longer body than either the duplex or BTW version and the latch has an additional latch actuator arm that is designed to assist in plugging as well to

3. Small Form Factor Fiber Optic Connectors 75

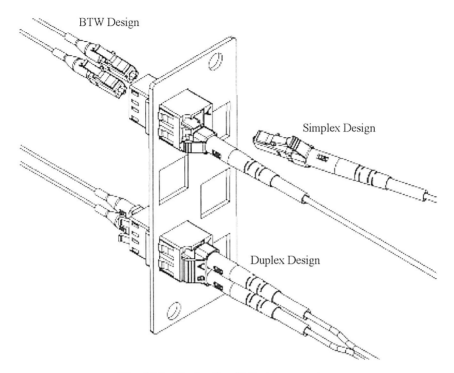

Fig. 3.15 The family of LC-style connectors.

prevent snagging in the field. The BTW connector is the smallest of the LC family and designed as a field or board mountable connector using 900-um buffered fiber and in some cases has a slightly extended latch for extraction purposes. The duplex version of this connector has a modified body to accept the duplexing clip that joins the two connector bodies together and actuates the two latches as one (Fig 3.15). Finally, even the duplex clip itself has variations depending on the vendor. In some cases the duplex clip is a solid one-piece design and must be placed on the cable prior to connectorization, while other designs have slots built into each side to allow the clip to be installed after connectorization. In conclusion, all LC connectors are not created equal and depending on style and manufacturer's preference there may be attributes that make one connector more suitable for a specific application than another.

The LC duplex connector incorporates two round ceramic ferrules with outer diameters of 1.25 mm and a duplex pitch of 6.25 mm. Alignment of these ferrules is achieved through the traditional couplers and bores using precision ceramic split or solid sleeves. In an attempt to improve the optical performance to better than 0.10 db at these interfaces most of the ferrule and backbone assemblies are designed to allow the cable manufacturer to tune them. Tuning of the LC connector simply consists of rotating the ferrule to one of four available positions dictated by the backbone design. The concept is basically to align the concentricity offset of each ferrule to a single quadrant at 12:00; in effect, if all the cores are slightly offset in the same direction, the probability of a core-to-core alignment is increased and optimum performance can be achieved. Although this concept has its merits, it is yet another costly step in the manufacturing process and in the case where a tuned connector is mated with an untuned connector, the performance increase may not be realized.

Typically the LC duplex connectors are terminated onto a new reduced-size zipcord referred to as mini-zip; however, as the product matures and the applications expand it may be found on a number of different cordages. The mini-zip cord is one of the smallest in the industry with an outer diameter of 1.6 mm compared with the standard zipcord for SC style product of 3.0-mm. Although this cable has passed industry standard testing there are some issues raised by the cable manufacturers concerning the ability of the 900-um fibers to move freely inside 1.6-mm jacket and others involving the overall crimped pull strengths. For these reasons some end users and cable manufactures are opting for a larger 2.0-mm, 2.4-mm or even the standard 3.0-mm zipcord. In applications where the fiber is either protected within a wall outlet or cabinet the BTW connector is used and terminated directly onto the 900-um buffers with no jacket protection.

The factory termination of the LC cable assemblies is very similar to other ceramic-based ferrules using the standard pot and polish processes with a few minor differences. The one-piece design of the connector minimizes production handling and helps to increase process yields when compared with other SFF and standard connector types. Because of the smaller diameter ferrule, the polishing times for an LC ferrule may be slightly lower than the standard 2.5-mm connectors, but the real production advantage is realized in the increased number of connectors that can be polished at one time in a mass polisher. For the reasons mentioned above and the fact that the process is familiar to most manufacturers, the LC connector may be considered one of the easiest SFF connectors to factory terminate.

Field termination of the LC connector has typically been accomplished through the standard pot and polish techniques using the BTW connector. However, a pre-polished, crimp and cleave connector is also available. The LCQuick Light field-mountable BTW-style connector made by Lucent Technologies is a one-piece design with a factory polished ferrule and an internal cleaved fiber stub. Unlike other pre-polished SFF connectors previously discussed, the LCQuick light secures the inserted field cleaved fiber to a factory polished stub by crimping or collapsing the metallic entry tube onto the buffered portion. This is accomplished by using a special crimp tool designed not to damage the fibers. However, this means the installer has but one chance for a good connection. The LCQuick light is designed specifically for use in protected environments such as cabinets and wall outlets and has no provision for outer jacket or Kevlar protection.

3.6. Other Types of SFF Connectors

Following a renewed interest in fiber optic cabling and transceiver footprint reduction, there have been many types of small form factor optical interfaces proposed. Some of these are no longer widely used; for example, the mini-MT connector was originally proposed as a 2-fiber version of the MTP connector, but was largely supplanted by the MT-RJ design. In this chapter, we have concentrated on optical interfaces for next-generation transceivers and cable plant infrastructures, and have limited our treatment so as not to include the MTP/MPO connectors (see Vol. 1, Chapter 2), the SMC parallel optical connector (see Vol. 1, Chapter 2), or other emerging multi-fiber interfaces for Infiniband. However, there are some other SFF interfaces currently in use besides the 4 major types discussed earlier; we will briefly describe them here.

The Fiber-Jack (FJ) is the non-trademarked name for the Opti-Jack connector developed by Panduit Corporation. As shown in Fig. 3.16, this connector incorporates two industry standard SC duplex ceramic ferrules, each 2.5 mm diameter. However, the spacing between ferrules has been

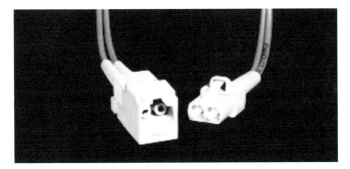

Fig. 3.16 Fiber-Jack.

reduced from 1.27 mm (0.5 inches) as in a standard SC Duplex connector to only 0.63 mm (0.25 inches). The ferrules are independently spring loaded, and are aligned in a receptacle by standard split-sleeve mechanical techniques. While this simplifies the connector design, it also means that the connector must incur the full cost of 2 ceramic ferrules; how this will eventually compare with other SFF ferrule costs has yet to be determined. The connector latch is modeled after the industry standard RJ-45 wall jack, and has found many initial applications in building wiring to wall outlets. The connector is available in both single-mode and multimode versions, which preserve the TIA industry standard color coding on the plug body and the termination cap on the jack. In addition, the FJ connector is available with color coding to identify different networks, applications, areas of the building, or portions of the cable infrastructure, to facilitate network adminstration. Following the category 5 wiring conventions, the FJ housing and plugs may be color coded in black, blue, gray, orange, red, beige, or white. The FJ was also among the first SFF connectors to be specified by a TIA FOCIS document for use with plastic optical fiber; because it employs standard ceramic ferrules, the FJ can accommodate plastic fiber with an outer diameter up to 1 mm. The FJ supports standard duplex jumper cables, couplers, and adapters; while FJ transceivers are not widely available, some development work in this area has been reported.

Another type of SFF connector, developed by NTT to serve both as a standard fiber optic patch cable and as an optical backplane interface, is the multi-termination unibody or MU connector. This connector is also available from various sources under slightly different names; for example, the version manufactured by Sanwa Corporation is called the SMU. As shown in Fig. 3.17, the basic MU connector is a simplex design measuring 6.6 mm wide and 4.4 mm high, with a center-to-center spacing of 4.5 mm in duplex or multi-fiber applications. It has been standardized by IEC 61754-6, "Interface standards type MU connector family," in 1997; other standard bodies including JIS and IEEE 1355 Heterogeneous Interconnect (a future bus architecture) have endorsed the MU as well. A backplane version of the MU is available, which measures 13 mm wide and 42 mm high. Its small size is achieved by using a ceramic ferrule 1.25 mm in diameter, roughly half the size of a standard SC connector ferrule. Consequently, a different type of physical contact polishing was developed to accommodate the smaller ferrules. Smaller diameter zirconia split-sleeves have also been developed to support duplex couplers, adapters, and similar jumper cable applications. A self-retention mechanism is employed, similar in design to a miniature push-pull SC latch; indeed, the MU is sometimes referred to as

3. Small Form Factor Fiber Optic Connectors 79

Fig. 3.17   MU.

a "mini-SC" connector because of the similarities in look and feel. This is consistent with the intended applications, as it permits blind mating of the connector in a printed circuit board backplane more readily than an RJ-45 style latch. The MU has found some applications as a front panel patch cable on optical switches or multiplexers that have a large amount of fiber connections; these interfaces are expected to be incorporated into optical backplanes on next generation equipment. The plug and jack structure of MU connectors can be easily cascaded to produce multi-fiber parallel interface designs, it has been suggested that future designs using the MU could accommodate hundreds or even thousands of fibers when used in this configuration. Transceivers for the MU interface are not yet widely available, although some development work is under way in this area.

3.7.   Transceivers

As noted earlier, optical transceivers are widely available from multiple vendors for the MT-RJ and LC interfaces, and to a lesser degree for the VF-45 and FJ interfaces. While there is some ongoing development of transceivers for the other SFF interfaces, they are not widely available at this time. The industry trend has been to adopt the MT-RJ interface for low data rate (below 1 Gbit/s), multimode applications, and the LC for high data rate (above 1 Gbit/s) applications, both single-mode and multimode (mainly using SX transceivers). Transceiver development has been facilitated by ad hoc industry standards or multi-source agreements (MSAs) which govern transceiver package dimensions, electrical interfaces and host board layouts, card bezel design, mechanical specifications (including insertion, extraction, and retention forces), and transceiver labeling. The original MSA for SFF transceivers was supported by 15 companies including Agilent, IBM, Lucent, Siemens/Infineon, Amp/Tyco, and others. It defined a pin through hole device with 2 rows of 5 pins each, using the signal definitions listed in Table 3.2.

Recently, a second MSA has been approved by the member companies, which defines a pluggable transceiver that mates with a surface mountable card receptacle. These small form factor pluggable (SFP) transceivers make it possible to change the optical interface at the last step of card manufacturing, or even in the field, to accommodate different connector interfaces or a mix of SX and LX transceivers. This should make it easier to adjust optical interface characteristics on future system designs, in much the same way that the GBIC transceiver did for the SC duplex interface (in fact, the SFP is sometimes known as a "mini-GBIC"). The SFP has twenty signal connections and provides three additional functions in addition to the original 10 SFF signal pins; these new functions include module definition pins which specify a serial ID indicating the type of transceiver function (such as LX vs SX transmitters), a data rate select function (such as 1 Gbit/s vs 2 Gbit/s), and a transmitter fault signal. The signal definitions are provided in Table 3.2.

A metal receptacle, sometimes called a cage or a garage, is surface mounted to the printed circuit board to accept the pluggable transceivers. In addition to providing easy replacement and reconfiguration of the transceiver interface, this offers several other advantages. The transceiver is often the only pin through hole component on a modern card design; the SFP cage allows the elimination of extra manufacturing processing steps and potentially reduces cost. By removing the optical components from the soldering process, the SFP should provide improved reliability of the optics, and permit the use of higher soldering temperatures (this may be important as future lead-free solders with higher process temperatures are required by environmental regulations).

3.8. SFF Comparison

Conventional duplex fiber optic connectors, such as the SC duplex defined by the ANSI Fibre Channel Standard [1], achieve the required alignment tolerances by threading each optical fiber through a precision ceramic ferrule. The ferrules have an outer diameter of 2.5 mm, and the resulting

Table 3.2 Comparison of SFF Connector Features

	LC	MT-RJ	SC-DC[viii]	VF-45[ix]
Fiber spacing	6.25 mm	0.75 mm	0.75 mm	4.5 mm
# of ferrules	2	1	1	0
Ferrule material	Ceramic	Plastic	Plastic	None
Alignment	Bore and ferrule	Pin and ferrule	Rail and ferrule	V-groove
Ferrule size	ϕ 1.25 mm	2.5 mm $\times$ 4.4 mm	ϕ 2.5 mm	None
Trx opening: (width $\times$ height $\times$ length)	11.1 mm $\times$ 5.7 mm $\times$ 14.6 mm	7.2 mm $\times$ 5.7 mm $\times$ 14 mm	11 mm $\times$ 7.5 mm $\times$ 12.7 mm	12.1 mm $\times$ 8 mm $\times$ 21 mm
Fiber cable	duplex	duplex or ribbon	duplex or ribbon	GGP polymer coated
Field term: Plug	pot and polish	pre-polished stub	pre-polished stub	Not Available
Field term: Socket	plug + coupler	plug + coupler and socket	plug + coupler	cleave and polish socket
Latch	RJ-top 2 latch coupled	RJ-top latch	SC push pull	RJ-top latch

fiber-to-fiber spacing (or pitch) of a duplex connector is approximately 12.5 mm. Because the outer diameter of an optical fiber is only 125 μm, it should be possible to design a significantly smaller optical connector. Smaller connectors with fewer precision parts could dramatically reduce manufacturing costs and have the potential to open up new applications such as fiber to the desktop. Smaller connectors and transceivers would also permit more ports to be added to enterprise servers, fiber optic switches, and communications equipment without increasing the size and cost of these devices[2]. Recently, a new class of small form factor (SFF) fiber optic connectors has been introduced with the goal of reducing the size of a fiber optic connector to one-half that of an SC Duplex connector while maintaining or reducing the cost[4], namely the LC[5], MT-RJ[6], SC-DC[7], and VF-45[8].

Table 3.2 gives a comparison of the different features of the four major SFF connectors. A brief description of each connector and its alignment method is given, followed by a discussion of the distinguishing characteristics and their impact on the connector and transceiver.

There are several different design approaches to reducing the dimensions of a fiber optic connector. One approach is to use a single ferrule with multiple fibers; this is the concept behind the SC-DC and MT-RJ connectors. The SC-DC (dual connect) and SC-QC (quad connect) use a standard SC connector body and latching mechanism with an offset key, but a new round plastic ferrule design that incorporates either 2 fibers (750-mm pitch) or 4 fibers (250-mm pitch) in a linear array. Alignment is provided by semicircular grooves in the sides of the SC-DC ferrule, which mate with corresponding ribs in the receptacle. This connector has been used by IBM Global Services as part of the Fiber Quick Connect system; it is currently limited to applications in patch panels and the cable infrastructure. The most radical, and innovative, approach for a smaller connector is to eliminate ferrules altogether; this is the case for the VF-45 connector. In this connector, a pair of optical fibers is aligned using injection-molded thermoplastic v-grooves; the fibers are cantilevered in free space on 4.5-mm pitch, and protected by the connector outer body. When plugged into a receptacle, the fibers bend slightly in order to achieve physical contact; better performance is achieved when using optical fibers that have a special strength coating in addition to the outer jacket. The MT-RJ connector uses the same rectangular plastic ferrule concept as the multifiber MTP connector, with 2 fibers on 750-mm pitch and a latching mechanism based on the RJ-45 connector. Alignment in this case is provided by a pair of metal guide pins in the connector, which mate with a corresponding pair of holes in the receptacle; this feature makes the MT-RJ the only small form factor connector with distinct male and female connector ends. A more evolutionary approach to designing SFF connectors involves simply shrinking the standard SC duplex connector, maintaining a single fiber in each of the ceramic ferrules and using conventional alignment techniques applied to the ferrules. The LC connector uses this approach and shrinks the ferrules to 1.25 mm in diameter with a fiber pitch of 6.25 mm (duplex). LC is the only small form factor connector that can be either simplex or duplex.

In a comparative analysis of the different connectors the first feature with striking differences is the fiber pitch. The connectors can be broken up into two classes: small (0.75 mm) and large (>4.5 mm) fiber pitch. The small pitch presents challenges to the transceiver design for cross talk and space transforms for use with optoelectronic devices packaged in standard f 5.4 mm TO cans. Suppliers are also breaking away from the hermetic TO can to non-hermetic silicon optical bench (SiOB) technology to address these packaging concerns. Small pitch connectors such as the MT-RJ also have both fibers in a single ferrule while large pitch connectors such as the LC have the fibers in separate ferrules (or without ferrules in separate v-grooves as in the VF-45). The number of ferrules is significant because the ferrules are precision-made parts (mm tolerances) and are traditionally the most costly part in the connector. Ferrule material as well as the quantity of material is an issue. The plastic ferrule connectors should have a cost advantage over the ceramic ones and this advantage should increase as the manufacturing volume increases. However, the ceramic ferrule technology is more mature and the prices are currently falling due to market pressures. New lower cost glass ceramic ferrules are currently being produced. The plastic ferrules are newer and are currently controlled by a limited number of suppliers, resulting in current prices being as much as twice the cost of two of the standard ceramic ferrules. The plastic ferrules are both made with the same glass-filled, thermoset material, which must be transfer molded. Transfer molding is generally slower than injection molding, but more accurate. Some ferrule suppliers are beginning development of low-cost injection-molded ferrules.

The alignment schemes vary for the connector types and this reflects on the complexity of the transceiver package design. For the LC connector each ferrule requires a precision bore (5 mm/<1 mm tolerance for multimode/single-mode fibers) on the transceiver, which may add cost to the transceiver. The MT-RJ single ferrule connector requires two precision pins (0.25 mm tolerance) placed at a precision (3 mm tolerance) separation, which may eliminate any cost advantage to the transceiver from having one fewer bore than the LC. The alignment process in the assembly of the transceiver optical coupler(s) can also be a source of differentiation for the small- and large-pitched connectors. Connectors with a single ferrule may have a potential cost advantage in that a single optical alignment may be possible. With a small-pitched single ferrule connector such as the MT-RJ, it may be possible to reduce the number of parts in the transceiver by placing both transmitter and receiver in a single package and aligning them simultaneously with a single X,Y,Z,q alignment. Large pitch connectors such as the LC do not have the fibers placed accurately with respect to each other and therefore require two separate packages for transmitter and receiver and two separate alignments (X,Y,Z). It should be recognized that the alignments may in some cases be reduced to passive alignment, which may reduce the cost benefit of having one ferrule.

The transceiver opening is a major concern because of the small form factor transceiver dimensional constraints of 14 mm width, 9.8 mm height, and 31 mm length. All SFF transceivers must comply with an industry consortium multiple source agreement (MSA), which governs the maximum outside dimensions of the transceiver body and the size of a transceiver opening in a card bezel. The MT-RJ connector requires the least volume in the transceiver; use of the space around the connector is questionable so a key dimension is the length into the transceiver. MT-RJ and LC are very close in length of their transceivers (VF-45 can be longer). The single ferrule connector has an added space burden. Part of the transceiver volume may be necessary to space apart the devices from the small fiber pitch to avoid electrical and optical cross talk in silicon optical bench packages and to accommodate devices packaged in conventional TO cans.

Field termination can be an important issue for some applications. Ceramic ferrule connectors such as the LC have standard pot and polish field termination kits so the time and cost to terminate the connectors should not change from standard SC duplex. The single-ferrule MT-RJ field termination is accomplished using a pre-polished fiber stub connector with index matching gel. This approach should offer installation time at least equivalent to the SC duplex, and potentially shorter.

High mechanical accuracy of the ferrule alignment surfaces relative to the fiber position is a necessary condition for a connector that uses a ferrule. Generally, suppliers of the MT-RJ have begun to offer slight amounts of fiber protrusion in order to ensure physical contact between two ferrules in a duplex coupler or between the ferrule and transceiver. Some designs of the MT-RJ duplex coupler have encountered a stubbing problem when the connector pins did not properly align with holes in the receptacle. Early coupler designs attempted to address this problem by including alignment tabs or cross-hairs inside the coupler, designed to provide coarse alignment of the ferrules; subsequent design refinements of the MT-RJ connector have shown that this feature may not be necessary to produce adequate performance. The correct amount of protrusion and the acceptable effects of environment and plugging on these protrusions remains an ongoing development item for the industry.

Although the SFF connectors are tested for resistance to both axial and off-axis pull forces, the computer equipment that uses these connectors must still provide some form of strain relief to prevent cable pull forces from being transmitted back to the optical transceivers. This is particularly true for connectors such as the LC and MT-RJ, which only provide latching mechanisms on one side of the connector body; if the cable assembly is pulled in a preferential direction opposite to the latch, excessive losses may result (SFF connectors are generally more susceptible to pull forces because of their reduced size). Some current designs for fiber cable strain relief are based on serpentine (S-shaped) grooves in a plastic housing, which are capable of absorbing very high forces from fiber cables. However, these current designs are not well suited to more than a dozen fibers at a time; the greatly increased density provided by SFF packaging may require a new form of optical cable strain relief. It is desirable for the new strain relief to also be more user-friendly; the current design requires fiber to be manually placed around grooves and tabs, and is rather cumbersome. As a result, it is not used in real-world applications as widely as might be desired. Furthermore, the older strain relief was designed for a larger cable outer jacket and would not accommodate SFF cables as well. In new computer environments that employ many SFF transceivers on a single-channel card, it is particularly desirable to avoid any failure mechanisms due to mechanical damage on the connectors, as this requires replacement of the entire card and disruption of a large number of channels. Although there are industry standard FOTPs that address many of the connector functional requirements, some important application tests do not have a corresponding FOTP procedure (for example, random connection loss, insertion/withdrawal force testing, temperature cycle, and off axis pull testing). In addition, comparison of different connector types is not always possible, since some of the FOTPs are not applicable to connectors with male and female ends (such as MT-RJ) or without a duplex coupler (such as VF-45). These additional tests simulate common stresses that are found when connectors are used in the field under raised floors, in racks of equipement, and even on the wall outlets. The limits of these tests are set by the application specifications; performance that is adequate for fiber-to-the-desktop, for example, may not be adequate for a mainframe or supercomputer environment.

The intent of many of the SFF connectors was to replace the SC Duplex and thus to have the same or better quality and performance as the SC Duplex but half the size. Thus it is reasonable to use the SC Duplex as a benchmark against which to measure the new SFF connectors. Recent independent testing has shown that the SFF interfaces are still evolving, and performance is likely to continue improving over the coming years; however, at present some versions of the LC and MT-RJ have demonstrated equivalent performance to SC duplex connectors. As the assembly procedure has matured and become more standardized, the variability across suppliers has decreased, at least for the larger or first-tier suppliers. The improved performance of LC connectors, for example, may be partially attributed to a novel polishing technique for the smaller ceramic ferrules, the use of low eccentricity standard bulk fiber, and a new manufacturing process of "tuning" the ferrules by rotating them at small angular increments to obtain the orientation with minimal eccentricity prior to assembling the ferrule into the cable connector. In addition, fine tuning of the MT-RJ design has taken place, including changes in the connector body outer dimensions which facilitate better engagement with the RJ-45 latch. A small chamfer has been added around the ferrule alignment holes; this avoids chipping of the hole edge as the guide pins are inserted, and makes for a more repeatable connection with reduced plug force. Since the MT-RJ employs a novel rectangular ferrule, polishing techniques also have needed to be refined. A conventional ferrule polishing machine intended for round endfaces and symmetrical, cylindrical ferrules may not provide uniform polish on the MT-RJ; in fact, it can polish the endface with an angular bias along one axis. It was originally thought that this would not affect connector performance, since the fiber ptich is so small that only a relatively small area of the ferrule needed to have a high degree of polish; however, it has subsequently been shown that polishing is critical for the ferrule area near the alignment pins as well. Improper handling can result in a ferrule endface polished with multiple compound angles, which interferes with the guide pin alignment and resulting optical performance. More recent polishing jigs developed specifically for the MT-RJ have reduced or eliminated this problem.

Axial pull requirements are also a very interesting area. Some SFF connectors such as the SC-DC and VF-45 connectors are designed to disengage from their receptacles under an applied pull force above 45 N; this is to prevent damage to the connector, and induce an obvious optical failure in the link (it is problematic to diagnose a link failure if the connector remains engaged but exhibits high optical losses under stress). Another design approach used by LC and MT-RJ is to retain the connector in the receptacle under much higher forces without excessive optical loss. The magnitude of the pull force is a requirement that will vary depending on the application; however, it is essential for all the applications that the cable unplug or mechanically fail before the loss increases. End users should be cautioned that despite improvements in standardized manufacturing processes, there is still a broad range of connector performance currently available from second- and third-tier suppliers. Differences in assembly procedures result in different performance, thus representing low cost/performance options arising from connector assembly procedures. The connectors continue to undergo minor design revisions to improve performance and tighten up the assembly procedures across multiple suppliers. With properly chosen suppliers, however, some SFF connectors have matured to the point where they can be considered adequate replacements for SC duplex.

References

1. ANSI Fibre Channel - physical and signaling interface (FC-PH) X3.230 rev. 4.3 (1994).
2. Trewhella, J., C. DeCusatis, and J. Fox. July 2000. "Performance comparison of small form factor fiber optic connectors." *IEEE Transactions on Advanced Packaging* **23**(2): 188–196.
3. DeCusatis, C. To be published. "Small form factor fiber optics for enterprise computing applications." IBM Journ. Research and Development.
4. Schwantes, C. Oct. 1998. "Small-form factors herald the next generation of optical components." *Lightwave* pp. 65–68.
5. Shahid, M. A., et al. "Small and Efficient Connector System." *Proc. 49th Electronic Components and Technology Conference*, San Diego, Calif., June 1999 and TIA FOICS 10.

6. Tamaki, Y., et al. "Compact and Durable MT-RJ Connector." *Proc. 49th Electronic Components and Technology Conference*, San Diego, Calif., June 1999 and TIA FOICS 12.
7. Wagner, K. "SC-DC/SC-QC connector." *Optical Engineering* **37**(12):3129 and TIA FOICS 11.
8. Selli, R., et al. "A novel v-groove based interconnect technology." *Optical Engineering* **37**(12):3134 and TIA FOICS 7.

Chapter 4 | Specialty Fiber Optic Cables

Casimer DeCusatis
IBM Corporation, Poughkeepsie, New York 12601

John Fox
ComputerCrafts, Inc., Hawthorne, New Jersey 07507

4.1. Introduction

In this chapter, we will describe different types of fiber optic cable that have been developed for a wide range of applications, including enhanced distance, optical amplification and attenuation, dispersion and polarization management, and other areas. Future optical fiber designs, still several years away from commercial use, will also be discussed. While it isn't possible to provide a comprehensive list of every type of specialty fiber currently available, we will describe the most common types and their uses. To begin, we need to review the manufacturing process for the more common graded index single-mode and multimode fiber types, which have been discussed in previous chapters.

4.2. Fabrication of Conventional Fiber Cables

The fabrication of conventional low loss silica fiber optic cables [1] involves precision control of the glass composition to both control impurities and to ensure accurate refractive index profiles. High purity materials and well-controlled tolerances are important to the manufacturing process. The desired refractive index profile is first fabricated in a large glass preform, typically several centimeters in diameter and about a meter long, which maintains the relative dimensions and doping profiles for the core and cladding. The glass is selectively doped to provide the desired index gradients. This preform, or boule, is later heated in an electric resistance furnace until it reaches its melting point over the entire cross-section; it can take up to an hour to establish uniform heating of the preform. Thin glass fibers are then drawn upward from the preform in a drawing tower, as illustrated in Fig. 4.1a; the fiber cools and solidifies very quickly, within a few centimeters of the furnace. The pulling force controls the rate of fiber production, and hence the fiber diameter, which is monitored by a laser interferometer. Bare glass fiber is then drawn through a vat of polymer and receives a protective coating extruded over it to a diameter of about 250 microns; this must be done as soon as possible after drawing to avoid water contamination in the fiber. Finally, the fibers are spooled evenly onto a mandrel about 20 cm in diameter, to avoid microbending. If the preform is uniformly heated (and therefore has a uniform viscosity) then the cross-section and index profile of the drawn fiber will be exactly the same as in the preform. In this manner, fibers with very complex refractive index profiles can be produced. In order to facilitate easy manufacturing of the glass fibers, it is easier to use fairly large preforms and doping methods with a fast deposition rate. The primary technology used in fiber preform manufacturing is chemical vapor deposition (CVD), in which submicron silica particles are produced through one or both of the following chemical reactions, carried out at temperatures of around 1800–2000 C:

$$\begin{aligned} SiCl_4 + O_2 &\longrightarrow SiO_2 + 2\,Cl_2 \\ SiCl_4 + 2H_2O &\longrightarrow SiO_2 + HCL \end{aligned} \tag{4.1}$$

This deposition produces a high purity silica soot which is then sintered to form optical quality glass. There are two basic manufacturing techniques commonly used. In the so-called "inside process," a rotating silica substrate tube is subjected to an internal flow of reactive gasses. There are two variations on this approach, modified chemical vapor deposition (MCVD) and plasma-assisted chemical vapor deposition (PCVD). In both cases, layers of material are successively deposited, controlling the composition at each step, in order to reach the desired refractive index. MCVD accomplishes this deposition by application of a heat source, such as a torch, over a small area on the outside of the silica tube. This heat is necessary for sintering the deposited SiO_2 and for the oxidation reactions shown in the equation. Submicron particles are deposited at the leading edge of the heat source; as the heat moves over these particles, they are sintered into a layered,

4. Specialty Fiber Optic Cables   91

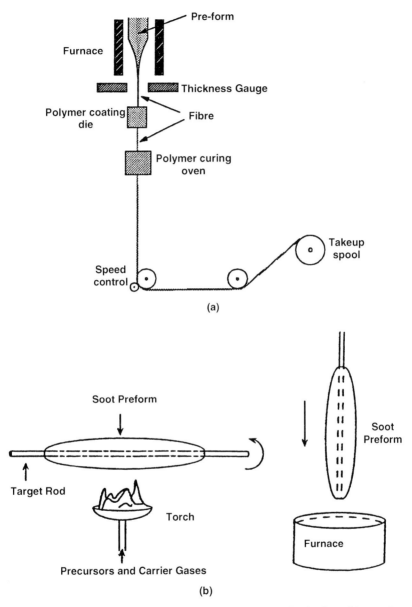

Fig. 4.1 (a) Typical optical fiber drawing tower. (b) Preform fabrication with a torch and carrier gasses.

glassy deposit. This requires fairly precise control over the temperature gradients in the tube, but has the advantage of accomplishing the sintering and deposition in one step. MCVD accounts for a large portion of the fiber produced today, especially in Europe and America. By contrast, the PCVD process provides the necessary energy for the chemical reactions by direct RF excitation of a microwave-generated plasma. Because the microwave field can be moved very quickly along the tube (since it heats the plasma directly, not the silica tube itself) it is possible to traverse the tube thousands of times and deposit very thin layers at each pass, which makes for very precise control of the preform index profile. A separate step is then required for sintering of the glass. In both cases, the preforms require a final heating to around 2150 C in a furnace to collapse the preform into a state from which glass is ready to be drawn. All inside vapor deposition (IVD) processes require a tube to be used as a preform; minor flaws in the tube can induce corresponding dips and peaks in the fiber index profile. This can be a problem in wavelength multiplexing systems, which may require only a few parts per million tolerance on the optical components in a link. This also affects the fiber's suitability for transmission of high data rates; the use of legacy fiber for Gigabit Ethernet links requires careful evaluation of these distortions in the core of IVD fibers.

In the so-called "outside process," a rotating, thin cylindrical mandrel is used as the substrate for subsequent CVD; the mandrel is then removed before the preform is sintered. An external torch fed by carrier gasses is used to supply the chemical components for the reaction, as well as to provide the necessary heat for the reaction to occur. Two outside processes have been widely used — the outside vapor deposition (OVD) and the vapor axial deposition (VAD) methods (today, Corning exclusively uses the OVD process for standard commercial fiber). Much of the control in these techniques lies in the construction of the torch. For example, OVD is basically a flame hydrolysis process in which the torch consists of discrete holes in a pattern of concentric rings which each provide a different constitutent element for the chemical reactions; a stream of oxygen is used between successive rings to act as a shield between the different chemicals. The torch is moved back and forth along the rotating preform and the dopants in the flame are dynamically controlled to generate the desired index profile, as illustrated in Fig. 4.1b OVD is not widely used any more, although it was among the first processes developed, because of its high cost (preforms are limited in size, as it is a batch process) and technical problems such as difficulty in removing all the water (OH groups) from the formed glass, a tendency for the fibers to have a large depression in refractive index near the core.

The VAD process is similar in concept, using a set of concentric annular apertures in the torch; in this case, the preform is pulled slowly across the stationary torch. It was found that by mixing $GeCL_4$ as a dopant into the $SiCl_4$-O_2 feed, the proportion of germania (GeO_2) deposited with the silica varies with the temperature of the flame; if a wide flame is used, the temperature gradient produces a graded portion of germania deposit. This process is still used, particularly in Japan, for commercial fiber production.

Silica glass has absorption bands in both the ultraviolet (UV) and mid infrared (IR) wavelength ranges, which provides a fundamental limit to the attenuation that can be achieved. This occurs despite the fact that the Raleigh scattering contribution decreases inversly as the fourth power of the wavelength, and the UV Urbach absorption edge decreases even faster with increasing wavelength. The infrared absorption increases at long wavelengths, becoming dominant at wavelengths greater than about 1.6 microns, which results in the minimal loss for wavelengths around 1.55 microns. Note that for many years, optical fiber attenuation was limited by a strong hydroxide (OH) absorption band near wavelengths of 1.4 microns; this has been steadily reduced over time with improved fabrication methods, until the loss minimum near 1.55 microns was brought close to the Raleigh limit. The best dopants for altering the refractive index of silica glass are those that provide a weak index change without inducing a large shift in the ultraviolet (UV) absorption edge. These include GeO_2 and P_2O_5 (which increase the refractive index in the fiber core), B_2O_3 and SiF_4 (which decrease the refractive index in the cladding). The chemical reactions for the general process often use high vapor pressure liquids such as $GeCl_4$, $POCl_3$, or SIF_4 along with oxygen as a carrier gas; these reactions are well documented [2–4]. It is more difficult to introduce more exotic dopants, such as the rare earth elements used in optical fiber amplifiers, and there is not a single widely used technique at this time. There are also other preform fabrication and fiber drawing techniques that are not generally used for telecommunication grade optical fiber, but which can be important for other material systems and applications. These include bulk casting of the preform and the non-CVD method of "rod and tube" casting (in which the core and cladding are cast separately and combined in a final melting step). There are also preform-free drawing techniques such as the "double crucible" method, in which the core and cladding are formed separately in a pair of platinum crucibles and combined in the drawing process itself. This method was important in the past, but is not widely used today because it does not provide the same precision control as the drawing tower process.

There has been a great deal of research into fiber materials with better transparency in the infrared. Although none of these materials has yet proven to be a serious competitor with doped silica, they may prove useful for other applications that do not have the strict requirements of telecommunication systems, such as CO_2 laser transmission, medical applications, or remote sensing and imaging [5]. For example, bulk infrared optics and fibers can be made from sulfide, selenide, and telluride glasses; collectivly known as chalcogenide fibers, they exhibit transmission loss on the order of 1 dB/meter in the wavelength range 5–7 microns [6]. Another example is the heavy-metal fluroide fibers [7], which hold some interest for communication systems because their theoretical limit for Rayleigh scattering is much lower than for silica (this is due to a high-energy ultraviolet absorption edge, and better infrared transparency). However, excess absorption has proven difficult to reduce, and current state-of-the-art fluoride fibers continue to exhibit losses near 1 dB/km, well above the Rayleigh limit. Fabrication of short lengths of fluroide fiber has been somewhat more successful in transmitting longer wavelengths, with reported losses as low as 0.025 dB/km at wavelengths of 2.55 microns [8]. The residual loss mechanisms in longer fibers remain the subject of ongoing research, and are thought to be due to extrinsic impurities or defects such as platinum particles from the fabrication crucibles, bubbles in the core preform and at the core-cladding boundry, and fluoride microcrystals.

Many types of optical fiber are available for different environments, including undersea cables, outdoors (with integrated strength members to facilitate hanging cables from telephone poles), and indoors. The properties of optical fiber cables are governed by various industry standards as noted in Chapter 1, including G.652 (with a maximum bandwidth of 50 GHz) and G.655 (with a maximum bandwidth of several THz). As discussed in Chapter 7, fiber jackets are typically rated as either riser, plenum, low smoke, low/zero halogen, or a dual-rated combination of the above; the IEC 1034 specification provides the dual-rated jacket specifications, while the zero halogen jackets are typically free of chlorine, fluroine, and bromides. New cable jacket types are also emerging from the national fire and electrical codes, such as the "limited combustibility" designation for some types of plenum cables. Some cable types are rated for use under a computer room raised floor, others for installation via air blowers in plenum ducts. As discussed in Vol. 1 Chapter 8, many types of multifiber connectors, structured cabling systems, and cable pullers, conduits, and patch panels are available. Special designs of conventional optical connectors are available, such as the so-called "elite MT" connectors, a customized version of the 12 fiber MT ferrule that is sorted to guarantee less than 0.35 dB maximum loss per fiber. New types of multifiber connectors are also being developed using two-dimensional ferrules to accommodate 24 or more fibers in a single plugging operation. In addition, many companies manufacture custom optical fibers to a user-specified refractive index profile, doping, core or cladding geometry, numerical aperture, cutoff wavelength, or other characteristics. Short sections of optical fiber may be packaged as loopbacks or wrap plugs for transceiver testing, while long haul spools may require special shielding for mechanical or structural reasons. Specialty fibers with coatings to increase their mechanical performance under bending are used today in small form factor VF-45 style connectors; these are described in more detail in Chapter 3. Furthermore, there are many types of fiber optic connectors available, including so-called "no polish" field installable connectors intended for quick and low-cost installation by untrained personnel, physical and non-physical contact connectors, flat ferrules and angle polished ferrules for low back reflectance, connectors with built-in variable attenuators or mode scramblers, metallized fibers that are soldered into place, and many, many other variants. While it isn't practical to give a comprehensive list of every specialty fiber type in this chapter, we will provide an overview of some major fiber types that are commonly encountered, as well as a synopsis of emerging fiber types that may become important in the future.

4.2.1. OPTICAL CABLE, COUPLERS, AND SPLITTERS

There are many different types of fiber optic cable; as shown in Fig. 4.2, it is possible to package multiple fibers into a single unibody cable or into a ribbon or zipcord structure. Bundles of fibers whose ends are bound together, ground, and polished can form flexible light pipes. Of course, it is possible to bundle the fibers in such a way that there is no fixed relationship between the location of an input fiber and an output fiber; the principle purpose of such structures is to conduct light from one location to another, for illumination as an example; these are sometimes referred to as incoherent bundles [9], although they have little to do with optical coherence theory. A more interesting case is when the fibers are carefully arranged so that they occupy the same relative positions at both ends of the bundle; such bundles are said to be coherent. A coherent bundle of single-mode fiber is capable of conducting a high-quality image even when the bundle is made highly flexible; such fiber arrays have many applications in remote vision systems, and are used in fiber optic endoscopes for medical applications.

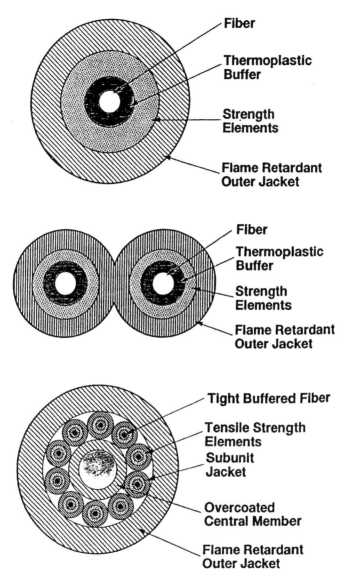

Fig. 4.2 Different types of fiber optic cable cross-sections.

Not all fiber arrays are made flexible; fused, rigid bundles or mosaics can be used to replace low-resolution sheet glass in cathode ray tubes. Mosaics consisting of hundreds to millions of individual fibers with their claddings fused together have mechanical properties very much like homogeneous glass [10]. Another common application of mosaics is as a field flattener.

If the image formed by a lens system falls on a curved surface, it is often desirable to reshape it into a plane, for example to match a photographic film plate. A mosaic can be ground and polished on one end surface to correspond with the contours of the image, and on the other surface to match the configuration of the detector. Similarly, a sheet of fused tapered fibers can be used to either magnify an image or miniaturize an image, depending on whether the light enters the smaller or larger end of the fibers.

Many simple devices such as fiber optic splitters, couplers, and combiners have been manufactured; the most common techniques include fiber tapering and fusion splicing [11–14], etching [15], and polishing [16–18]. Other fabrication techniques can also be used, including micro-optics and integrated optical components; however, optical fiber devices are particularly useful because they can be inserted into existing networks as just another piece of cable. One of the most common devices is a tapered fiber optic power splitter, often implemented in single-mode fiber [19]. In this process, two glass fibers with their protective jackets removed are brought close together and parallel to each other, then fused and stretched using a torch or similar heat source. Light that is initially launched into only one fiber will be partially coupled into the adjacent fiber as it propagates through the tapered region. Light propagating in the single-mode fiber is not confined to the core but extends into the surrounding cladding. In the case of a fiber taper it has been shown [20] that light propagating through the input fiber core is initially transferred to the cladding interface as it enters the tapered region, then to the core-cladding mode of the adjacent fiber. The light transfers back to the core modes as it exits the tapered region. This is known as a cladding mode coupling device. Light that is transferred to a higher order mode of the core-cladding structure is readily stripped away by the higher refractive index of the fiber coating, resulting in excess attenuation. The simplest case of light coupling from the cladding of one fiber into another through a fused taper can be described to a good approximation by the scalar wave equation and first-order perturbation theory [21]; if light is propagating along the z axis, then the exchange of optical power, P, is given by

$$P = \sin^2(kz) \tag{4.2}$$

where z is propagation distance and k is a complex function of the optical wavelength, refractive indices of the core and cladding, material properties, and the overlap distance between the two fibers [22]. Although this is only an approximation and neglects higher order terms, it does reflect the sinusoidal dependence of coupled power on wavelength and the dependence of power transfer on cladding diameter and other effects. Tapered couplers can be used to separate wavelengths using this dependence; by proper choice of the device length and taper ratio, two wavelengths can be made to emerge from two different output ports. Some applications include filters for wavelength division multiplexing (WDM) systems, or multiplexing signal and pump beams in an erbium-doped fiber amplifier. In some cases, such as optical power splitters, it is more desirable to remove the dependence of coupled power on wavelength; acromatic couplers can be fabricated by using two fibers with different propagation constants. These are known as dissimilar fibers; in most cases, fibers are made dissimilar by changing their cladding diameters or cladding indices. In this case, the preceding equation for coupled power must be modified and the power vs. distance is not simply sinusoidal, but becomes much more complex [22].

Other approaches are also possible, such as tapering the device such that the modes expand well beyond the cladding boundaries [23], or encapsulating the fibers in a third material with a different refractive index [24–25]. Often, it is desirable to taper multiple fibers together so that an input signal is split between many output fibers. Typically, a single input is split into 2^M outputs (i.e., 2, 4, 8, or 16), where the configuration of fibers in the tapered region affects the output power distribution; care must be taken to achieve a uniform optical power distribution among the output fibers [26]. The optical power coupled from one fiber into another can also be changed by bending the tapered device at its midpoint; this frustrates coupled power transfer. For example, displacing one end of a 1 cm long taper by only 1 mm can change the coupled power by over 30 dB [11]. Applications for this effect include variable optical attenuators and optical switches.

4.3. Fiber Transport Services

Given the many different types of fiber optic data links in a modern enterprise data center, the design of an optical cable infrastructure that will accommodate both current and future needs has become increasingly complicated. For example, IBM Site and Connectivity Services has developed structured cabling systems to support multi-gigabit cable plants. In this section, we briefly describe several recent innovations in fiber optic cable and connector technology for the IBM structured cabling solution, known as Fiber Transport Services (FTS) or Fiber Quick Connect (FQC).

A central concept of FTS is the use of multifiber trunks, rather than collections of 2-fiber jumper cables, to interconnect the various elements of a large data center [27–30]. FTS provides up to 144 fibers in a common trunk, which greatly simplifies cable management and reduces installation time. Cable congestion has become a significant problem in large data centers, with up to 256 ESCON channels on a large director or host processor. With the introduction of smaller, air-cooled CMOS-based processors and the extended distance provided by optical fiber attachments, it is increasingly common for data processing equipment to be rearranged and moved to different locations, sometimes on a daily basis. It can be time consuming to reroute 256 individual jumper cables without making any connection errors or accidentally damaging the cables. To relieve this problem, this year FTS and S/390 have introduced the Fiber Quick Connect system for multifiber trunks. The trunks are terminated with a special 12-fiber optical connector known as a Multifiber Termination Push-on (MTP) connector, as shown in Fig. 4.3. Each MTP contains 12 fibers or 6 duplex channels in a connector smaller than most duplex connections in use today (barely 0.5 inches wide). In this way, a 72-fiber trunk cable can be terminated with 6 MTP connectors; relocating a 256-channel ESCON director now requires only re-plugging 43 connections. Trunk cables terminated with multiple MTP connectors are available in 4 versions, either 12-fiber/6 channels, 36-fiber/18 channels, 72-fiber/36 channels, or 144-fiber/72 channels. Optical alignment is facilitated by a pair of metal guide pins in the ferrule of a male MTP connector, which mate with corresponding holes in the female MTP connector. Under the covers of a director or enterprise host processor, the MTP connectors attach to a coupler bracket (similar to a miniature patch panel); from there, a cable harness fans out each MTP into 6 duplex connectors which mate with the fiber optic transceivers as shown in Fig. 4.4. Since the qualification of the cable harness, under the covers patch panel, and trunk cable strain relief for FTS are all done in collaboration with the mainframe server development organization, the FTS solution functions as an integral part of the applications.

At the other end of the FTS trunk, individual fiber channels are fanned out at a patch panel or main distribution facility (MDF), where duplex fiber connectors are used to re-configure individual channels to different destinations. These fanouts are available for different fiber optic connector types, although ESCON and Subscriber Connection (SC) duplex are most common for multimode and SC duplex for single-mode. Fanning out the duplex fiber connections at an MDF also offers the advantage of being

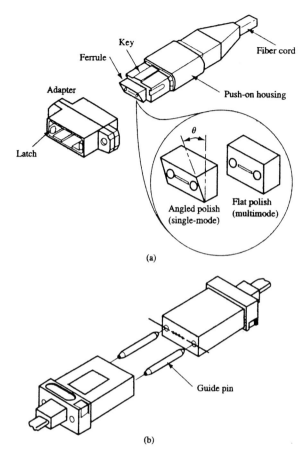

Fig. 4.3 (a) MTP connector with flat or angle polished ferrule; (b) detail of alignment pins.

able to arrange the MDF connections in consecutive order of the channel identifiers on the host machine, greatly simplifying link reconfigurations. As the size of the servers has been reduced and the number of channels has increased, the size of the MDF soon became a limiting factor in many installations. In order to keep the MDF from occupying more floor space than the processors, a more dense optical connector technology was required for the Fiber Quick Connect system. To meet this need, IBM Global Services has adopted a new small form factor fiber optic connector as the preferred interconnect for multimode patch panels, the SC-DC, which is further described in Chapter 3. Structured cabling solutions similar to this are available from other companies as well; they may include overhead or

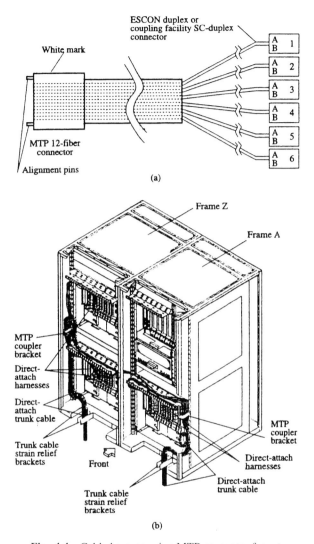

Fig. 4.4 Cable harness using MTP connector fanout.

underfloor cable trays and raceways, as well as cabinets or rack-mounted enclosures (standard units are compatible with either 19- or 23-inch-wide equipment racks, with heights between 1 and 7 U[1] tall).

[1] The Electronics Industry Association (EIA) defines a standard height of 1 U as equivalent to 1.75 inches for a data center equipment rack.

4.3.1. ATTENUATED CABLES FOR WDM AND CABLE TV

Some applications require optical attenuators to control or limit the optical power on a fiber link. One example is wavelength division multiplexing (WDM) equipment, as described in Chapter 5, which uses fixed attenuators to allow a common transceiver to interoperate with many different physical layers. Another is the fiber links in the cable television industry.

Fixed attenuators can be expensive, and must be incorporated into the design of the cable system; this means that duplex cables cannot be used if attenuation is required in only one path of a duplex fiber link. Separating duplex connectors defeats the keying that prevents the connector from being improperly inserted into a receptacle or transceiver. The attenuators also provide an extra connection point in the link, which must be cleaned and may be susceptible to mechanical vibration that will tend to dislodge connectors in data communication products. Instead of using fixed, pluggable attenuators, it is possible to manufacture in-line optical attenuators as part of the fiber optic cable assembly. Several approaches can be used. For multimode attenuators, a short piece of single-mode fiber can be spliced into the cable; by controlling the alignment between the single-mode fiber stub and the multimode cable on either side, as well as the length of the stub, various levels of attenuation can be achieved. The stub may be actively aligned during cable maufacturing, then protected by an external sheath or package to protect it from mechanical shock, vibration, and cable flexing. Similar effects can be achieved by deliberate misalignment of a multimode or single-mode fiber fusion splice. In most cases, arbitrary attenuation values from 0.5 dB to over 20 dB can be realized with a tolerance of less than 0.5 dB. The resulting attenuated cables are quite robust, and in many cases they serve a dual purpose, since the application would have required jumper cables anyway to adapt to different styles of optical connectors or connect with subtended equipment.

For other applications in which optical power must be controlled, specialty fibers are available with high attenuation that is flat over a certain spectral region. With so much design effort directed toward reducing the fiber attenuation to improve link budgets and distances, it can be easy to forget that optical fiber can just as easily be designed for high attenuation. Using the same precision-controlled manufacturing techniques that consistently yield low loss fiber, it is possible to dope the fiber in such a way that very consistent, high attenuation is provided over a wide range of operating wavelengths. This can be an advantage in designing attenuators for WDM systems. As with the offset spliced cable attenuators, these fibers tend to be more reliable and robust than conventional airgap attenuators. They have the additional advantage that a controlled amount of attenuation can be selected by simply cutting and splicing a desired length of the fiber. Short sections of high attenuation fiber are also being integrated in some types of fiber optic components, and are finding applications in optical test and measurement systems. Typical fibers are available with attenuations ranging from 0.5 dB/meter to 30 dB/meter in 0.5 dB increments, or from 0.25 dB/cm to 25 dB/cm in 0.25 dB increments.

4.3.2. ENCAPSULATED FIBER AND FLEX CIRCUITS

Recently, there has been increasing interest in using optical interconnections within the backplanes of computer systems, or for connections between equipment frames or racks (see Chapter 6). Various types of embedded surface waveguides have been proposed, which could be patterned using standard photolithographic techniques to produce an optical wiring plane on a standard multilayer printed circuit board or within a multichip module. However, many of these proposals require special materials or handling procedures, and it is difficult to make waveguide materials capable of withstanding high processing temperatures involved in circuit board rework without becoming opaque. Consequently, there have been several proposals and commercial product offerings dealing with embedding optical fibers into printed circuit boards. One approach is to embed conventional fibers into a flexible polymer laminate; in this way, optical circuit boards can be designed using existing computer drafting techniques, with the fibers being treated as if they were wires on a separate circuit board plane. The fibers can be terminated with standard optical connectors, or special blind-mateable multifiber backplane connectors, which can co-exist with copper interconnects. This has the advantage of reducing the size and complexity of optical interconnect designs, making them less susceptible to mechanical or thermal degradation, as well as allowing implementation of time domain filters and optical delay lines. One example is the OptiFlex[2] material developed by Lucent; acrylate or polyimide-coated fibers can be embedded in a 0.5-mm-thick polymer up to 75 square cm in area [31]; the resulting material sheet can withstand operating temperatures from -20 to $+85$ C, and accommodates all standard types of single-mode and multimode fibers.

[2]OptiFlex is a trademark of Lucent Corporation.

4.3.3. NEXT-GENERATION MULTIMODE FIBER

Conventional datacom links use single-mode fiber for long-distance, high-speed links and multimode fiber for shorter links. The recent deployment of short wave (780–850 nm) laser transceivers has made it possible to use multimode fiber at data rates of 1.25 Gbit/s at distances of a few hundred meters. Because the cost of short wave transceivers is presently lower than for long wave transceivers, there remains some question as to the preferred fiber to install for some applications, and the best mixture of 62.5-micron and 50-micron multimode fiber. The IEEE has recently recommended using 62.5-micron multimode fiber in building backbones for distances up to 100 meters, and 50-micron fiber for distances between 100 and 300 meters. Care must be taken not to mix different types of multimode fiber in the same cable plant, as the resulting mismatch in core size and numerical aperture creates high losses. This can make it difficult to administer a mixed cable plant, as there is no industry standard connector keying to prevent misplugging different types of multimode fiber into the wrong location. In general, 50-micron fiber has been widely deployed in Europe and Japan, while North America has primarily used 62.5-micron multimode fiber.

Because there is such a large amount of legacy multimode fiber installed, new industry standards have attempted to accommodate multimode fiber even at higher data rates. Much of the interest in 50-micron fiber has been a result of its higher bandwidth and longer distances that can be achieved for shortwave laser links. While the idea of backward compatibility works reasonably well up to 1 Gbit/s (distances of a few hundred meters can be achieved), it begins to break down at higher data rates when the achievable distance is reduced even further. Designing a future-proof cable infrastructure under these conditions becomes increasingly difficult; at some point, new fiber needs to replace the legacy multimode fiber. An alternative to single-mode is the emerging "next-generation" multimode fiber, a 50-micron fiber optimized for 850 nm transmission, which can achieve distances up to at least 300 meters at 10 Gbit/s data rates. This would allow less expensive VCSEL lasers to support 10 Gigabit Ethernet, Fibre Channel, and telecom data rates, rather than more expensive long-wave lasers over single-mode fiber. The cost tradeoff at a system level is less clear at this point, but this concept has been jointly presented to various standards bodies by Corning and Lucent. An example is the Systimax LazrSPEED[3] fiber recently introduced by Lucent; backward compatible with existing

---
[3] Systimax and LazrSPEED are trademarks of Lucent Corporation.

4. Specialty Fiber Optic Cables

multimode systems and requiring no special installation tools or skills, this fiber uses a green jacket to distinguish it from existing multimode (orange), single-mode (yellow), and dispersion-managed (purple) fiber cables. Attenuation is about 3.5 dB/km at 850 nm and 1.5 dB/km at 1300 nm; bandwidth is 2200 MHz-km at 850 nm (500 MHz-km overfilled) and 500 MHz-km at 1300 nm (no change when overfilled). Another example is the Corning InfiniCore[4] fiber; the CL 1000 line consists of 62.5-micron fiber made with an outside vapor deposition process that achieves 500 meter distances at 850 nm and 1 km at 1300 nm. Similarly, the CL 2000 line of 50-micron fiber supports 600 meter distance at 850 nm, and 2 km at 1300 nm. Additional details on modal dispersion in multimode fiber can be found in the Telecommunication Industry Association (TIA) task group on modal dependencies of bandwidth (TIA FO-2.2).

4.3.4. OPTICAL MODE CONDITIONERS

Because of the bandwidth limitations of multimode optical fiber, future multi-gigabit fiber optic interconnects will be based on single-mode fiber cables. For this reason, most new fiber installations include at least some single-mode fiber in the cable infrastructure. However, many applications continue to use multimode fiber extensively; a recent survey of building premise cable installers reported that most LAN infrastructures currently installed are composed of about 90% Multimode fiber [32]. As the fiber cable plant is upgraded to support higher data rates on single-mode fiber, we must also provide a migration path that continues to reuse the installed multimode cable plant for as long as possible. The need to migrate from multimode to single-mode fiber affects many important datacom applications [33]:

- I/O applications currently using multimode fiber for ESCON will need to migrate the cable plant to single-mode fiber in order to take full advantage of the higher bandwidth of FICON links. Future FICON enhancements that extend this protocol to multi-gigabit data rates will also require single-mode fiber.
- Networking applications such as ATM have traditionally used different adapter cards to support multimode and single-mode fiber. Gigabit Ethernet standard (IEEE 802.3z) is the first industry standard to propose the use of both fiber types with the same adapter card.

[4]InfiniCore CL 1000 and CL 2000 are trademarks of Corning Corporation.

- Parallel Sysplex links were originally offered as either 50 Mbyte/s data rates over multimode fiber or 100 Mbyte/s data rates over single-mode fiber. With the announcement of more recent servers, support for multimode fiber has been withdrawn as a standard feature and is now available only on special request. There is a need to support 100 Mbyte/s adapter cards over installed multimode fiber to facilitiate migration of those customers who have been using the 50 Mbyte/s option.

In order to address these concerns, special fiber optic adapter cables have been developed, known as mode conditioning patch cables (MCP). This cable contains both single-mode and multimode fibers, and should be inserted on both ends of a link to interface between a single-mode adapter card and a multimode cable plant. This allows the maximum achievable distance for multimode fiber (550 meters) and enables some applications to continue using the installed multimode cable plant. The MCPs for parallel sysplex links, Gigabit Ethernet, Fibre Channel, and many other applications are available today.

Next, let us describe the technical issues associated with this approach. The bandwidth of an optical fiber is typically measured using an over-filled launch condition, which results in equal optical power being launched into all fiber modes [32]. This is also known as a mode scrambled launch, and is approximately equivalent to the conditions achieved when using a Lambertian source such as an LED. By contrast, laser sources being more highly collimated tend to produce an under-filled launch condition; this can result in either larger or smaller effective bandwidth relative to an over-filled launch, and is sensitive to small changes in the fiber's refractive index profile. As discovered in recent gigabit link tests [34], bandwidth measured using over-filled launch conditions is not always a good indication of link performance for laser applications over multimode fiber. As illustrated in Fig. 4.5, when a fast rise time laser pulse is applied to multimode fiber,

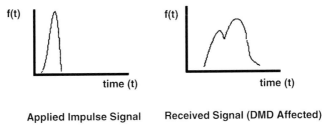

Fig. 4.5 Effect of DMD on signals passing through multimode fiber.

significant pulse broadening occurs due to the difference in propagation times of different modes within the fiber. This pulse broadening is known as differential mode delay (DMD); it is observed as an additional contribution to timing jitter (measured in ps/m) and can be large enough to render a gigabit link inoperable. DMD values are unique to the modal weighting of a source, the modal delay and mode group separation properties of the fiber, mode-specific attenuation in the fiber, and the launch conditions of the test. DMD is made worse by the excitation of relatively few modes with similar power levels in widely spaced mode groups and a high percentage of modal power concentrated in lower order modes. The impact of DMD increases with link length. There is, unfortunately, not a simple relationship between the industry specified over-fill launch measured bandwidths of the fiber and the effective bandwidth due to DMD.

The radial over-fill launch method was developed as a way to establish consistent and repeatable modal bandwidth measurement of a given fiber coupled with a given source [34]. A radial over-fill launch is obtained when a laser spot is projected onto the core of the multimode fiber, symmetric about the core center with the optic axis of the source and fiber aligned; the laser spot must be larger than the fiber core, and the laser divergence angle must be less than the fiber's numerical aperture. When these conditions are satisfied, the worst case modal bandwidth of the link is taken to be the worse of the over-fill and radial over-fill launch condition measurements (although for most applications, the radial over-fill launch will be the worst case). There is a good correlation between the radial over-fill launch bandwidth and the DMD limited bandwidth of a fiber; thus, high-speed laser links implemented over multimode fiber will likely experience bandwidth values closer to the radial over-fill launch method rather than the more commonly specified over-fill launch method.

To allow for laser transmitters to operate at gigabit rates over multimode fiber without being unduly limited by DMD, a special type of fiber optic jumper cable was developed to "condition" the laser launch and obtain an effective bandwidth closer to that measured by the over-fill launch method. The intent is to excite a large number of modes in the fiber, weighted in the mode groups that are highly excited by over-fill launch conditions, and to avoid exciting widely separated mode groups with similar power levels. This is accomplished by launching the laser light into a conventional single-mode fiber, then coupling into a multimode fiber that is off-center relative to the single-mode core, as shown in Fig. 4.6. There are two ways in which the offset launch can be introduced. One version requires manufacturing a splice between the single-mode and multimode fiber with a

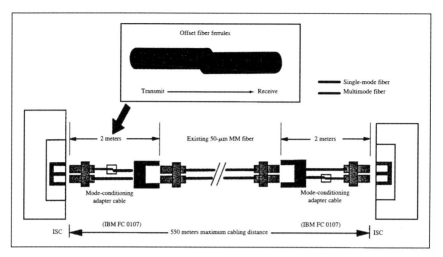

Fig. 4.6 Off-center ferrule design for mode conditioning patch cables.

controlled amount of lateral offset between the fiber cores. A tolerance analysis of this approach revealed that some installations could experience unacceptable variability in the splice elements, resulting in poor alignment and ineffective mode conditioning. For this and other reasons, the preferred embodiment uses standard ceramic ferrule technology with an offset in the ferrule alignment. Different offsets are required for 50.0- and 62.5-micron multimode fiber cores. Evaluations conducted by the Gigabit Ethernet Task Force, Modal Bandwidth Investigation Group, have verified that single-mode to 62.5-micron multimode MCPs with lateral offsets in the 17–23 micron range can achieve an effective modal bandwidth equivalent to the over-fill launch method across 99% of the installed multimode fiber infrastructure. Similar work has shown that single-mode to 50 micron multimode offset launch cables with lateral offsets in the 10–16 micron range will achieve similar results.

The MCP is illustrated in Fig. 4.7; its form factor is similar to a standard 2 meter jumper cable, except that it contains both single-mode and multimode fibers and includes a small package for the offset ferrules near one end. During the manufacturing process, the offset ferrules are actively aligned and then sealed with a potting compound to provide thermal and mechanical stability. The active alignment apparatus is shown in Fig. 4.8; a wide field charge-coupled device (CCD) camera is used to measure the two-dimensional spatial distribution of optical power at the output of the MCP. Typical results of this measurement are shown in Fig. 4.9(a) and (b);

4. Specialty Fiber Optic Cables   109

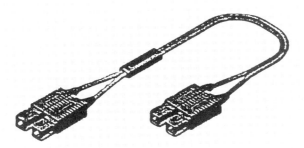

Fig. 4.7 Mode conditioning patch cable (MCP). The small box just behind one connector contains the offset ferrules; the transmit fiber from the optical transceiver is singlemode, all other fibers are multimode.

the first plot illustrates the optical power distribution for a long wavelength laser source coupled directly into single-mode fiber, and the second plot shows the same laser launched into an MCP and then into a 3-meter multimode jumper cable. It can be seen from these figures that the MCP-conditioned launch provides a uniform distribution of optical power among all the modes of the multimode fiber, and that the MCP-conditioned launch is virtually indistinguishable from the laser launch into single-mode fiber. Once the ferrules have been aligned and sealed into their optimal position, a simple assembly loss-type measurement can be used to evaluate the MCP performance, rather than measuring the complete two-dimensional coupled power profile. As shown in Fig. 4.10, there is a good correlation between the connection loss of the offset ferrules and the coupled power ratio. Note that alternate designs for MCPs have also been proposed, in which the offset launch condition is replaced by special optics that convert the laser

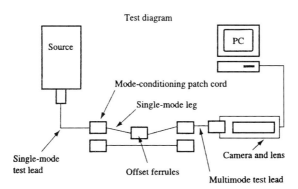

Fig. 4.8 Schematic of alignment apparatus for MCP cable manufacturing.

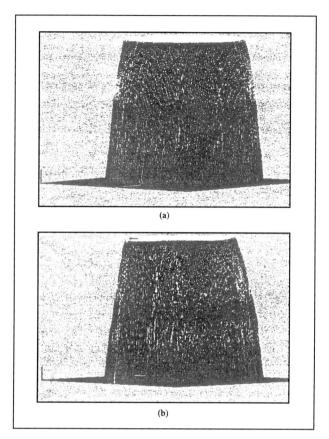

Fig. 4.9 MCP optical power profiles (a) single-mode fiber (b) multimode fiber.

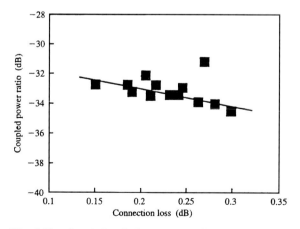

Fig. 4.10 Coupled optical power vs MCP connection loss.

spot into a donut-shaped launch. This acts to minimize power in the fundamental mode of the fiber, to achieve the same effect as an offset launch. MCPs have been tested under a variety of stressful conditions [32, 33].

4.4. Polarization Controlling Fibers

In the design of fiber optic systems it is important to know how many modes can propagate in the fiber, the phase constants of the different modes, and their spatial profiles. To do this we need to solve the wave equation for a particular fiber geometry as described in Chapter 1. The solution depends on the specific refractive index profile of the fiber. For the case of step index fiber profiles, a complete set of analytical solutions exists [35]; these can be grouped into three different types of modes depending on the direction of the electric field vector relative to the direction of propagation. They are called transverse electric (TE), transverse magnetic (TM), and hybrid modes. The hybrid modes can be further separated into two classes depending on whether the electric field, E, or magnetic field, H is larger in the transverse direction; these are called EH and HE modes, respectively. In practice, the refractive index difference between the core and cladding of an optical fiber is so small (about 0.002 to 0.009) that most of these modes are degenerate and it is sufficient to use a single notation for all modes, called the linearly polarized or LP notation. An LP mode is denoted by 2 subscripts, which refer to the radial and azimuthal zeros of the particular mode; for example, the fundamental mode is the LP_{01} mode. This is the only mode that will propagate in a single-mode fiber.

The cylindrical symmetry of an optical fiber leads to a natural decoupling of the radial and tangential components of the electric field vector; hence, standard single-mode fiber does not maintain the polarization state of the light when it is launched. However, these two polarizations are nearly degenerate, and a fiber with circular symmetry is most often described in terms of orthogonal linear polarizations. This near-degeneracy of the two polarization modes is easily broken by any imperfections in the cylindrical symmetric geometry of the fiber, including mechanical stresses on the fibers. These effects can either be introduced intentionally during the fiber manufacturing process, or they may arise inadvertently after the optical fiber has been installed. The effect is known as birefringence; it results in two orthogonally polarized modes with slightly different propagation constants (note that the two modes need not be linearly polarized, and in general they will be elliptical polarizations). Because each mode experiences a slightly different refractive index, the modes will drift in phase relative to each other; at any point in time, the light in the fiber exists in a state of polarization that is a superposition of the two orthogonally polarized modes. Birefringence of a fiber may be specified as the difference in refractive index between the two modes of propagation. The net polarization evolves as the light propagates through various states of ellipticity and orientation; after some distance, the two modes will differ in phase by a multiple of 2π, resulting in a state of polarization identical to that at the fiber input. This characteristic length is known as the beat length, and is a measure of the intrinsic material birefringence in the fiber; the time delay between the two modes is called polarization dispersion, and it can impact the performance of communication links in a manner similar to intermodal dispersion [9]. For example, if the time delay is less than the coherence time of the light source, the light in the fiber remains coherent and fully polarized. For sources of wide spectral width, however, this condition is reversed and light emerges from the fiber in a partially polarized and unpolarized state (the orthogonal polarizations have little or no statistical correlation). Links producing an unpolarized output can experience a 3 dB power penalty when passing through a polarizing optical element at the output of the fiber.

A stable polarization state can be ensured by deliberately introducing birefringence into an optical fiber; this is known as polarization preserving fiber or polarization maintaining fiber (PMF). Fibers with an asymmetric core profile will be strongly birefringent, having a different refractive index and group velocity for the two orthogonal polarizations (this is sometimes known as loss discrimination between modes). Such fibers are useful in some types of systems that require control of the transmitted light polarization. There are many possible core configurations as shown in Fig. 4.11; for example, elliptical cores provide a simple form of PMF by using very

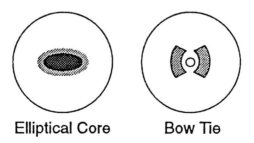

Fig. 4.11 Cross sections of polarization maintaining fibers.

4. Specialty Fiber Optic Cables   113

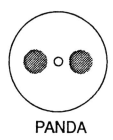

Fig. 4.12 Cross-section of PANDA fiber, showing mechanical members that apply strain to the fiber.

high levels of dopant in the core. These so-called "high birefringence" fibers also experience high attenuation because of the elevated dopant levels in the core. A double-core geometry (not shown in the figure) will also introduce a large birefringence. Another approach is to create mechanical stress within the fiber, such as in the bow tie configuration, by inserting stress-inducing members near the fiber core. Note that in all of these examples, polarization is only preserved if the initial signal is polarized along one of the preferred directions in the PMF; otherwise, the polarization of the signal will continue to drift as light propagates along the fiber.

Another example is the Polarization Maintaining and Absorption Reducing (PANDA) fiber; the areas highlighted in Fig. 4.12 show parts of the fiber core doped to create an area with a different coefficient of expansion than that of the cladding. In manufacturing, as this fiber cools, stresses are set up due to this difference, which in turn modifies the refractive index without requiring high levels of dopants in the core. These stress-applying members are present along the entire length of the fiber; such fiber is commercially available.

There are other ways to make PMF, although they are not widely used in commercial products [36]. For example, so-called "low birefringence" fibers can be made by very carefully controlling the fiber profile, since there is no reason for power to couple between orthogonally polarized modes if there are no irregularities in a perfectly circular fiber. Another way is to twist the fiber during manufacturing, or deliberately make the core off-center, so that the two polarization modes become circular in opposite directions and power coupling cannot take place. Yet another variation is called "spun fiber"; a PANDA fiber pre-form is spun while the fiber is being drawn, producing a full revolution about every 5 mm. Spun fiber has no polarization dependence at all, but is very difficult to make successfully at lengths much beyond about 200 meters and is very expensive; consequently it is not commonly used for communication systems.

Polarization effects can manifest themselves in a number of ways. For example, standard erbium-doped fiber amplifiers (EDFAs) exhibit two forms of birefringence, which are usually considered trivial but may build up in systems with many optical amplifiers. First, polarization dependent loss (PDL) refers to the fact that most EDFAs exhibit higher gain for one polarization state than for the orthogonal state. Because the arriving signal is composed of a superposition of states, the gain changes slowly (over a timescale of minutes); however, the amplified spontaneous emission noise is unpolarized and experiences a fixed gain. Hence, there is variation in the signal-to-noise ratio over time. Note that this effect accumulates as the root mean square of all the amps in a chain, rather than as a straight summation. Second, polarization dependent gain (PDG) is a saturation effect in the EDFA itself; the amplifier exhibits a higher gain in the polarization state orthogonal to that of the signal. One way to combat these effects is by polarization scrambling of the input signals. Another is to design a polarization-insensitive EDFA; this can be done by using a circulator to direct light into a Faraday polarization rotating mirror such that the light makes two trips through the EDFA. The PDL and PDG effects introduced on one polarization state in the first pass through the EDFA are also induced in the second polarization state during the second pass through the EDFA, such that the emerging light has uniform gain across both polarization states.

Recently, new types of polarization maintaining fibers have been announced for high capacity fiber systems. The Lucent TruePhase family of polarization-maintaining fibers is offered in application-specific wavelengths. A new nonzero-dispersion fiber from Pirelli, the Advanced FreeLight fiber features a reduction of 50% in its polarization mode dispersion ratio over previously available fibers. As this is a rapidly evolving field, we can expect many new fiber types to be introduced in the coming years with improved properties.

4.5. Dispersion Controlling Fibers

As discussed in Chapter 1, multimode optical fibers are subject to modal dispersion, while both multimode and single-mode fibers experience a combination of material (or chromatic) dispersion and waveguide dispersion. It was also noted that chromatic and waveguide dispersion have opposite signs

4. Specialty Fiber Optic Cables

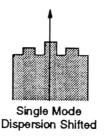

Fig. 4.13 Index profile of dispersion-shifted fiber.

so they may cancel each other out; this is why conventional silica fiber has a dispersion minima around 1300-nm wavelength. We may group together the collective effects of all these factors under the term group velocity dispersion (GVD). Standard single-mode fiber can exhibit either normal or anomalous dispersion. Under normal dispersion, long wavelengths have a higher group velocity than short wavelengths; if a wide spectrum of light is launched into this fiber, the red wavelengths will emerge first, followed by the blue wavelengths (this is also known as a positive frequency chirp or up-chirp). For anomalous dispersion, the situation is reversed; short wavelengths travel faster than long wavelengths, and blue light will emerge from the fiber before red light (this is called a negative frequency chirp or down-chirp).[5] Because modulation of a signal necessarily increases its bandwidth, and all practical light sources have some finite spectral width, dispersion effects occur in all communication systems.

As shown earlier, a typical silica optical fiber has an attenuation minimum around wavelengths of 1.55 microns, and a zero dispersion point at 1.3 microns. This represented a fundamental tradeoff in fiber link design, depending on whether the optical links were intended to be loss-limited or dispersion-limited; shorter distance data communication systems such as ESCON typically chose to operate at 1.3 microns, while links designed for long-haul communications were designed around 1.55 microns. Both the loss minimum and material dispersion are inherent physical properties of the silica fiber materials. However, waveguide dispersion can be affected by the refractive index profile design [36]. The profile shown in Fig. 4.13 has been successfully used to shift the zero dispersion point 1.55 microns; this is known as dispersion shifted fiber (DSF). Conventional fiber that has

---

[5] Note that while this terminology is consistent with most other reference books, in some engineering texts the meaning is reversed; the definition given here for normal dispertion is called anomalous, and the definition given here for anomalous is called normal.

not been treated in this manner is called non-zero dispersion shifted fiber (NZDSF). Currently available, DSF has a number of practical problems. It is more prone to some forms of signal nonlinearity, especially because its slightly smaller mode field diameter concentrates electromagnetic field more strongly in the core. WDM systems can also experience strong interchannel interference, or so-called near-end crosstalk (NEXT). For example, nonlinear effects such as four-wave mixing (FWM), also known as four-photon mixing, can be a serious design issue for WDM systems (see Chapter 5). FWM is strongly influenced by the wavelength channel spacing and by the fiber dispersion. In order for FWM to occur, each channel must stay in phase with its adjacent channel for a considerable distance. Thus, if fiber dispersion is high (as with standard NDSF in the 1550 nm band, typically around 17 ps-nm-km) the effects of FWM are minimal for channel spacings greater than about 25 Ghz (if channel spacing is reduced below about 15 GHz, the effect of FWM can be severe even on standard fiber). If DSF is used (dispersion less than 1 ps-nm-km), then FWM effects are maximized; the effect causes degradation at channel spacings of less than 80 GHz, and at a channel spacing of 25 GHz around 80% of the optical energy in the original two signals will be transferred into either sum or difference frequencies. So, one might ask why all fiber isn't dispersion-shifted to take advantage of the minimal dispersion properties; part of the answer is that DSF significantly increases problems like FWM. Other nonlinear effects, such as stimulated Raman scattering (see Chapter 7) are also worse when operating over DSF.

In order to address these problems, there are new types of fiber available, which essentially guarantee a certain level of dispersion (around 4 ps-nm-km), although these are currently very expensive. Other variants known as dispersion optimized fiber are also available, with a refractive index profile as shown in Fig. 4.14. This guarantees around 4 ps-nm-km dispersion in the

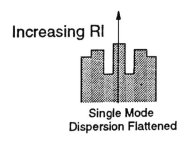

Fig. 4.14 Index profile of dispersion-flattened fiber.

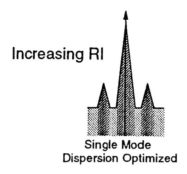

Fig. 4.15 Index profile of dispersion-optimized fiber.

1530–1570 nm wavelength range; it is available under various brand names, including Tru-Wave[6] fiber from AT&T and so-called SMF-LS fibers from Corning.

In addition, more sophisticated dispersion compensation can be achieved by adding several core and cladding layers to the fiber design. The index profile shown in Fig. 4.15 is quite complex, but it is possible to realize dispersion less than 3 ps-nm-km over the entire wavelength range 1300 nm–1700 nm using this approach. This is known as dispersion-flattened fiber; it was intended to allow users to easily migrate from 1300-nm systems to 1550-nm systems without changing the installed fiber. This fiber also has potential applications to broadband WDM applications, for which fiber dispersion must be kept as uniform as possible over a range of operating wavelengths. The principle drawback is its high loss, around 2 dB/km, which prevents general use in the wide area network.

It is also possible to construct a fiber index profile for which the total dispersion is over 100 ps-nm-km in the opposite direction to the material dispersion; this can be used to reverse the effects of conventional fiber dispersion. So-called dispersion compensating fiber (DCF) is commercially available with an attenuation of around 0.5 dB/km. DCF has a much narrower core than standard single-mode fiber, which accentuates nonlinear effects; it is also typically birefringent and suffers from polarization mode dispersion.

Different fiber designs intended for use in WDM environments have recently been introduced. For example, Corning has recently introduced its large effective area fiber, known as LEAF, for use in the 1550-nm window (both C-band and L-band wavelengths) at data rates up to 10 Gbit/s

---

[6]Tru-Wave is a trademark of AT&T.

and beyond. The LEAF fiber is a single-mode, non-zero dispersion-shifted fiber with a larger effective area in which optical power can be transmitted; typically this fiber can accommodate 2 dB more than conventional fibers without introducing nonlinear effects that can arise because of high power levels in the core, especially at the high power levels associated with multi-wavelength WDM systems. LEAF fiber has enhanced bandwidth, and effectively quadruples the information-carrying capacity of the fiber. This has made LEAF fiber particularly well-suited to long distance carriers and datacom service providers. A variation on this technology is the so-called MetroCor[7] fiber from Corning, a single-mode NZDSF compatible with industry standard G.655, which is designed to handle both C-band and L-band transmission in metropolitan area networks (1280 to 1625 nm). Similarly, the Allwave[8] fiber from Lucent provides a 50% larger spectrum than conventional fiber, lowering the attenuation between the 1300-nm and 1550-nm windows while maintaining low dispersion at 1300 nm. To protect against bending loss on single-mode WDM systems, special fiber is available such as the Blue Tiger[9] cables from Lucent. Identified by a blue cable jacket, this fiber is specially designed to support very tight bend radius applications (only 0.3 dB loss on a 10 mm bend, as compared with 1 to 1.5 dB for a conventional fiber under these conditions). Because a sharp fiber bend can induce enough loss to force WDM equipment to protection switch, this type of fiber may be important as WDM finds increasing applications in the metropolitan-area network.

Dispersion-flattened fibers have recently been investigated in soliton propagation systems for long distance communication without optical repeaters. We mention this for completeness, since datacom systems typically do not use soliton transmission; this is presently reserved for long haul telecommunication links. The nonzero dispersion slope in the fiber causes different wavelengths to experience different average dispersions; this can be a significant limitation in classical soliton long distance WDM transmission. By combining different types of commercially available fiber with different signs of dispersion and dispersion slopes, it is possible to design paths with essentially zero dispersion slope. For example, this was demonstrated in a recent experiment in which almost flat average dispersion ($D = 0.3$ ps/nm-km) was achieved by combining standard, dispersion compensated, and Tru-Wave fibers; this enabled soliton transmission

---

[7] MetroCor is a trademark of Corning Corporation.
[8] Allwave is a trademark of Lucent.
[9] Blue Tiger is a trademark of Lucent.

of twenty-seven WDM channels, each carrying 10 Gbit/s, over more than 9000 km. Soliton experiments may lead to other types of specialty fibers; for example, adiabatic soliton compression can be obtained by using a fiber whose dispersion slowly decreases with distance (so-called dispersion tapered fiber).

4.6. Photosensitive Fibers

Many types of glass are sensitive to ultraviolet light, which can induce a permanent change in their refractive index. These are known as photosensitive or photorefractive materials; in this case, the refractive index profile of the fiber can be changed by light that either propagates along the fiber length or illuminates an unjacketed fiber from the side. Many different types of materials can be used for this effect [37]; standard glass fiber can be made photosensitive by doping it with hydrogen, for example, while other fiber types do not require the hydrogenation process. If the fiber is illuminated through a transmission mask, or by an interference pattern created by two light beams, then the photorefractive effect can be used to write a diffraction grating into the fiber index variations. Recently, a class of devices known as in-line fiber Bragg gratings (FBGs) have been developed which hold great promise for many applications such as optical filters, wavelength add/drop multiplexers, and dispersion compensators. Writing a fiber Bragg grating requires a larger full width at half maximum response due to saturation of the index change. Due to coupling of radiation modes, undesirable side lobes of the grating can also be created at shorter wavelengths; however, this can be controlled with more recently developed types of fiber and grating writing techniques. For example, a grating stronger than 30 dB can be written in a few minutes using specialty fibers, with sidelobes kept below 0.1 dB.

Photosensitive fibers can also be used to write chirped fiber Bragg gratings for dispersion compensation. Recently, it has been demonstrated that the delay curve of grating-based dispersion compensators can be designed for a nearly perfectly linear response. This has caused increased interest in the use of FBGs for dispersion compensation in WDM systems. The FBG does have a certain amount of intrinsic polarization mode dispersion, but this can be overcome by coupling them with a suitable PMD compensator. This type of dispersion compensation is one of the many alternatives to dispersion shifted, flattened, and compensated fibers discussed earlier; with

...d interest in high bit rates (10 to 40 Gbit/s and beyond) over ...distances (hundreds of km), the market for dispersion compensa-...owing rapidly, and there will likely be many types of specialized fiber designed for different applications.

4.7. Plastic Optical Fiber

In an effort to reduce cost, many of the metal and ceramic components in optical fiber connectors are being replaced by plastic or polymer materials. Eliminating metal from fiber optic connectors has the advantage of improving electromagnetic radiation susceptibility, since the other components of a typical fiber cable are non-conductive. For example, radiation is generated by electronic equipment, which can escape from the receptacle of a fiber optic transceiver; metallic elements in the fiber connector act like antennas to re-radiate these emissions, possibly causing interference with either the original circuit board or with nearby unshielded electrical cables. In this manner, a product with many transceivers such as a fiber optic switch may pass electromagnetic noise emission testing with no fiber cables attached, but may fail if fiber cables with metal elements are plugged into the transceiver ports.

While ferrules in most optical connectors are traditionally made of ceramic, there are only a few companies in the world that manufacture these ferrules to the precision tolerances required. This can lead to capacity problems when there is a high demand for optical fiber; it also means that multimode ferrules may become scarce, as production lines migrate to higher volume (and higher margin) single-mode parts. With improvements in the single-mode ferrule manufacturing process, yields are high and there is very little product that can be recycled into multimode ferrules with their less restrictive tolerances. As a result, many companies are investigating plastic or polymer ferrules to address high-volume requirements and to reduce the cost of optical connector manufacturing. New polishing techniques need to be developed for plastic ferrules; there are also concerns with reliability and damage to the ferrule after hundreds of mating cycles. Plastic ferrules are just emerging as a viable alternative to ceramics, and are expected to play an increasingly important role in future cable systems.

This effort begs the question of whether the fiber itself could be replaced by plastic, at least for applications that can use transmission wavelengths not highly attenuated by the new material. Although most optical fibers are made of doped silica glass, plastic optical fibers are also available; these are commonly used in applications that do not require long transmission distances, such as medical instrumentation, automobile and aircraft control systems, and consumer electronics. In fact, short pieces of plastic fiber with large cores (about 900 microns) have already been used in optical loopbacks and wrap plugs, due to their low cost and ease of alignment with both multimode and single-mode transceivers. The short lengths required for a loopback mean that attenuation is not an issue, and in some cases the high attenuation of short plastic fibers is an advantage because it prevents saturation of the transceiver. However, there is some interest in using plastic fiber for very low cost data communication links, especially for the small office/home office (SOHO) environment. The combination of simplified alignment with optical sources and detectors (due to the large core diameter of plastic fibers) as well as the low cost of visible optical sources makes plastic optical fiber cost competitive for some applications. Plastic fibers are also very easy to connectorize; with minimal training and very simple tools, an amateur can connectorize bare plastic fiber in a few minutes. By contrast, glass fiber requires highly trained technicians and expensive equipment; this is a major difference, and one of the reasons for interest in plastic fiber for the do-it-yourself installations of homes and small offices; a significant inhibitor to wider use of optical links in this environment is the high cost of installation compared with the more reasonable cost of hardware and raw materials. Plastic fiber links are also used with visible light sources around 570 to 650 nm wavelength; this makes alignment of the fibers easier to perform, and high-power visible sources are readily available at low cost. A major drawback to plastic fiber is the difficulty in creating fusion splices with acceptable attenuation; typical splice losses are about 5 dB.

Both step index and graded index plastic fiber are available, although only step index is considered a commercial product at this time. While there are many potential plastics that could be considered, the most commonly used is Poly(Methyl MethylAcrylate) or PMMA. Doping the PMMA fibers can dramatically change their transmission properties; for example, Fig. 4.16 shows typical plastic fiber attenuation vs. wavelength for both standard PMMA fiber and a more recently introduced fiber in which deuterium replaces hydrogen in some parts of the polymer molecules. Attenuation is very high compared with glass fiber at all wavelengths; transmission

Table 4.1 **Typical Specifications of Plastic Optical Fiber**

Core Diameter	980 microns
Cladding Diameter	1000 microns (1 mm)
Jacket Diameter	2.2 mm
Attenuation (at 850 nm)	<18 dB per 100 meters (180 dB/km)
Numerical Aperture	0.30
Bandwidth (at 100 meters)	Step index: 125 MHz Graded Index: 500 MHz

windows using visible light near 570 nm and 650 nm are feasible for short distances (up to 100 meters). At this distance, step index fiber has a bandwidth of about 125 MHz, while for graded index fibers bandwidths of better than 500 MHz have been reported [9]. Table 4.1 gives some typical properties of plastic optical fiber; note the large core diameter, which is 100 times bigger than single-mode fiber. Although plastic fiber has been adopted outside the United State, most notably in Japan, there are not yet established standards for its use; some proposals suggest that plastic fiber could be implemented for links up to 50 meters and 50 to 100 Mbit/s data rates. Various types of plastic fiber and connectors have been proposed

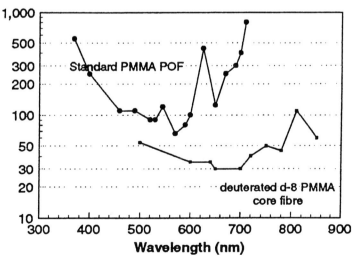

Fig. 4.16 Attenuation vs Wavelength for different types of plastic optical fiber.

4. Specialty Fiber Optic Cables

Table 4.2 **Properties of Typical HPCF Fiber**

Core Diameter	200 microns
Cladding Diameter	225 microns
Buffer Diameter	500 microns
Jacket Diameter	2.2 mm
Attenuation (at 650 nm)	0.8 dB/100 meters
Numerical Aperture	0.3 max.
Bend Radius	2 cm, loss of 0.05 dB
Bandwidth	10 MHz-km

for standardization by ATM and other groups. One example is the so-called Lucina[10] graded index plastic optical fiber developed by Asahi Glass Company; it is made of a transparent fluorpolymer, CYTOP, which is proprietary to Asahi Glass at this time. Available in both single-mode and multimode versions, including up to 200-micron core diameters, this fiber can potentially support over 1 Gbit/s up to 500 meters; attenuation is about 50 dB/km at 850 nm wavelength, bandwidth is around 200–300 MHz-km.

Other combinations of fiber materials are the subject of ongoing research, in particular Hard Polymer Clad (glass) Fiber or HPCF. This uses a relatively thick glass fiber core with a step-index hard plastic cladding; it is claimed to offer significantly less attenuation than conventional plastic fiber at around the same cost. It is also thinner than POF and consequently suffers from less modal dispersion. Although there are no general standards yet for HPCF fiber, the specification in Table 4.2 has been approved by the ATM Forum for use at 155 Mbit/s up to 100 meters. Since the fiber is using wavelengths around 650 nm, attenuation is fairly high; this wavelength is used to keep HPCF compatible with plastic fiber links.

4.8. Optical Amplifiers

In order to increase the maximum transmission distance of a loss limited optical link, various types of optical amplifiers have been proposed. This topic can easily consume an entire series of books, and many excellent references are available [38, 39]; we will note here only a few important points related to the design of such systems. Some types of optical amplifiers are

---

[10]Lucina is a trademark of Ashai Glass Corporation.

based on semiconductor devices, which act much like optically pumped lasers in the fiber link. Similarly, optical amps made by selectively doping fibers can be thought of as an optically pumped light source, and such fibers have been proposed for just such applications. Some types of rare earth elements have transition bands that correspond to the near infrared optical communication spectrum. When inserted into a glass matrix, these materials can absorb light from a pumping wavelength, storing energy that can subsequently be used to amplify incident light. The most common type used today is erbium-doped fiber amplifiers (EDFAs), which operate in the 1550-nm band, and are transparent to protocol, bit rate, and bit format.

In order to amplify signals at other wavelengths, including the 1300-nm band commonly used for data communication, other rare earth dopants may be used. Because glass is amorphous rather than crystalline, it offers a large gain spectrum when doped with a rare earth element; this can be shared equally between many different wavelength channels, making doped fiber a good candidate for WDM systems. Praseodymium (Pr) doped amplifiers are one solution; this is based on implanting Pr^{3+} ions in the fiber material. Silica glass cannot be used for this type of amplifier; instead, Fluorozirconate or ZBLAN glasses are used with a very narrow (2-micron diameter) core to concentrate the pump light. This creates significant losses when attempting to couple into standard optical fiber links. These amps can be practically pumped at wavelengths of either 1017 nm (using a semiconductor InGaAs laser) or 1047 nm (using a Nd:YLF crystal laser). Recently, there has been a good deal of research into new material systems using Pr-doped Chalcogenide fibers with much higher gains (around 24 dB). Another possible dopant for the 1300-nm band is Neodymium (Nd), which can amplify over wavelength ranges of 1310 to 1360 in ZBLAN glass and 1360 to 1400 nm in silica. Efficient pump wavelengths for Nd-doped amplifiers are either 795 nm or 810 nm. Recent research has also been done in amplifiers using plastic fiber [40], especially as part of compact integrated optical systems, or as part of a hybrid glass–plastic long-haul communication system (for longer distance applications using plastic fiber, it may be easier to use silica links rather than to develop a plastic fiber amplifier). The main interest in this area lies in the fact that the gain medium can be an organic compound, which is introduced into the plastic at a relatively low temperature; this is much easier than the process of doping standard glass or ZBLAN with rare earth elements. Organic dyes cannot be used as a gain medium in glass fiber, since they tend to break down around silica's melting point of 2000°C. One example is a plastic fiber amplifier using Rhodamine B doped PMMA, with a reported gain window between 610 and 640 nm, a pump efficiency of 33%, and a gain of 24 dB [40].

One way to get higher pump power in a rare-earth-doped fiber is by using broad area laser diodes or arrays. Although these devices emit very high pump power, they cannot be efficiently coupled into the single-mode fiber used for rare-earth-doped amplifiers. An alternative is to combine a multi-mode pump and a single-mode signal into one fiber, the so-called double-clad optical amplifier. The single-mode core is placed inside the mulimode pump region; if the pump light overlaps with the core, it will be absorbed over a certain section of the fiber. To optimize this overlap, the circular symmetry of the fiber must be broken using either an off-center core or a pump region that is triangular, rectangular, polygonal, or D-shaped. Absorption into the doped core must be made very strong in order to reach the bleaching power (gain region) with relatively low pump power densities. One method of achieving this is co-doping — a first rare-earth dopant is used as an absorber that in turn transfers its energy to a second dopant, which then provides gain in the desired wavelength band. A common example is the use of ytterbium in combination with erbium in a double-clad fiber amplifier or fiber booster.

4.9. Futures

In this chapter, we have discussed many types of optical fibers and cable assemblies in use today. We conclude with a short description of the new types of optical fibers currently under development; while these fibers are not yet commercially available, they are promising for long-term applications in optical data communications.

4.9.1. PHOTONIC CRYSTAL FIBERS

Another type of optical fiber that holds promise for future applications is the so-called photonic crystal fibers, also known as microstructured fibers, photonic bandgap fibers, or holey fibers. This technology was first demonstrated by Phillip Russel at the University of Bath, England, in 1997. Rather than guiding light through silica using the principle of total internal reflection, these fibers consist of a microstructure cross-section along their length (long capillaries filled with air). A typical configuration consists of air holes surrounding a silica core as shown in Fig. 4.17. The fiber is made by assembling a small (7-mm diameter) bundle of silica capillary tubes surrounding a solid silica rod in a hexagonal packing arrangement. This structure is first drawn into a 1-mm diameter strand; then it is inserted into an 8-m diameter borosilicate jacket and drawn to a diameter of 260 microns. The fiber is then covered with a protective polymer coating. Light is guided in the hollow fiber core using the same quantum confinement principles of a photonic bandgap structure in semiconductor materials. There are two basic types of photonic crystal fibers, namely high-index fibers (using a solid silica core and micro-pores in the cladding) and low-index fibers (using a hollow core or a core with micro-pores). The micro-holes in the fiber are typically about 1.9 microns diameter on a 3.9 micron spacing. This fiber has several unique properties; for example, in theory there is no minimum propagation distance before the light becomes single-moded (this has led to the design of "endlessly single-mode" fibers with 6-micron core and 125-micron acrylate cladding). Bending losses should be negligible, even for a few centimeters bend radius.

Although bandgap effects are present in these fibers, it has also been found that light is guided by a secondary mechanism, namely the difference in average refractive index between the core and cladding. The large index difference between glass and air allows for much stronger confinement of the modes than with conventional fibers. Because the light is effectively guided through air, very low transmission losses are predicted; current fibers exhibit around 0.2 dB/meter at wavelengths between 600 and 1550 nm. Group velocity dispersion of about 50 picoseconds/nm/km was measured between wavelengths of 1480 nm–1590 nm using a 6.5-meter long sample of fiber. These fibers also offer very small effective mode field area, and a selectable zero dispersion point; together, these features mean that the fiber can experience nonlinear effects over very short lengths (less than 1 meter) as opposed to the minimum interaction lengths required in silica fibers for effects such as four-wave mixing. This has potential applications to wavelength conversion, dispersion compensation, soliton generation, and white-light fiber lasers. These fibers also facilitate the construction of double-core or multi-core fibers, with selective optical coupling between modes; these are most often used for optical remote sensors, but may find applications in multichannel communication systems as well. The fibers can also have doped cores, and even use ytterbium doping to build photonic crystal fiber lasers. They have applications as polarization-preserving fibers or mode conditioners/converters. Furthermore, different

4. Specialty Fiber Optic Cables 127

Endlessly Single-Mode Photonic Crystal Fibre

Contrary to standard optical fibres, this remarkably new photonic crystal fibre is **always single-moded**. And bending losses are negligible. Even turns with a few centimetre diameter, has no visible impact on the guided mode.

The fibre has a polymer coating, which makes it robust and easy to handle. Furthermore, the core-region has a diameter of approximately 6 microns, allowing **low-loss splicing to standard optical single-mode fibres**.

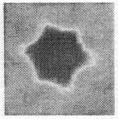

Near-field picture of guided mode at 633 nm wavelength

Close-up of fibre core-region

▸ **> 10 m lengths available**
▸ **Very low bending losses**

Type:	Endlessly single-mode
Loss:	< 0.2 dB/m @ 600 – 1550 nm
Material:	Pure silica
Core diameter:	6 μm
Cladding diameter:	125 μm
Coating diameter:	250 μm
Coating material:	Acrylate
Air hole diameter:	≅ 1.9 μm
Air hole distance:	≅ 3.9 μm

▸ **Easy coupling to standard fibre**
▸ **Large core SM fibre for visible light**

www.crystal-fibre.com   contact@crystal-fibre.com

Fig. 4.17   Cross-section of photonic bandgap fiber.

wavelengths in the fiber experience a different number of holes and hence a different refractive index; having an effective index strongly dependent on wavelength opens up possible applications in wavelength multiplexing (the fiber could be used as part of a laser system to generate a wide band of wavelengths). The area in which modes propagate is very large, and can be controlled over three orders of magnitude; if the effective mode field diameter is made very large, higher optical power can be transmitted without nonlinear effects. Recently, it was shown that the fiber can be designed with anomalous dispersion at short wavelengths (780 nm), opening up possible applications in dispersion management; soliton transmission at short wavelengths may even be possible. However, photonic crystal fiber is still in the research phase and considerable work remains to be done before it is widely available as a commercial product; researchers at the Technical University of Denmark and NKT Group have collaborated to form Crystal Fibre company, while researchers in the United Kingdom have started Blaze Photonics, to further development of this technology.

4.9.2. *FREE SPACE OPTICAL LINKS*

For some applications, optical fiber may not be necessary at all; in cases where it is difficult or expensive to install fibers, free space optical communication may provide an alternative. For example, free space optics have been proposed for communication between buildings in areas where fiber cannot be readily installed underground or between tall office buildings in metropolitan areas [41], between ships at dock and their mooring station, or to bridge the last mile between private homes and a service provider network [42]. Most of these systems consist of some form of telescope optics to collimate a tight beam for long-distance transmission, or to receive the dispersed light at a remote location. As one might expect, these approaches must deal with pointing accuracy and do not work as well under adverse weather conditions, since the infrared light signal is strongly absorbed by moisture in the atmosphere. On a smaller scale, free space optical interconnects have also been proposed for intra-machine communications, using micro-optics and lenses to collimate and focus light beams within a computer system or between adjacent equipment racks. While many such systems have been proposed, commercial products for use within a computer system are not yet available.

4.9.3. OMNIDIRECTIONAL FIBERS

In 1998, researchers at MIT developed a new type of mirror that reflects light from all angles and polarizations, just like metallic mirrors, but can also be as low loss as dielectric mirrors. Prior to this breakthrough, it was widely thought that a mirror could only be designed to offer high reflectivity for light with certain polarizations and angles of incidence. There are many potential applications for this so-called "perfect mirror" technology, one of which is fabricating the new mirrors into a tube with a hollow air-filled core to create an omnidirectional optical waveguide. Because this approach guides light through air rather than through silica glass, it should theoretically result in lower attenuation per km, as well as increased optical power per channel, polarization insensitivity, and improved dispersion characteristics. Also, unlike conventional waveguides, which are subject to loss at tight bend radius, the omnidirectional waveguide can make very small bends without incurring optical loss. This could enable new applications for integrated optics by allowing the miniaturization of many optical components. It has also been suggested that the monodirectional reflective material could be fabricated into a conventional coaxial cable structure with a metal or dielectric core, capable of performing as either an electrical or optical cable.

4.9.4. SUPERLUMINAL WAVEGUIDES

Finally, researchers at NEC corporation have recently published [43] the first evidence for optical signals propagating faster than the speed of light in a vacuum. It's a well-known principle of physics and Fourier optics that a pulsed signal (finite duration in time) can be created by adding up an infinite number of waves with infinite duration at different frequencies. The shorter the desired pulse, the larger the bandwidth of frequencies that must be used. All pulses of light or microwaves are thus formed by a packet of waves, each of which has a different frequency, amplitude, and phase. There is a distinction between the speeds of the individual frequency components, called the phase velocity, and the speed at which the wave packet propagates, called the group velocity. Conventional physics allows the phase velocity to exceed the speed of light, c, although the group velocity (the speed at which information is actually conveyed) must be less than c. It's possible to have a medium in which the phase and group velocities are not only different, but in opposite directions (one positive, the other negative). While the idea of some wave components moving backward while the wave packet only moves forward is non-intuitive, these so-called backward propagating modes are not new, and have been observed under many circumstances. In a medium with negative group velocity or so-called anomalous dispersion, it is possible in theory to propagate optical pulses faster than c. There are some materials that exhibit negative group velocities, typically near an atomic absorption frequency; unfortunately, this region also corresponds to very high optical attenuation and nonlinearities, which have made many prior experiments inconclusive. However, recent experiments have shown evidence of microwave pulses propagating about 7% faster than c over distances of about 1 meter — about ten times the wavelength of the radiation itself. Subsequently, researchers at NEC have demonstrated [43] superluminal propagation of optical signals, again over distances much larger than the wavelength of light. Working with a medium that exhibits gain, or optical amplification, near the region of negative group velocity, a narrow optical pulse (3.7 microseconds) was launched into a chamber 6 cm long that was filled with excited atomic cesium vapor. If the chamber had been filled with vacuum, the optical pulse should have taken about 0.02 nanoseconds to pass through; instead, the leading edge of the optical pulse appeared at the chamber's exit 62 nanoseconds before the bulk of the pulse entered the chamber. In other words, the wave packet had traveled nearly 20 meters away from the chamber exit before the incoming pulse entered. The optical pulse was shifted forward in time by about 1.7% of its width, giving a group velocity inside the chamber of about $-c/330$. The optical pulse emerging from the chamber was identical in shape to the one that entered. The mechanism for this is not completely understood, although it does not appear to violate the causality principle since the leading edge of the input pulse contains all the information necessary to reconstruct the peak of the output pulse, so the entire pulse does not need to enter the chamber before it exists at the opposite side. The output pulse may be just a clever interference effect, shaping newly created photons from the gain region into an exact copy of the incident light pulse. The original incoming pulse is subsequently canceled out by a backward propagating mode that travels from the chamber exit to its entrance with a phase velocity about 330 times faster than c. While we have little hope of developing a superluminal optical data link any time soon, research like this into the basic physics of optical propagation may lead to very practical benefits in future communication systems.

References

1. Nolan, D. 2000. "Tapered fiber couplers, mux and demux." Chapter 8 in *Handbook of Optics* vol. III and IV. Optical Society of America.
2. Hill, K. 2000. "Fiber bragg gratings." Chapter 9 in *Handbook of Optics* vol. III and IV, Optical Society of America.
3. Adams, M.J. 1981. *An Introduction to Optical Waveguides.* Chichester, England: John Wiley and Sons.
4. Carlisle, A.W. 1985. "Small size high performance lightguide connectors for LANs." *Proc. Opt. Fiber Comm.* Paper TUQ 18: 74–75.
5. Miller, S.E. and A.G. Chynoweth. 1979. *Optical Fiber Telecommunications*, New York: Academic Press.
6. Nishii, J. et.al., 1992. "Recent advances and trends in chalcogenide glass fiber technology: a review." *Journ. Non-crystalline Solids* 140: 199–208.
7. Sanghera, J.S., B.B. Harbison, and I.D. Agrawal. 1992. "Challenges in obtaining low loss fluoride glass fibers," *Journ. Non-crystalline Solids* 140: 146–149.
8. Takahashi, S. 1992. "Prospects for ultra-low loss using fluoride glass optical fiber," *Journ. Non-Crystalline Solids* 140: 172–178.
9. DeCusatis, C. 2001."Design and engineering of fiber optic systems." In *The Optical Engineer's Desk Reference*, E. Wolfe, ed. Optical Society of America.
10. Hect, E., and A. Zajac. 1979. *Optics.* New York: Addison Wesley.
11. Nolan, D. 2000. "Tapered fiber couplers, mux and demux." Chapter 8 in *Handbook of Optics* vol. IV. OSA Press.
12. Ozeki, T., and B.S. Kawaski. 1976. "New star coupler compatible with singe multimode fiber links." *Elec. Lett.* 12: 151–152.
13. Kawaski, B.S., and K.O. Hill. 1977. "Low loss access coupler for multimode optical fiber distribution networks." *App. Optcs* 16: 1794–1795.
14. Rawson, G.E., and M.D. Bailey. 1975. "Bitaper star couplers with up to 100 fiber channels." *Elec. Lett.* 15: 432–433.
15. Sheem, S.K., and T.G. Giallorenzi. 1979. "Singlemode fiber optical power divided; encapsulated teching technique." *Opt. Lett.* 4: 31.
16. Tsujimoto, Y., et al. 1978. "Fabrication of low loss 3 dB couplers with multimode optical fibers." *Elec. Lett.* 14: 157–158.
17. Bergh, R.A., G. Kotler, and H.J. Shaw. 1980. "Singlemode fiber optic directional coupler." *Elec. Lett.* 16: 260–261.
18. Parriaux, O., S. Gidon, and A. Kuznetsov. 1981. "Distributed coupler on polished singlemode fiber." *App. Opt.* 20: 2420–2423.
19. Kawaski, B.S., K.O. Hill, and R.G. Lamont. 1981. "Biconical-tapered singlemode fiber coupler." *Opt. Lett.* 6: 327.
20. Lamont, R.G., D.C. Johnson, and K.O. Hill. 1984. *App. Opt.* 24: 327–332.
21. Snyder, A., and J.D. Love. 1983. *Optical Waveguide Theory.* New York: Chapman and Hall.

22. Brown, T. 2000. "Optical fibers and fiber optic communications." Chapter 1 in *Handbook of Optics* vol. III and IV. Optical Society of America.
23. Weidman, D.L. December 1993. "Achromat overclad coupler." US Patent no. 5,268,979.
24. Truesdale, C.M., and D.A. Nolan. 1986. "Core-clad mode coupling in a new three-index structure." European Conference On Optical Communications, Barcelona, Spain.
25. Keck, D.B., A.J. Morrow, D. A. Nolan, and D. A. Thompson. 1989. *Journ. Lightwave Tech.* 7: 1623–1633.
26. Miller, W.J., D.A. Nolan, and G.E. Williams. April 1991. "Method of making a 1 X N coupler." U.S. Patent no. 5,017,206.
27. "Fiber Transport Services Physical and Configuration Planning." 1998. (IBM document number GA22-7234), IBM Corp., Mechanicsburg, Pa.
28. "Planning for Fiber Optic Channel Links." 1993. (IBM document number GA23-0367), IBM Corp., Mechanicsburg, Pa.
29. "Maintenance Information for Fiber Optic Channel Links." 1993. (IBM document number SY27-2597), IBM Corp., Mechanicsburg, Pa.
30. DeCusatis, C. December 1998. "Fiber Optic Data Communication: Overview and Future Directions," *Optical Engineering,* special issue on Optical Data Communication.
31. See Lucent product information at www.lucent.com.
32. Giles, T., J. Fox, and A. MacGregor. 1998. "Bandwidth reduction in gigabit ethernet transmission over multimode fiber and recovery through laser transmitter mode conditioning." *Opt. Eng.* 37: 3156–3161.
33. DeCusatis, C., D. Stigliani, W. Mostowy, M. Lewis, D. Petersen, and N. Dhondy. Sept./Nov. 1999. "Fiber optic interconnects for the IBM Generation 5 parallel enterprise server." *IBM Journal of Research and Development.* **43**(5/6): 807–828.
34. Giles, C.R. 1997. "Lightwave applications of fibre Bragg gratings." *IEEE Journ. Lightwave Tech.* **15**(8): 1391–1404.
35. Okoshi, T. 1982. *Optical Fibers.* New York: Academic Press.
36. Dutton, H. 1999. *Optical Communications.* New York: Academic Press.
37. Buck, J. 2000. "Nonlinear effects in optical fibers." Chapter 3 in *Handbook of Optics* vol. III and IV, Optical Society of America.
38. Delavaux, J-M P. 1995. "Multi-stage erbium doped fiber amplifier designs." *IEEE Journ. Lightwave Tech.* **13**(5): 703–720.
39. Olsen, N.A. 1989. "Lightwave systems with optical amplifiers." *IEEE Journ. Lightwave Tech.* **13**(7): 1071–1082.
40. Peng, G.D., P.L. Chu, Z. Xiong, T. Whitbread, and R.P. Chaplin. 1996. "Broadband tunable optical amplification in Rhodamine B-doped step-index polymer fibre." *Opt. Comm.* 129: 353–357.

41. Kim, I., R. Steiger, J. Koontz, C. Moursund, M. Barclay, P. Adhikari, J. Schuster, E. Korevaar, R. Ruigrok, and C. DeCusatis. 1988. "Wireless optical transmission of fast ethernet, FDDI, ATM, and ESCON protocol data using the TerraLink laser communication system." *Opt. Eng.* 37: 3143–3156.
42. Humbert, P., and W. Weller. Dec. 2000. "Mesh algorithms enable the free space laser revolution." *Lightwave* **17**(13): 170–178.
43. Wang, L.J., A. Kuzmich, and A. Dogariu. July 20, 2000. "Gain-assisted superluminal light propagation." *Nature* 406: 277–279.

Chapter 5 | Optical Wavelength Division Multiplexing for Data Communication Networks

Casimer DeCusatis

IBM Corporation, Poughkeepsie, New York 12601

5.1. Introduction and Background

Although fiber optic communication systems have seen widespread commercial use for some time, in recent years there has been increasing use of fiber in computer and data communication networks, as compared with their applications in voice and telephone communications. There are several unique requirements that distinguish the sub-field of optical data communication. Datacom systems must maintain very low bit error rates, typically between 10^{-12} and 10^{-15}, since the consequences of a single bit error can be very serious in a computer system; by contrast, background static in voice communications, such as cellular phones, can often be tolerated by the listener. Optical data links also face a tradeoff between optical power and unrepeated distance. Computer applications such as distributed computing or real-time remote backup of data for disaster recovery require fiber optic links with relatively long unrepeated distances (10–50 km). At the same time, because computer equipment is often located in areas with unrestricted access, the optical links must comply with international laser safety regulations, which limit the transmitter output power [1]. Telecom links may stretch hundreds of kilometers or more using low jitter signal regenerators; by contrast, most datacom systems cannot use mid-span repeaters in long links. This makes link budget analysis a critical element for the datacom system designer; it also implies that there are a much larger number of datacom transceivers per km of fiber than in telecom applications, making the datacom market very sensitive to the cost of optoelectronics. This drives a further tradeoff between low-cost and high-reliability components. Datacom systems require rugged components because data centers require continuous availability of the computer applications, but they are a hostile environment for optical devices; connectors are not cleaned regularly, cable reconfigurations are frequent, cable strain relief and bend radius are often not managed properly, and transceivers must withstand large numbers of reconnections, high pull forces on fiber cables, and other environmental stress.

Despite these challenges, the use of optical networking has brought about some convergence between telecommunication and data communication, at least at the physical layer. The need for ever-increasing bandwidth, data rates, and distance have made fiber optic links an integral part of computer system architectures over the past 10 years, particularly in high-performance applications, which have traditionally driven the need for higher bandwidth and consequently been the first computer applications to employ new technologies. A typical example is the increasing use of fiber optic data links on the IBM System/390 Enterprise Servers (mainframe computers). This continues to be an active area of development; mainframes and large servers, long recognized for their high security, reliability, and continuous availability, remain the data repository for major Fortune 1000 companies and contain an estimated 70% of the world's data. We can categorize the applications driving high-end optical data links into three areas:

(1) input/output (I/O) devices (such as tape or magnetic disk storage)
(2) clustered parallel computer processors (such as the IBM Parallel Sysplex architecture)
(3) inter-networking in the local, metropolitan, or wide area network (LAN/WAN/MAN)

The bandwidth requirements for networking environments using asynchronous transfer mode (ATM) over synchronous optical network (SONET) or similar protocols are well known [1]. However, Internet traffic has recently been growing at a rate of about 150% per year, and the amount of data traffic carried on the public telephone network now exceeds the amount of voice traffic by many estimates. This traffic has somewhat different properties than voice traffic; for example, the average connection distance for

Internet traffic is about 3000 km, or about 5 times larger than for voice traffic. This implies that future optical transport networks (OTNs) can reduce costs by using recently developed technologies such as long-range optical transmission, amplification, and switching. Internet traffic also tends to come in bursts, with a longer average connection duration than voice traffic, which makes capacity planning more challenging. However, telecommunications remains an important area for fiber optic networking. Virtually all telephone systems, including wireless and satellite communications, rely on a backbone of fiber optic networks. Most voice traffic is carried over protocols such as ATM/SONET; switching equipment designed for these protocols provides features such as fast protection switching so that calls are uninterrupted in the event of a fiber break or equipment failure. They also help ensure high quality of service (QOS) by providing error-free transmission of voice signals, sometimes by retiming them for increased fidelity. However, these features have until recently been available only on very expensive telecom switching equipment designed for a phone company's central office, and have applied to ATM/SONET protocols only. Users were forced to provide a separate overlay network for data communications, often at high cost, which did not offer the same features for voice and data. This required the construction of protocol-specific overlay networks, as shown in Fig. 5.1, which are expensive and do not scale well. Recent DWDM equipment provides QOS functions such as protection switching across all protocols; this allows users to save on leased optical fiber cost by combining voice and data over a single network, and for the first time makes it cost effective to build so-called virtual private networks (VPNs) to handle all the communication requirements of a single organization. Furthermore, DWDM can be used by telecom and datacom service providers as the basis for a service offering. Some telecom companies currently lease networking services in the same way that they lease telephone lines for voice traffic; this is a first step toward making bandwidth a commodity and enabling so-called all-optical networks (AON). The ability to manage an entire network from one location at the service provider central office is critical, as well as the ability to manage remotely (for example, one customer has products installed in Phoenix, Arizona, which are monitored through their call center in Minneapolis, Minnesota). DWDM devices also allow telecom carriers to migrate from their legacy voice-only, SONET-based equipment to new, more cost efficient network types, without sacrificing performance.

Transport networks in the MAN continue to evolve at a rapid pace. Datacom applications such as virtual private networks, Internet Protocol (IP) over WDM, electronic commerce, and the exponential growth of the

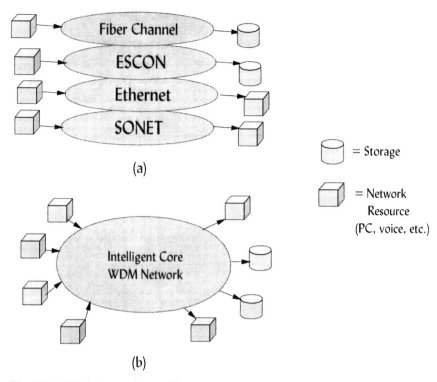

Fig. 5.1 (a) Existing service specific overlay network (b) future service transparent optical network with DWDM core.

Internet and Web continue to drive the need for increased bandwidth in the MAN (indeed, growth of the "optical internet" is among the largest driving factors in the deployment of WDM solutions). In this environment, the amount of installed fiber is being consumed quickly by new service offerings, and many networks face congested routes or fiber exhaust problems which limit their scalability and constrain new service offerings. Installation of new fiber, upgrading to higher line rates, or adding new terminal equipment as a workaround to fiber exhaust can be prohibitively expensive and may disrupt existing services. As fiber becomes scarce, there is increasing pressure to decommission legacy services in order to free fiber for higher data rate systems; a migration path is needed to maintain use of existing critical services and protect the installed service base, while maximizing use of the existing fiber plant. Network topologies within the MAN also continue to evolve; there is strong interest in sharing data among many widely distributed locations, and converging the data network with voice, video, and other services. Because the underlying protocols for ESCON and Parallel

Sysplex were designed for point-to-point systems, and switches to enable fabrics of FICON and Fibre Channel devices are not yet commercially available, new MAN topologies are being driven by the capability of next-generation DWDM systems to support more than point-to-point configurations. While first-generation DWDM systems were strictly point-to-point, next-generation systems include self-healing hubbed ring and meshed ring configurations.

Large servers require dedicated storage area networks (SANs) to interconnect with various types of magnetic disk, tape, and optical storage devices, printers, and other equipment; the dramatic increase in use of fiber optics for these applications has led to the term "Optical Data Center" (ODC). A SAN is a network of computers and storage devices (magnetic disk and tape, optical disk, printers, etc.) interconnected by a switching device to provide continuous access to all the data in the event that any single communication path fails. Examples of SANs include networks of ESCON, FICON, Fibre Channel, or Gigabit Ethernet devices, which can require tens to hundreds of fiber optic links. For example, large organizations such as airlines, banks, credit card companies, the federal government, and others are currently using huge SANs with many petabytes of storage; in the near future, exabyte SANs will likely emerge. A closely related concept is network attached storage (NAS), which refers to any type of storage device that is not locally connected to a server but attached only via a network interface. The cost and manageability of these systems is greatly improved by using wavelength multiplexing technology to reduce the number of fibers required; because it is protocol independent, it also provides the ability to work with different communication protocols and equipment from many different companies. Large SANs also benefit from the ability to construct protocol independent DWDM rings with QOS features such as "1 + 1" fast protection switching on individual channels for all protocols, not just those used for telecom voice communication. New applications such as data mining and requirements to archive data in real time at a remote location for disaster recovery with minimal service interruptions continue to drive bandwidth demands in this area. This has led to the emergence of client-server-based networks employing either circuit or packet switching, and the emergence of network-centric computing models in which a high bandwidth, open protocol network is the most critical resource in a computer system, surpassing even the processor speed in its importance to overall performance.

The recent trend toward clustered, parallel processors to enhance performance has also driven the requirement for high bandwidth fiber optic coupling links between computers. For example, large water-cooled mainframe computers using bipolar silicon processors are being replaced by smaller, air-cooled servers using complimentary metal oxide semiconductor (CMOS) processors [1, 2]. These new processors can far surpass the performance of older systems because of their ability to couple together many central processing units (CPUs) in parallel. There are many examples of parallel coupled processor architectures, as discussed in Vol. 1, Chapter 7; one important example is a Parallel Sysplex, an ad hoc standard developed by IBM, which allows mainframe computers to be clustered together using optical fiber links, and to work in parallel as if they were a single system. In this manner, the advantages of parallel processing can be used to increase the performance of the computer system without increasing the speed of the microprocessors, and to increase system reliability to better than 99.999%. A Parallel Sysplex requires a minimum of 30 to 60 duplex fiber optic channels for a small installation, and larger installations can require hundreds of channels. For disaster recovery purposes, the elements of a Parallel Sysplex are often split among physical locations up to 40 km or more apart; this is known as a Geographically Dispersed Parallel Sysplex (GDPS). The high cost of leased optical fiber over these distances ($300 per mile per month for 1 channel) makes it cost prohibitive for many users to implement Parallel Sysplex without using wavelength multiplexing. However, a Parallel Sysplex has unique requirements for the fiber optic channels and the wavelength multiplexer. In particular, a Sysplex requires some links (known as InterSystem Channels) to support the ANSI Open Fiber Control (OFC) protocols; this protocol specifies point-to-point channels only, and does not describe how to include repeaters or multiplexers in the link without violating the proper channel operation, or how to extend an OFC channel beyond distances of 20 km as required for many Sysplexes. Furthermore, a Sysplex requires that the computer's clock be distributed to remote locations up to 40 km apart, which causes a variety of latency and timing concerns when a multiplexer is included in the link. For these and other technical reasons, very few WDM solutions are able to support GDPS.

The combined effects of these application areas has made high-end computer systems a near term application for multi-terabit communication networks incorporating wavelength division multiplexing. In the remainder of this chapter, we will first describe the dominant approaches to data storage and coupled Parallel Sysplex processors using optical fiber connectivity and datacom protocols. We will then discuss how dense wavelength division multiplexing (DWDM) is being used in large data communication systems, including specific examples of current and next-generation

DWDM devices and systems. Technical requirements, network management and security, fault tolerant systems, new network topologies, and the role of time division multiplexing in the network will be presented. Finally, we describe future directions for this technology.

5.2. Wavelength Multiplexing

Multiplexing wavelengths is a way to take advantage of the high bandwidth of fiber optic cables without requiring extremely high modulation rates at the transceiver. With an available bandwidth of about 25 THz, a single optical fiber could carry all the telephone traffic in the United States on the busiest day of the year (recently, Mother's Day has been slightly exceeded by Valentine's Day). This technology represents an estimated $1.6 B market with over 50% annual growth; wavelength multiplexing systems may be classified according to their wavelength spacing and number of channels as follows [3]:

- Coarse WDM systems typically employed only 2–3 wavelengths widely spaced, for example 1300 nm and 1550 nm. Applications of this technology in data communications are limited, although recently coarse WDM systems with 4 to 8 channels have been used in small networks.
- Wide Spectrum WDM (WWDM) systems can support up to 16 channels, using wavelengths that are spaced relatively far apart; there is no standardized wavelength spacing currently defined for such systems, although spacing of 1 to 30 nm has been employed. These systems are meant to serve as a low-cost alternative to dense wavelength division multiplexing (DWDM) for applications that do not require large numbers of channels on a single fiber path, and are being considered as an option for the emerging 10 Gbit/s Ethernet standard [4]. These are also known as Sparse WDM (SWDM) systems.
- Dense WDM systems (DWDM) employed wavelengths spaced much closer together, typically following multiples of the International Telecommunications Union (ITU) industry standard grid [5] with wavelengths near 1550 nm and a minimum wavelength spacing of 0.8 nm (100 GHz). This may be further subdivided as follows:
- First-generation DWDM systems typically employed up to 8 full duplex channels multiplexed into a single duplex channel.

5. Optical Wavelength Division Multiplexing

- Second-generation DWDM systems employ up to 16 channels.
- Third-generation DWDM systems employ up to 32 channels; this is the largest system currently in commercial production for data communication applications.
- Fourth-generation or Ultra-dense WDM are expected to employ 40 channels or more and may deviate from the current ITU grid wavelengths; channel spacing as small as 0.4 nm (50 GHz) have been proposed [6]. These systems are not yet commercially available, and are the subject of much ongoing research.

Normally, one communication channel requires two optical fibers, one to transmit and the other to receive data; a multiplexer provides the means to run many independent data or voice channels over a single pair of fibers. This device takes advantage of the fact that different wavelengths of light will not interfere with each other when they are carried over the same optical fiber; this principle is known as wavelength division multiplexing (WDM). The concept is similar to frequency multiplexing used by FM radio, except that the carrier "frequencies" are in the optical portion of the spectrum (around 1550 nanometers wavelength, or 2×10^{14} Hertz). Thus, by placing each data channel on a different wavelength (frequency) of light, it is possible to send many channels of data over the same fiber. More data channels can be carried if the wavelengths are spaced closer together; this is known as dense wavelength division multiplexing (DWDM). Following standards set by the International Telecommunications Union (ITU) [5], the wavelength spacing for DWDM products is a minimum of 0.8 nm, or about 100 GHz; in practice, many products use a slightly broader spacing such as 1.6 nanometers, or about 200 GigaHertz, to simplify the design and lower overall product cost. A list of ITU grid standard wavelengths for DWDM is shown in Table 5.1. The concept of combining multiple data channels over a common fiber (physical media) is illustrated in Fig. 5.2. Note that the process is in principle protocol independent; it provides a selection of fiber optic interfaces to attach any type of voice or data communication channel. Input data channels are converted from optical to electrical signals, routed to an appropriate output port, converted into optical DWDM signals, and then combined into a single channel. The wavelengths may be combined in many ways; for example, a diffraction grating or prism may be used. Both of these components act as dispersive optical elements for the wavelengths of interest; they can separate or re-combine different wavelengths of light. The prism or grating can be packaged with fiber optic pigtails

Table 5.1 **ITU Grid Standard Wavelengths for Dense WDM Systems. C-band (or blue band) extends from about 1528.77 nm (196.1 THz) to 1556.31 nm (191.4 THz), and L-band (or red band) extends from 1565.48 nm (191.5 THz) to 1605.73 nm (186.7 THz). The minimum channel spacing is 0.8 nm (100 GHz) Anchored to a 193.1 THz reference (after ITU standard G.MCS Annex A of COM15-R 67-E)**

Wavelength (nm)	Frequency (THz)
1528.77	196.1
1529.55	196.0
1530.33	195.9
1531.12	195.8
1531.90	195.7
1532.68	195.6
1533.47	195.5
1534.25	195.4
1535.04	195.3
1535.82	195.2
1536.61	195.1
1537.40	195.0
1538.19	194.9
1538.98	194.8
1539.77	194.7
1540.56	194.6
1541.35	194.5
1542.14	194.4
1542.94	194.3
1543.73	194.2
1544.53	194.1
1545.32	194.0
1546.12	193.9
1546.92	193.8
1547.72	193.7
1548.51	193.6
1549.32	193.5
1550.12	193.4
1550.92	193.3
1551.72	193.2
1552.52	193.1
1553.33	193.0
1554.13	192.9
1554.94	192.8
1555.75	192.7

Table 5.1   continued

Wavelength (nm)	Frequency (THz)
1556.55	192.6
1557.36	192.5
1558.17	192.4
1558.98	192.3
1559.79	192.2
1560.61	192.1
1561.42	192.0
1562.23	191.9
1563.05	191.8
1563.86	191.7
1564.67	191.6
1565.48	191.5
1566.31	191.4
1567.13	191.3
1567.94	191.2
1568.77	191.1
1569.59	191.0
1570.42	190.9
1571.24	190.8
1572.06	190.7
1572.05	190.6
1573.71	190.5
1574.54	190.4
1575.37	190.3
1576.19	190.2
1577.03	190.1
1577.85	190.0
1578.69	189.9
1579.51	189.8
1580.35	189.7
1581.18	189.6
1582.02	189.5
1582.85	189.4
1583.69	189.3
1584.52	189.2
1585.36	189.1
1586.19	189.0
1587.04	188.9
1587.88	188.8

Table 5.1 continued

Wavelength (nm)	Frequency (THz)
1588.73	188.7
1589.56	188.6
1590.41	188.5
1591.25	188.4
1592.10	188.3
1592.94	188.2
1593.80	188.1
1594.64	188.0
1595.49	187.9
1596.34	187.8
1597.19	187.7
1598.04	187.6
1598.89	187.5
1599.74	187.4
1600.60	187.3
1601.45	187.2
1602.31	187.1
1603.16	187.0
1604.02	186.9
1604.88	186.8
1605.73	186.7

and integrated optical lenses to focus the light from multiple optical fibers into a single optical fiber; demultiplexing reverses the process. The grating or prism can be quite small, and may be suitable for integration within a coarse WDM transceiver package. Such components are typically fabricated from a glass material with low coefficient of thermal expansion, since the diffraction properties change with temperature. For coarse WDM applications, this is not an issue because of the relatively wide spacing between wavelengths. For DWDM systems, the optical components must often be temperature compensated using heat sinks or thermoelectric coolers to maintain good wavelength separation over a wide range of ambient operating temperatures. Because of the precision tolerances required to fabricate these parts and the complex systems to keep them protected and temperature stable, this solution can become expensive, especially in a network with many add/drop locations (at least one grating or prism would be required for each add/drop location).

Dense Wavelength Division Multiplexing (DWDM)

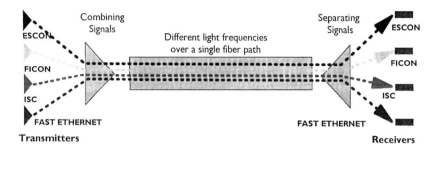

Fig. 5.2 Multiple data channels and protocols transmitted over a single optical fiber by WDM.

Another form of multiplexing element is the array waveguide grating (AWG); this is basically just a group of optical fibers or waveguides of slightly different lengths, which controls the optical delay of different wavelengths and thereby acts like a diffraction grating. AWGs can be fabricated as integrated optical devices either in glass or silicon substrates, which are thermally stable and offer proven reliability. These devices also provide low insertion loss, low polarization sensitivity, narrow, accurate wavelength channel spacing, and do not require hermetic packaging; one example of a commercial product is the Lucent Lightby 40 channel AWG mux/demux. Recently, the technology for fabricating in-fiber Bragg gratings has shown promise as an alternative approach; this will be discussed later in the chapter. Another promising new technology announced recently by Mitel Corporation's Semiconductor division implements an Echelle grating by etching deep, vertical grooves into a silica substrate; this process is not only compatible with standard silicon processing technology, which may offer the ability to integrate it with other semiconductor WDM components, but also offers a footprint for a 40-channel grating up to 5 times smaller than most commercially available devices.

Dielectric Add/Drop Filter

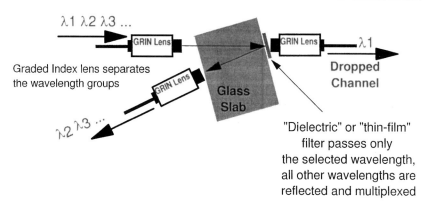

- This device is bi-directional, it operates in the opposite direction as an Add filter.

Fig. 5.3 Dielectric thin film optical filters for wavelength multiplexing.

However, in many commercially available DWDM devices a more common approach is to use thin film interference filters on a glass or other transparent substrate, as illustrated in Fig. 5.3. These multilayer filters can selectively pass or block a narrow range of wavelengths; by pigtailing optical fibers to the filters, it is possible to either combine many wavelengths into a single channel or split apart individual wavelengths from a common fiber. Note that many filters may be required to accommodate a system with a large number of wavelengths, and each filter has some insertion and absorption loss associated with it; this can affect the link budget in a large WDM network. For example, a typical 4-channel thin film add/drop filter can have as much as 3–4 dB loss for wavelengths that are not added or dropped; a cascade of many such filters can reduce the effective link budget and distance of a network by 10–20 km or more. The combined wavelengths are carried over a single pair of fibers; another multiplexer at the far end of the fiber link reverses the process and provides the original data streams.

There are many important characteristics to consider when designing a DWDM system. One of the most obvious design points is the largest total number of channels (largest total amount of data) supported over the multiplexed fiber optic network. Typically, one wavelength is required to support a data stream; duplex data streams may require two different wavelengths in

5. Optical Wavelength Division Multiplexing

each direction or may use the same wavelength for bi-directional transmission. As we will discuss shortly, additional channel capacity may be added to the network using a combination of WDM and other features, including TDM and wavelength reuse. Wavelength reuse refers to the product's ability to reuse the same wavelength channel for communication between multiple locations; this increases the number of channels in the network. For example, consider a ring with 3 different locations A, B, and C. Without wavelength reuse, the network would require one wavelength to communicate between sites A and C, and another wavelength to communicate between sites B and C, or two wavelength channels total. With wavelength reuse, the first and second sites may communicate over one wavelength, then the second and third sites may reuse the same wavelength to communicate rather than requiring a new wavelength. Thus, the second wavelength is now available to carry other traffic in the network. Wavelength reuse is desirable because it allows the product to increase the number of physical locations or data channels supported on a ring without increasing the number of wavelengths required; a tradeoff is that systems with wavelength reuse cannot offer protection switching on the reused channels.

Time Division Multiplexing (TDM) is another way in which some WDM products increase the number of channels on the network. Multiple data streams share a common fiber path by dividing it into time slots, which are then interleaved onto the fiber as illustrated in Fig. 5.4. TDM acts as a

Time Division Multiplexing (TDM)

▶ Different Data Streams with the Same Protocol can share a Single Physical Channel

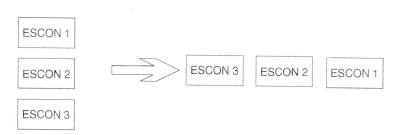

Fig. 5.4 Example of time division multiplexing 3 ESCON data frames.

front-end for WDM by combining several low data rate channels into a higher data rate channel; because the higher data rate channel only requires a single wavelength of the WDM, this method provides for increased numbers of low-speed channels. As an example, if the maximum data rate on a WDM channel is 1 Gbit/s then it should be possible to TDM up to 4 channels over this wavelength, each with a bit rate of 200 Mbit/s, and still have some margin for channel overhead and other features. The TDM function may be offered as part of a separate product, such as a data switch, that interoperates with the wavelength multiplexer; preferably, it would be integrated into the WDM design. Some products only support TDM for selected telecommunications protocols such as SONET; in fact, the telecom protocols are designed to function in a TDM-only network, and can easily be concatenated at successively faster data rates. However, since WDM technology can be made protocol independent, it is desirable for the TDM to also be bit-rate and protocol-independent, or at least be able to accommodate other than SONET-based protocols. This is sometimes referred to as being "frequency agile." Note that while a pure TDM network requires that the maximum bit rate continue to increase in order to support more traffic, a WDM network does not require the individual channel bit rates to increase. Depending on the type of network being used, a hybrid TDM and WDM solution may offer the best overall cost performance; however, TDM alone does not scale as well as WDM. A comparison of the two multiplexing approaches is given in Fig. 5.5.

Another way to measure the capacity of the multiplexer is by its maximum bandwidth, which refers to the product of the maximum number of channels and the maximum data rate per channel. For example, a product that supports up to 16 channels, each with a maximum data rate of 1.25 Gigabit per second (Gbps), has a maximum bandwidth of 20 Gbps. Note that the best way to measure bandwidth is in terms of the protocol-independent channels supported on the device; some multiplexers may offer very large bandwidth, but only when carrying well-behaved protocols such as SONET or SDH, not when fully configured with a mixture of datacom and telecom protocol adapters. Another way to measure the multiplexer's performance is in terms of the maximum number of protocol-independent, full-duplex wavelength channels that can be reduced to a single channel using wavelength multiplexing only. Some products offer either greater or fewer numbers of channels when used with options such as a fiber optic switch.

Relative cost scaling for different multiplexing technologies

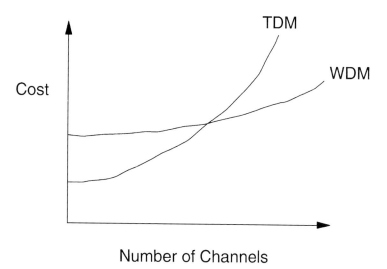

Fig. 5.5 Comparison of TDM and WDM scaling potential.

5.2.1. WDM DESIGN CONSIDERATIONS

Another important consideration in the design of WDM equipment is the number of multiplexing stages (or cards) required. It is desirable to have the smallest number of cards supporting a full range of datacom and telecom protocols. Generally speaking, a WDM device contains 2 optical interfaces, one for attachment of input or client signals (which may be protocol specific) and one for attachment of the WDM signals. Each client interface may require a unique adapter card; for example, some protocols require a physical layer that is based on an LED transmitter operating over 62.5-micron multimode fiber, others use short-wavelength lasers with 50-micron multimode fiber, and still others require long wavelength lasers with single-mode fiber. Likewise, each channel on the WDM interface uses a different wavelength laser transmitter tuned to an ITU grid wavelength, and therefore requires a unique adapter card. Some designs place these 2 interfaces on a single card, which means that more cards are required to support the system; as an example, a product with 16 wavelength channels may require

16 cards to support ESCON, 16 more to support ATM, and in general to support N channels with M protocols would require $N*M$ cards. Typically $N = 16$ to 32 channels and $M = 10$ to 15 protocols, so this translates into greater total cost for a large system, greater cost in tracking more part numbers and carrying more spare cards in inventory, and possibly lower reliability (since the card with both features can be quite complex).

This common card is sometimes known as a transponder, especially if it offers only optical input and output interfaces. An advantage of the transponder design is that all connections to the product are made with optical fiber; the backplane does not carry high-speed signals, and upgrades to the system can be made more easily by swapping adapter cards. However, a cross-connected high-bandwidth backplane is still a desirable feature to avoid backplane bus congestion at higher data rates and larger channel counts; furthermore, it provides the possibility of extending the backplane into rack-to-rack type interconnections using parallel optical interconnects. Another possible design point places the client interface and WDM interface on separate cards, and uses a common card on the client side to support many protocol types, so only 1 card for each protocol is required. Continuing our example, a product with 16 channels may support 10 different protocols; using the first design point discussed above this requires 160 cards, while the second design point requires only 16 cards, a significant simplification and cost savings. The tradeoff with this second design point is that more total cards may be required to populate the system, since 2 cards are required for every transponder, or in other words a larger footprint for the same number of wavelength channels. In practice, a common client interface card may be configurable to support many different protocols by using optical adapters and attenuators at the interface, and making features such as retiming programmable for different data rates. This means that significantly less than 1 card per protocol is required; a maximum of 2–5 cards should be able to support the full range of networking protocols listed in Table 5.2. Note that WDM devices that do not offer native attachment of all protocols may require separate optical patch panels, strain relief, or other protection for the optical fibers; this may require additional installation space and cost.

Features such as adapters or patch panels that must be field installed also implies that a particular configuration cannot be fully tested before it reaches the end user location; field-installed components also tend to have lower reliability than factory built and installed components. For example, if a product supports ESCON protocols, a user should be able to plug an

Table 5.2 Protocols Supported by WDM, Including Native Physical Layer Specifications and Attachment Distances; MM = multimode fiber, SM = single-mode fiber, TX = transmitter output, RX = receiver input, LX = long wavelength transmitter, SX = short wavelength transmitter

Protocol type	Physical layer specification	Native attach distance
ESCON/SBCON MM and Sysplex Timer MM	TX: -15 to -20.5 RX: -14 to -29	3 km
ESCON/SBCON SM	TX: -3 to -8 RX: -3 to -28	20 km
FICON SM	TX: -3 to -8.5 RX: -3 to -22	10 km
ATM 155 MM	TX: -14 to -19 RX: -14 to -30	2 km
ATM 155 SM	TX: -8 to -15 RX: -8 to -32.5	10 km
FDDI MM	TX: -14 to -19 RX: -14 to -31.8	2 km
Gigabit Ethernet LX SM	TX: -14 to -20 RX: -17 to -31	5 km
Gig. EN SX MM (850 nm)	TX: -4 to -10 RX: -17 to -31	550 meters
HiPerLinks for Parallel Sysplex and GDPS	TX: -3 to -11 RX: -3 to -20	10 km

ESCON duplex connector directly into the product as delivered, without requiring adapters for the optical connectors. Note that some ad hoc industry standards such as the Parallel Sysplex architecture for coupling mainframe computers have very specific performance requirements, and may not be supported on all DWDM platforms; recently published technical data from a refereed journal is a good reference to determine which protocols have been tested in a given application; there may also be other network design considerations, such as configuring a total network solution in which the properties of the subtended equipment are as well understood as those of the DWDM solution.

There are several emerging technologies that may help address these design points in the near future. One example is wavelength tunable or agile lasers, whose output wavelength can be adjusted to cover several possible wavelengths on the ITU grid. This would mean that fewer long wavelength adapter cards or transponders would be required; even if the wavelength agile laser could be tuned over only 2 ITU grid wavelengths, this would still cut the total number of cards in half. Another potentially useful technology is pluggable optical transceivers, such as the gigabit interface converter (GBIC) package or emerging pluggable small form factor transceivers. This could mean that an adapter card would be upgradable in the field, or could be more easily repaired by simply changing the optical interface; this reduces the requirement to keep large numbers of cards as field spares in case of failure. Also, a pluggable interface may be able to support the full physical layer of some protocols, including the maximum distance, without the need for optical attenuators, patch panels, or other connections; the tradeoff for this native attachment is that changing the card protocol would require swapping the optical transceiver on the adapter card.

Using currently available technology, it is much easier to implement ITU grid lasers and optical amplifiers at C-band wavelengths than at L-band wavelengths. Hence, some implementations use tighter wavelength spacing in order to fit 32 or more channels in C-band; the scalability of this approach remains open to question. Other approaches available today make use of L-band wavelength and larger inter-wavelength spacing, and can be more easily scaled to larger wavelength counts in the future. Performance of the multiplexer's optical transfer function (OTF), or characterization of the allowable optical power ripple as a function of wavelength, is an important parameter in both C-band and L-band systems.

5.2.2. NETWORK TOPOLOGIES

Conventional SONET networks are designed for the WAN and are based on reconfigurable ring topologies, while most datacom networks function as switched networks in the LAN and point-to-point in the WAN or MAN. There has been a great deal of work done on optimizing nationwide WANs for performance and scalability, and interfacing them with suitable LAN and MAN topologies; WDM plays a key role at all three network levels, and various traffic engineering approaches will be discussed later in this chapter. Despite the protocol independent nature of WDM technology, many WDM products targeted at telecom carriers or local exchange carriers (LECs) were designed to carry only SONET or SDH compatible traffic. Prior to the introduction of WDM, it was not possible to run other protocols over a ring unless they were compatible with SONET frames; WDM has made it possible to construct new types of protocol-independent network topologies. For the first time, datacom protocols such as ESCON may be configured into WDM rings, including hubbed rings (a central node communicating with multiple remote nodes), dual hubbed rings (the same as a single hub ring except that the hub is mirrored into another backup location), meshed rings (any-to-any or peer-to-peer communication between nodes on a ring), and linear optical add/drop multiplexing (OADM) or so-called "opened rings" (point-to-point systems with add/drop of channels at intermediate points along the link in addition to the endpoints). These topologies are illustrated in Fig. 5.6. Note that since the DWDM network is protocol independent, care must be taken to construct networks that are functionally compatible with the attached equipment; as an example, it is possible to build a DWDM ring with attached Fibre Channel equipment that does not comply with recommended configurations such as Fibre Channel Arbitrated Loop. Other network implementations are also possible; for example, some metro WDM equipment offers a 2-tier ring consisting of a dual fiber ring and a separate, dedicated fiber link between each node on the ring to facilitate network management and configuration flexibility (also known as a "dual homing" architecture).

5.2.3. DISTANCE AND REPEATERS

It is desirable to support the longest distance possible without repeaters between nodes in a WDM network. Note that the total supported distance for a WDM system may depend on the number of channels in use; adding more channels requires additional wavelength multiplexing stages, and the optical fibers can reduce the available link budget. The available distance is also a function of the network topology; WDM filters may need to be configured differently, depending on whether they form an optical seam (configuration that does not allow a set of wavelengths to propagate into the next stage of the network) or optical bypass (configuration that permits wavelengths to pass through into the rest of the network). Thus, the total distance and available link loss budget in a point-to-point network may be different from the

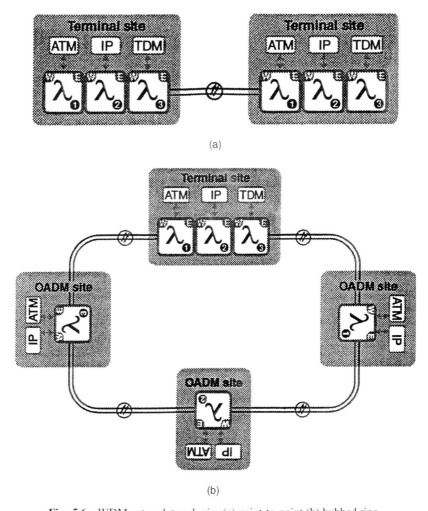

Fig. 5.6 WDM network topologies (a) point-to-point (b) hubbed ring.

distance in a ring network. The available distance is typically independent of data rate up to around 2 Gbit/s; at higher data rates, dispersion may limit the achievable distances. This should be kept in mind when installing a new WDM system that is planned to be upgraded to significantly higher data rates in the future. The maximum available distance and link loss may also be reduced if optional optical switches are included in the network for protection purposes. In some cases, it is possible to concatenate or cascade WDM networks together to achieve longer total distances; for example, by

5. Optical Wavelength Division Multiplexing   155

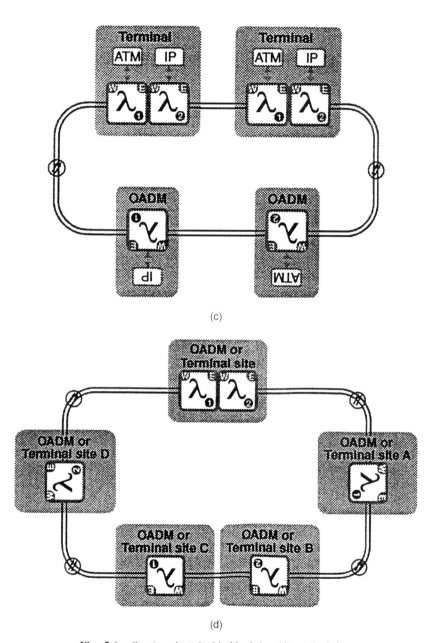

Fig. 5.6 *Continued* (c) dual hubbed ring (d) meshed ring.

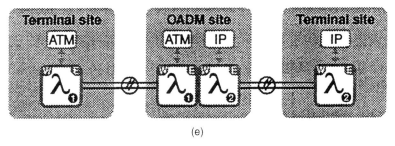

Fig. 5.6 *Continued* (e) linear optical add/drop multiplexer (OADM).

daisy chaining two point-to-point networks the effective distance can be doubled (if there is a suitable location in the middle of the link). Better performance in concatenated applications is usually achieved with channels that retime the data; data retiming is a desirable property, because it improves the signal fidelity and reduces noise and jitter. There are three levels of functionality, namely Retiming (removes timing jitter to improve clock recovery at the receiver), Reshaping (removes pulse shape distortion such as that caused by dispersion), and Regeneration (ensures the outgoing signal has sufficient power to reach its next destination). Devices that support only the first two are known as "2R" repeaters, while devices that support all three are called "3R" repeaters. Generally speaking, longer distances and better data fidelity are possible using 3R repeaters; however, this class of repeaters must be configured in either software, hardware, or both to recognize at least the data rate on the link. Care must be taken to keep the advantages of a protocol-independent design when configuring a 3R repeater.

Some WDM devices may also support longer distances using optical amplifiers in either pre-amp, post-amp, or mid-span configurations. Recently, record-breaking terabit per second point-to-point transmission in systems over more than 2000 km has been reported [7] based on dispersion-managed nonlinear transmission techniques, without the need for signal regeneration devices, using only linear optical amplifiers. However, if we extend our discussion to consider terrestrial photonic networks with dynamic routing capability, more sophisticated regeneration schemes could be necessary in order to compensate for signal quality discrepancies between high data rate WDM channels, which could be routed at different times over variable distances. Although classical optoelectronic regenerators as used in the modern telecommunication industry constitute an attractive solution to this problem for single-channel, single-data-rate transmission, it is not yet clear whether they could also serve as a cost-effective solution for a rapidly growing broadband WDM network. All-optical regenerators based on components that can process low as well as high bit rate signals could be an interesting alternative.

Signal degeneration in fiber systems arises from various sources, including amplified spontaneous emission (ASE) due to optical amplifiers, pulse spreading due to group velocity dispersion (GVD) (which can be corrected in principle through passive dispersion compensation schemes), polarization mode dispersion (PMD), and other nonlinear effects such as Kerr effect signal distortion and jitter for data rates above 10 Gbit/s. If timing jitter is negligible, simple amplification and reshaping processes are usually enough to maintain signal quality over long distances by preventing the accumulation of noise and distortion. Many repeaters convert the WDM optical signal to an electrical signal, then back into an optical signal for re-transmission; however, various schemes for all-optical 2R and 3R repeaters have been suggested. A 2R regenerator consists mainly of a linear amplifier (which may be an optical amplifier) followed by a data-driven nonlinear optical gate (NLOG), which modulates a low-noise continuous light source. If the gate transmission vs. signal intensity characteristics yield a thresholding or limiting behavior, then the signal extinction ratio can be improved and ASE noise can be partially reduced; in addition, accumulated signal chirp can be compensated. In some cases timing jitter is also a concern, for example, due to cross-phase modulation in WDM systems or pulse edge distortions due to the finite response times of nonlinear analog signal processing devices (wavelength converters). In these cases, 3R regeneration may be necessary. The basic structure of an optical 3R regenerator consists of an amplifier (which may be optical), a clock recovery function to provide a jitter-free short pulse clock signal, and a data-driven nonlinear optical gate that modulates this clock signal. The core function of the optical regenerators is thus a nonlinear gate featuring signal extinction ratio enhancement and noise reduction. Such gates can be either optical in nature (as in the case of proposed all-optical repeaters) or electronic (as in the case of hybrid optical-to-electrical conversion-based devices). Semiconductor-based gates are much more compact than fiber-based devices. It is also possible to further subdivide this class of semiconductor devices into passive devices, such as saturable absorbers, and active devices, such as semiconductor optical amplifiers that require an electrical power supply. For optical 3R regeneration, a synchronous jitter-free clock stream must be recovered from the incident signal [8]. Many solutions to this problem have been reported, and it would be beyond the scope of this chapter to describe them all.

Although 2R regenerators are attractive because of their relative simplicity, it is not clear whether they will be adequate above 10 Gbit/s data rates since this would require components with very short transition time responses.

5.2.4. LATENCY

DWDM devices also function as channel extenders, allowing many datacom protocols to reach previously impossible distances (50–100 km or more). Combined with optical amplifier technology, this has led some industry analysts to proclaim "the death of distance," meaning connection distances should no longer pose a serious limitation in optical network design. However, in many real-world applications, it is not sufficient to simply extend a physical connection; performance of the attached datacom equipment must also be considered. Latency, or propagation delay due to extended distances, remains a formidable problem for optical data communication. The effects of latency are often protocol specific or device specific. For example, using DWDM technology it is possible to extend an ESCON channel to well over 50 km. However, many ESCON control units and DASD are synchronous, and exhibit timing problems at distances beyond about 43 km. Some types of asynchronous DASD overcome this limitation; however, performance of the ESCON protocol also degrades with distance. Due to factors such as the buffer size on an ESCON channel interface card and the relatively large number of acknowledgments or handshakes required to complete a data block transfer (up to 6 or more), ESCON begins to exhibit performance drop at around 9 km on a typical channel, which grows progressively worse at longer distances. For example, at 23 km a typical ESCON channel has degraded from a maximum throughput of 17.5 MByte/s to about 10 MByte/s; if the application is a SAN trying to back up a petabyte database, the time required to complete a full backup operation increases significantly. This problem can be addressed to some degree by using more channels (possibly driving the need for a multiplexer to avoid fiber exhaust) or larger data block sizes, and is also somewhat application dependent. Other protocols, such as FICON, can be designed to perform much better at extended distances.

A further consideration is the performance of the attached computer equipment. For example, consider the effect of a 10-km duplex channel with 100 microseconds round-trip latency. A fast PC running at 500 MHz clock rate would expend 50,000 clock cycles waiting for the attached device to respond, while a mainframe executing 1000 MIPS could expend over

100,000 instruction cycles in the same amount of time. The effect this may have on the end user depends on factors such as the application software; we have noted in Vol. 1, Chapter 7 some of the effects this latency can have on the performance of a Parallel Sysplex under similar conditions. As another example, there is tremendous activity in burst mode routing and control traffic for next-generation Internet (NGI) applications. In many cases, it is desirable for data to be transported from one point in the network to one or more other points in the least possible time. For some applications the time sensitivity is so important that minimum delay is the overriding factor for all protocol and equipment design decisions. A number of schemes have been proposed to meet this requirement, including both signaling based and equipment intensive solutions employing network interface units (NIUs) and optical crossbar switches (OXBS). In order to appreciate the impetus for designing burst mode switching networks, it is useful to consider the delays that are encountered in wide area data networks such as the Internet. For example, it takes 20.5 ms for light to travel from San Francisco to New York City in a straight line through an optical fiber, without considering any intervening equipment or a realistic route. Many transmission protocols require that data packets traverse the network only after a circuit has been established; the setup phase in TCP/IP, for example, involves 3 network traversals (sending a SYN packet, responding with an ACK, and completing the procedure with another ACK) before any data packets can be sent. While this helps ensure reliable data transport, it also guarantees a minimum network delay of more than 80 ms before data can be received. For some applications this may be an unnecessary overhead to impose.

One way to mitigate the impact of network setup time on packet latency is to pipeline the signaling messages. Two methods for accomplishing these time savings have been proposed [9, 10] which allow the first packet delay to be reduced from 2 round trips to $1^{1}/_{2}$ round trips; these schemes are currently undergoing field trials. Another paradigm for providing a burst mode switching capability using conventional equipment is to send a signaling packet prior to the actual data packet transmission, and transmitting the data while the setup is in progress. Some WDM systems have proposed using this approach [11], using a dedicated wavelength for signal control information, a burst storage unit in order to buffer packets when necessary, and an implementation-friendly link scheduling algorithm to ensure efficient utilization of the WDM channels. Another alternative is commonly known as "spoofing" the channel; the attached channel extension equipment will be configured to send acknowledgments prior to actually delivering the data.

While this reduces latency and improves performance, it also makes the assumption that data can be delivered very reliably on the optical link; if there is a transmission problem after the attached channels have received their acknowledgment of successful delivery, the error recovery problem becomes quite difficult. Although this approach has been implemented in commercial devices, it has not been widely accepted because of the potential data integrity exposures inherent in the design.

In many optical networks, cross-connect switches are employed, which act like electronically reconfigurable patch panels. The cross-connect changes its switching state in response to external control information, such as from an outband signal in the data network. The alternative to signaled data transmission is data switching using header information. This is somewhat different from a typical ESCON or Gigabit Ethernet switch, which performs optical to electrical conversion, reads the data header, sets the switch accordingly, and then reconverts the data to the optical domain for transmission (sometimes retiming the signal to remove jitter in the process). An example of outband switching networks has been demonstrated [12]; in the approach, a data packet is sent simultaneously with a header that contains routing information. The headers are carried along with the data, but out of band, using subcarrier multiplexing or different wavelengths. At switch nodes inside the network, the optical signal is sampled just prior to entering a short fiber delay line. While most of the signal is being delayed, a fast packet processing engine determines the correct state of the switching fabric, based on the incoming header and a local forwarding table. The switch fabric is commanded to enter a net state just before the packet exits the delay line and enters the switch. This method provides the lowest possible latency for packet transmission and removes the task of pre-calculating the arrival time of burst mode data packets.

5.2.5. PROTECTION AND RESTORATION

Backup fiber protection or restoration refers to the multiplexer's ability to support a secondary fiber path for redundancy in case of a fiber break or equipment failure. The most desirable is so-called "1 + 1" SONET-type protection switching, the standard used by the telecom industry, in which the data is transmitted along both the primary and backup paths simultaneously, and the data switches from the primary to the backup path within 50 ms. (This is the SONET industry standard for voice communications; while it has been commonly adopted by protocol-independent WDM devices, the effects of switching time on the attached equipment depends on the application, as in the previous discussion of latency effects.) There are different meanings for protection, depending on whether the fiber itself or the fiber and electronics are redundant. In general, a fully protected system includes dual redundant cards and electrical paths for the data within the multiplexer, so that traffic is switched from the primary path to the secondary path not only if the fiber breaks, but also if a piece of equipment in the multiplexer fails. A less sophisticated but more cost-effective option is the so-called fiber trunk switch, which simply switches from the primary path to the backup path if the primary path breaks. Not all trunk switches monitor the backup path as $1 + 1$ switches do, so they can run the risk of switching traffic to a path that is not intact (some trunk switches provide a so-called "heartbeat" function, sending a light pulse down the backup path every second or so to establish that the backup link remains available). Trunk switches are also slower, typically taking from 100 ms to as much as 2 seconds to perform this switchover; this can be disruptive to data traffic. There are different types of switches; a unidirectional switch will only switch the broken fiber to its backup path (for example, if the transmit fiber breaks, then the receive fiber will not switch at the same time). By contrast, a bidirectional switch will move both the transmit and receive links if only one of the fibers is cut. Some types of datacom protocols can only function properly in a bidirectional switches environment because of timing dependencies on the attached equipment; the link may be required to maintain a constant delay for both the transmit and received signal in a synchronous computer system, for example. Some switches will toggle between the primary and secondary paths searching for a complete link, while others are non-revertive and will not return to the primary path once they have switched over. It is desirable to have the ability to switch a network on demand from the network management console, or to lock the data onto a single path and prevent switching (for example, during link maintenance).

Another desirable switching feature is the ability to protect individual channels on a per channel basis as the application requires; this is preferable to the "all or none" approach of trunk switches, which require that either all channels be protected, or no channels be protected. Hybrid schemes using both trunk switching and $1 + 1$ protection are generally not used, as they require some means of establishing a priority of which protection mechanism will switch first and they defeat the purpose of the lower cost trunk switch. Other features such as dual redundant power supplies and cooling units with concurrent maintenance (so-called "hot swappable" components, which can be replaced without powering down the device) should also be part of a high-reliability installation.

Another desirable property is self-healing, which means that in the event of a fiber break or equipment failure the surviving network will continue to operate uninterrupted. This may be accomplished by re-routing traffic around the failed link elements; some form of protection switching or by-pass switching can restore a network in this manner. In a larger network consisting of multiple cross-connects or add/drop multiplexers, there are two approaches to producing optical self-healing networks [13]. One is to configure a physical-mesh topology network with optical cross connects (OXCs), and the other is to configure a physical ring topology network with optical add/drop multiplexers (OADMs, or WDM ADMs). Ring topology based optical self-healing networks are generally preferred as a first step because of the lower cost of OADMs compared with OXCs; also the protection speed in a ring topology is much faster, and the OADM is more transparent to data rate and format. Eventually, in future photonic networks, multiple optical self-healing rings may be constructed with emerging large-scale OXCs. New architectures have also been proposed, such as bi-directional wavelength path-switched ring (BWPSR), which uses bi-directional wavelength-based protection and a wavelength-based protection trigger self-healing ring network [14].

5.2.6. NETWORK MANAGEMENT

Some DWDM devices offer minimal network management capabilities, limited to a bank of colored lamps on the front panel; others offer sophisticated IP management and are configured similar to a router or switch. IP devices normally require attached PCs for setup (defining IP addresses, etc.) and maintenance; some can have a "dumb terminal" attachment to an IP device which simplifies the setup, but offers no backup or redundancy if the IP site fails. Some devices offer minimal information about the network; others offer an in-band or out-band service channel that carries management traffic for the entire network. When using IP management, it is important to consider the number of network gateway devices (typically routers or switches) that may be attached to the DWDM product and used to send management data and alarms to a remote location. More gateways are desirable for greater flexibility in managing the product. Also, router protocols such as Open Shortest Path First (OSPF) and Border Gateway Protocol (BGP) are desirable because they are more flexible than "static" routers; OSPF can be used to dynamically route information to multiple destinations.

Many types of network management software are available — in datacom applications, these are often based on standard SNMP protocols supported by many applications such as HP Openview, CA Unicenter, and Tivoli Netview. IP network management can be somewhat complex, involving considerations such as the number of IP addresses, number of gateway elements, resolving address conflicts in an IP network, and others. Telecom environments will often use the TL-1 standard command codes, either menu-driven or from a command line interface, and may require other management features to support a legacy network environment; this may include CLEI codes compatible with the TERKS system used by the telecommunications industry and administered by Telcordia Corp., formerly Bell Labs [15]. User-friendly network management is important for large DWDM networks, and facilitates network troubleshooting and installation. Many systems require one or more personal computers to run network management applications; network management software may be provided with the DWDM, or may be required from another source. It is desirable to have the PC software pre-loaded to reduce time to installation (TTI), pre-commissioned and pre-provisioned with the user's configuration data, and tested prior to shipment to ensure it will arrive in working order. Finally, note that some protocol-specific WDM implementations also collect network traffic statistics, such as reporting the number of frames, SCSI read/write operations, Mbytes payload per frame, loss of sync conditions, and code violations (data errors). Optical management of WDM may also include various forms of monitoring the physical layer, including average optical power per channel and power spectral density, in order to proactively detect near end-of-life components or optimize performance in amplified WDM networks.

5.2.7. NONLINEAR EFFECTS AND OPTICAL AMPLIFIERS FOR WDM

The interaction of light with the optical fiber material is typically very small, particularly at low optical power levels. However, as the level of optical power in the fiber is increased nonlinear effects can become significant; this is especially important for long-distance fiber links, which provide the opportunity for smaller effects to build up over distance. Because WDM involves transmitting many optical signals over a common fiber, nonlinear

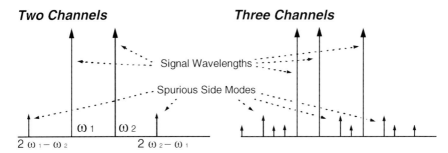

Fig. 5.7 Example of four wave mixing (FWM) in WDM networks with 2 channels and with 3 channels.

effects in the link can become significant. In an optical communication system, nonlinear effects can induce transmission errors that place fundamental limits on system performance, in much the same fashion as attenuation or dispersion effects. At the same time, some important components such as amplifiers for extended distance WDM systems rely on nonlinear effects for their operation.

One of the most common nonlinear interactions is known as "four wave mixing" (FWM), which occurs when two or more optical signals propagate in the same direction along a common single-mode fiber. As illustrated in Fig. 5.7, optical signals in the fiber can mix to produce new signals at wavelengths that are spaced at the same intervals as the original signals. The effect can also occur between three or more signals, making the overall effect quite complex. FWM increases exponentially with signal power, and becomes greater as the channel spacing is reduced; in particular, it is a concern with dense wavelength division multiplexing systems. If WDM channels are evenly spaced, then the spurious FWM signals will appear in adjacent wavelength channels and act as noise. One method of dealing with this problem is to space the channels unevenly to reduce the effect of added noise on adjacent channels; however, FWM still removes some optical power from the desired signal levels. Because FWM is caused by signals that remain in phase with each other over a significant propagation distance, the effect is stronger for lasers with a long coherence length. Also, FWM is strongly influenced by chromatic dispersion — because dispersion ensures that different signals do not stay in phase with each other for very long, it acts to reduce the effect of FWM.

Another nonlinear optical phenomena is known as frequency chirping, or inducing a linear frequency sweep in an optical pulse. Until now, we have assumed that the refractive index of the fiber core is a constant, independent of the optical power. Actually, sufficiently high optical power levels can affect the material properties of the glass and induce small changes in the refractive index. The frequency chirp is generated by self-phase modulation, which arises from the interaction of the propagating light and the intensity dependent portion of the fiber's refractive index [17]. Because these effects are caused by the propagating signal itself, they are known as carrier-induced phase modulation (CIP). The fiber's refractive index can be expressed as follows:

$$n_1 = n_{10} + n_{12}I(t) \tag{5.1}$$

where n_{10} is the refractive index under low optical power conditions (for this case $n_1 = n_{10}$), $I(t)$ is the intensity profile of the propagating light, and n_{12} is a positive material constant [17]. The propagation constant is then given by

$$k_1 = \omega_0 n_{10}/c + \Gamma I(t) \tag{5.2}$$

where the constant Γ represents a collection of terms [17]. The phase of the optical pulse now becomes

$$\psi = \omega_0 t - \omega_0 n_{10} z/c - \Gamma I(t) \tag{5.3}$$

where z is the propagation distance. The instantaneous frequency of the light is thus proportional to the negative time derivative of the intensity profile,

$$\omega_i = d\psi/dt = \omega_0 - \Gamma dI(t)/dt \tag{5.4}$$

and the properties of the resulting chirp depend on the time-varying light intensity; for many pulse shapes including Gaussian, this results in a nonuniform frequency chirp. Another property of interest is the group velocity dispersion (GVD); this is calculated by expanding the propagation constant, $k(\omega)$, about its center frequency and retaining the second-order derivative [18]. Optical fibers with positive GVD cause the frequency components to spread out as the light propagates along the fiber; by contrast, in a fiber with negative GVD the frequency components move closer together as the light propagates. The specific material properties of the fiber and the wavelength of light determine whether the GVD is positive or negative. The combined effects of self-phase modulation and GVD are called dispersive self-phase modulation; this effect has applications in optical pulse compression systems [18]. When there are multiple signals at different wavelengths in the same fiber, nonlinear phase modulation in one signal can induce phase modulation of the other signals. This is known as cross-phase modulation; in contrast with other nonlinear effects, it does not involve power transfer between signals. Cross-phase modulation can introduce asymmetric spectral broadening and distortion of the pulse shapes.

At higher optical power levels, nonlinear scattering may limit the behavior of a fiber optic link. The dominant effects are stimulated Raman and Brillouin scattering. When incident optical power exceeds a threshold value, significant amounts of light may be scattered from small imperfections in the fiber core or by mechanical (acoustic) vibrations in the transmission media. These vibrations can be caused by the high-intensity electromagnetic fields of light concentrated in the core of a single-mode fiber. Because the scattering process also involves the generation of photons, the scattered light can be frequency shifted [16]. Put another way, we can think of the high-intensity light as generating a regular pattern of very slight differences in the fiber refractive index; this creates a moving diffraction grating in the fiber core, and the scattered light from this grating is Doppler shifted in frequency by about 11 GHz. This effect is known as stimulated Brillouin scattering (SBS); under these conditions, the output light intensity becomes nonlinear as well. Stimulated Brillouin scattering will not occur below a critical optical power threshold, as discussed in Vol. 1, Chapter 7. Brillouin scattering has been observed in single-mode fibers at wavelengths greater than cutoff with optical power as low as 5 mW; it can be a serious problem in long-distance communication systems when the span between amplifiers is low and the bit rate is less than about 2 Gbit/s, in WDM systems up to about 10 Gbit/s when the spectral width of the signal is very narrow, or in remote pumping of some types of optical amplifiers [16]. In general, SBS is worse for narrow laser linewidths (and is generally not a problem for channel bandwidth greater than 100 MHz), wavelengths used in WDM (SBS is worse near 1550 nm than near 1300 nm), and higher signal power per unit area in the fiber core. The effects of these factors has been described in Vol. 1, Chapter 7. In cases where SBS could be a problem, the source linewidth can be intentionally broadened by using an external modulator or additional RF modulation on the laser injection current. However, this is a tradeoff against long-distance transmission, since broadening the linewidth also increases the effects of chromatic dispersion.

When the scattered light experiences frequency shifts outside the acoustic phonon range, due instead to modulation by impurities or molecular vibrations in the fiber core, the effect is known as stimulated Raman scattering (SRS). The mechanism is similar to SBS, and scattered light can occur in both the forward and backward directions along the fiber; the effect will not occur below a threshold optical power level as noted in Vol. 1, Chapter 7. As a rule of thumb, the optical power threshold for Raman scattering is about three times larger than for Brillouin scattering. Another good rule of thumb is that SRS can be kept to acceptable levels if the product of total power and total optical bandwidth is less than 500 GHz-W. This is quite a lot. For example, consider a 10-channel DWDM system with standard wavelength spacing of 1.6 nm (200 GHz). The bandwidth becomes $200 \times 10 = 2000$ GHz, so the total power in all 10 channels would be limited to 250 mW in this case (in most DWDM systems, each channel will be well below 10 mW for other reasons such as laser safety considerations). In single-mode fiber, typical thresholds for Brillouin scattering are about 10 mW and for Raman scattering about 35 mW; these effects rarely occur in multimode fiber, where the thresholds are about 150 mW and 450 mW, respectively. In general, the effect of SRS becomes greater as the signals are moved further apart in wavelength (within some limits); this introduces a tradeoff with FWM, which is reduced as the signal spacing increases.

Optical amplifiers can also be constructed using the principle of SRS; if a pump signal with relatively high power (half a watt or more) and a frequency 13.2 THz higher than the signal frequency is coupled into a sufficiently long length of fiber (about 1 km), then amplification of the signal will occur. Unfortunately, more efficient amplifiers require that the signal and pump wavelengths be spaced by almost exactly the Raman shift of 13.2 THz, otherwise the amplification effect is greatly reduced. It is not possible to build high-power lasers at arbitrary signal wavelengths; one possible solution is to build a pump laser at a convenient wavelength, then wavelength shift the signal by the desired amount [16]. However, another good alternative to SRS amplifiers is the widely used erbium-doped fiber amplifiers (EDFA). These allow the amplification of optical signals along their direction of travel in a fiber, without the need to convert back and forth from the electrical domain. While there are other types of optical amplifiers based on other rare earth elements such as praseodymium (Pd) or neodymium (Nd), and even some optical amplifiers based on semiconductor devices, the erbium-doped amplifiers are the most widely used because of their maturity and good performance at wavelengths of interest near 1550 nm.

An EDFA operates on the same principle as an optically pumped laser; it consists of a relatively short (about 10 meters) section of fiber doped with a controlled amount of erbium ions. When this fiber is pumped at high power (10 to 300 mW) with light at the proper wavelength (either 980 nm or 1480 nm) the erbium ions absorb the light and are excited to a higher energy state. Another incident photon around 1550 nm wavelength will cause stimulated emission of light at the same wavelength, phase, and direction of travel as the incident signal. EDFAs are often characterized by their gain coefficient, defined as the small signal gain divided by the pump power. As the input power is increased, the total gain of the EDFA will slowly decrease; at some point, the EDFA enters gain saturation, and further increases to the input power cease to result in any increase in output power. Because the EDFA does not distort the signal, unlike electronic amplifiers, it is often used in gain saturation. The gain curve of a typical EDFA as a function of wavelength is shown in Fig. 5.8; note that the gain at 1560 nm is about twice as large as the gain at 1540 nm. This can be a problem when operating WDM systems; some channels will be strongly amplified and dominate over other channels that are lost in the noise.

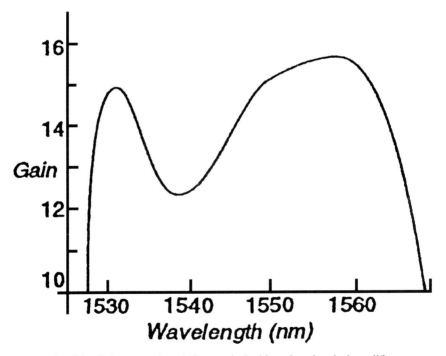

Fig. 5.8 Gain vs wavelength for a typical erbium-doped optical amplifier.

Furthermore, a significant complication with EDFAs is that their gain profile changes with input signal power levels; so, for example, in a WDM system the amplifier response may become nonuniform (different channels have different effective gain) when channels are added or dropped from the fiber. This requires some form of equalization to achieve a flat gain across all channels. There has been a great deal of research in this area; some proposals include adding an extra WDM channel locally at the EDFA to absorb excess power (gain clamping), and manipulating either the fiber doping or core structure. Another concern with EDFAs is that some of the excited erbium undergoes spontaneous emission, which can create light propagating in the same direction as the desired signal. This random light is amplified and acts as background noise on the fiber link; the effect is known as amplified spontaneous emission (ASE). Since ASE can be at the same wavelength as the desired signal, it may be difficult to filter out; furthermore, ASE accumulates in systems with multiple amplifier stages and is proportional to the amplifier gain.

Other nonlinear effects can be used to produce useful devices by changing the properties of the fiber itself; one of the most common examples is fiber Bragg gratings. When an optical fiber is exposed to ultraviolet light, the fiber's refractive index is changed; if the fiber is then heated or annealed for a few hours, the index changes can become permanent. This phenomena is called photosensitivity [19, 20]. The magnitude of the index change depends on many factors, including the irradiation wavelength, intensity, and total dose, the composition and doping of the fiber core, and any materials processing done either prior or subsequent to irradiation. In germanium-doped single-mode fibers, index differences between 10^{-3} and 10^{-5} have been obtained. Using this effect, periodic diffraction gratings can be written in the core of an optical fiber. This was first achieved by interference between light propagating along the fiber and its own reflection from the fiber endface [21]; this is known as the internal writing technique and the resulting gratings are known as Hill gratings. Another approach is the transverse holographic technique, in which the fiber is irradiated from the side by two beams which intersect at an angle within the fiber core. Gratings can also be written in the fiber core by irradiating the fiber through a phase mask with a periodic structure. These techniques can be used to write fiber Bragg gratings in the fiber core; such gratings reflect light in a narrow bandwidth centered around the Bragg wavelength, λ_B, which is given by

$$\lambda_B = 2N_{\text{eff}}\Lambda \quad (5.5)$$

where Λ is the spatial period, or pitch, of the periodic index variations and N_{eff} is the effective refractive index for light propagating in the fiber core. There are many applications for fiber Bragg gratings in optical communications and optical sensors, such as tapped optical delay lines, filters, multiplexers, optical strain gauges, and others (an extensive review is provided in ref. 22, 23). Fiber Bragg gratings function in reflection, while many applications require a transmission effect; this conversion is accomplished using an optical circulator, a Michaelson or Mach-Zender interferometer, or a Sagnac loop [24]. Fiber Bragg gratings can be used to multiplex and demultiplex wavelengths in a WDM system, or to fabricate add/drop filters within the optical fibers that offer very low insertion loss; they can also be used in various dispersion compensation schemes. These devices represent a promising new technology for future commercial WDM applications.

5.3. Commercial WDM Systems

Many commercial DWDM products in use today have been developed for the telecommunication and data communications market; there are also a number of testbeds and service trials underway, and new products or technologies are being proposed at a rapid pace. Some examples include the MultiWave WDM terminal from Ciena Corp., which accommodates up to 16 OC-48 channels and will soon be upgradable to more than 32 channels, and the WaveMux from Pirelli, which handles up to 10 OC-192 channels or 32 OC-48 channels. There is also a unique implementation of WDM using free-space optics, the Lucent OptiAir (a free space 4-channel wavelength multiplexed communication link, for applications where physical fiber connections are not practical, such as communication with ships at dock in a seaport). Several other corporations are also investigating applications of WDM, including the ESPRIT program in Europe, the Optical Network Technology Consortium, the All Optical Network Consortium, the MONET consortium led by AT&T, and others [25].

In particular, the MONET project is a multi-vendor government-sponsored network consortium and testbed for future WDM systems [26]. While its capabilities and applications continue to evolve, we can discuss a recent snapshot of the network as an example of how WDM is being evaluated for future networking applications. The MONET network in Washington, D.C. consists of a pair of two fiber rings denoted as the east and west ring. The east ring is provided and maintained by Lucent Technologies. This ring consists of two wavelength selective cross-connects located at the Laboratory for Telecommunication Sciences at the National Security Agency and the Naval Research Laboratory, one wavelength add/drop multiplexer at the National Aeronautics and Space Administration, and two wavelength amplifiers in the Bell Atlantic central office. The west ring is provided by Tellium and managed by Telcordia Technologies. This ring consists of three wavelength ADMs at the Defense Advanced Research Projects Agency/Information Systems Institute, Defense Information Systems Agency, and Defense Intelligence Agency, as well as four wavelength amplifiers in the Bell Atlantic central offices. Bell Atlantic provides the in-ground fiber infrastructure for the network. The whole network is controlled and managed by Telcordia Technologies' CORBA-based network control and management (NC&M) system, which runs on the ATM network at OC-3c rates over a supervisory wavelength channel at 1510 nm. The Lucent switching fabric is based on lithium niobate devices and is thus protocol transparent. The Tellium ADMs are based on wavelength transponders and transceivers using O/E/O conversion and an electrical switching fabric. Thus, the west ring is opaque and has wavelength interchange capability.

The west ring has 4 modes of transmission; OC-3, OC-12, OC-48, and optical signal (OSIG) mode, which is used to transmit non-SONET rate signals up to 2.5 Gbit/s without timing recovery during the O/E/O conversion. Many experiments have been conducted on this network; for example, recently it was used to test the long-distance transmission capabilities of Gigabit Ethernet in OSIG mode [26]. This was accomplished using two Pentium II workstations as test hosts running short wave Gigabit Ethernet links to a switch, which was in turn configured into two virtual LANs (VLANs) to prevent the hosts from bridging packets directly between them. Packets were thus forced over the Gigabit Ethernet LX ports into the WDM network. This work has demonstrated for the first time transport of Gigabit Ethernet packets directly over WDM for 1062 km; this suggests that this protocol, and probably also the Fibre Channel physical layer, can be used for WAN backbone transmission, bypassing the intermediate ATM/SONET protocol layers. Of course, performance issues remain to be resolved; using the jumbo frames defined in the Gigabit Ethernet standard, a maximum of about 9 Kbytes data block size is permitted, which is not large compared with Fibre Channel or ESCON block sizes. Thus, the role of Gigabit Ethernet in the SAN remains to be determined.

MONET is an example of an active WDM WAN employing a multivendor environment. Some of the other WDM networking efforts currently underway deal with passive optical networks (PON) [27]; other research programs are underway to investigate new architectures for client-independent OTNs including IP over DWDM; one example is the EURESCOM project P918 [28]. Similarly, commercial WDM products offer a wide range of options. Coarse WDM products may consist of individual transceivers or multichip modules with built-in optical multiplexing functions. Most DWDM products are based on a fundamental building block, such as a card cage, which can be daisy-chained or scaled to accommodate more wavelength channels. Some building blocks are mounted in a standard 19- or 23-inch equipment rack, which is assembled in the field; others are pre-packaged as standalone boxes, which vary from large systems to smaller packages (the size of a large PC); the physical size of a WDM system depends on many factors, including the choice of design point, transponders vs. separate client/network cards, use of small form factor optical transceivers, etc. In some cases, a smaller "satellite box" with only a few wavelengths or even a coarse WDM solution is deployed close to the end user premises, which feeds into a larger DWDM node some distance away on the WAN. Many different types of network management are offered, which may consist of a proprietary graphic user interface with protocol-specific performance monitoring to an industry standard SNMP approach or a TL-1 (per EIA standard 232) and CMISE interface which allows telco standard operations, administration, maintenance, and provisioning functions (OAM&P). Use of sub-rate TDM or CDMA, supported network topologies with different bandwidths depending on the configuration, number of channels over a single fiber or pair of fibers, and many other design choices contribute to the differences between commercially available WDM products. To further complicate matters, many WDM products are resold or rebranded by different companies offering a range of technical support and maintenance options.

Given the rapid growth of this technology, it is not possible to comprehensively list all of the commercial products being introduced, or all of the research efforts currently underway to support future products. A sample of selected WDM companies and WAN backbone carriers is given in Table 5.3. Product marketing literature and claims of supported features should always be examined with care; potential users should evaluate the most recent technical information from prospective companies to find the best solution for a given application. Instead of attempting to give a detailed technical description of all the major commercial WDM product offerings, we will select one representative product to illustrate first-generation and

5. Optical Wavelength Division Multiplexing

Table 5.3

(a) partial, non-comprehensive list of commercial WDM device and equipment providers; for latest information on resale agreements and brand names, contact the companies listed below. Many new companies enter the WDM industry each year; this is not intended to be a comprehensive list, and does not imply any endorsement of the product or companies listed below; this information is provided for reference purposes only. All brand names listed below are registered trademarks of their respective parent companies.

Adva Fiber Service Platform (also offered under various OEM and resale agreements over the past several years, including those with Canoga-Perkins (Lambda-Access or WA 8/16), Controlware, Hitachi, Centron, CNT or Computer Network Technologies (UltraNet Wave Multiplexer), Inrange, formerly General Signal Networks (OptiMux 9000), and Cisco (Metro 1500). The initial product offering was known as the OCM-8 (Optical Channel Multiplexer).

Alcatel Networks WDM platform (backbone and OXC equipment)

Astral Point ON 5000 optical access transmission system and Optical Services Architecture

Avanex PowerMux DWDM devices

BrightLink Networks (OXC equipment)

Centerpoint Broadband Technologies (telecommunications equipment)

Ciena LightWorks product line; MultiWave 1600, MultiWave Sentry 1600/4000, MultiWave CoreStream, MultiWave CoreDirector, MultiWave EdgeDirector

Cisco Metro 1500 product line, WaveMux 6400, TeraMux, Wavelength Router (ONS 15900 series)

Corvis CorWave product family; Optical Network Gateway, Optical Amplifier, Optical Routing Switch, CorManager system

Ericsson ERION (Ericsson Optical Networking) product line, including ERION Linear and FlexRing

Finisar OptiCity metro DWDM product line

Fujitsu (backbone and OXC equipment)

IBM 2029 Fiber Saver (replaces the IBM 9729 optical wavelength division multiplexer; the 2029 is part of a joint development agreement with Nortel Networks)

Juniper Networks (backbone equipment)

Kestrel Solutions TalonMX optical frequency division multiplexer

Lucent WaveStar product line; OLS 40G, OLS 80G, OLS 400G, TDM 2.5G, TDM 10G, ADM 16/1, DACS 4/4/1, DVS, Bandwidth Manager, LambdaRouter

Table 5.3 continued

LuxN WavStation product line and Multiplex Channel Module (MCM)

NEC (backbone and OXC equipment)

Nortel Networks OPTera Metro 5200 MultiService Platform, OPTera LH product line, and OPTera Connect

Osicom GigaMux (a subsidiary, Sorrento Networks, offers the EPC sub-rate TDM solution)

Pirelli TeraMux product line

Sycamore Networks SN 8000 Metro Core system, Intelligent Optical Network Node SN 8000, optical switch SN 16000

Tellabs Titan 6100 series WDM platform

Tellium Aurora 32 and 512 product lines (backbone and OXC equipment)

(b) major North American optical fiber network backbone carriers

AT&T
Nortel
MCI WorldCom
Sprint
Global Crossing
Williams
Qwest
IXC
Level 3
GTE
Enton

(c) major European optical fiber network backbone carriers

Interoute
GTS Group
Viatel
Teleglobe Communications
Energis
COLT
Carrier1

third-generation devices (fourth-generation ultra-dense WDM devices are not yet commercially available). For each of these typical devices, we will provide a detailed description of its components, packaging, and functionality; similar building blocks are used by other commercial products.

5.3.1. FIRST-GENERATION WDM: THE IBM 9729 OPTICAL WAVELENGTH DIVISION MULTIPLEXER

One example of a first-generation DWDM system is the IBM 9729 Optical Wavelength Division Multiplexer [1, 29, 30]. The first WDM product developed specifically for the datacom industry, it became available as a special request product in 1993 and was released as a commercial product in 1996. The device is shown in Fig. 5.9; it allows the transmission of up to 10 full duplex links (20 independent data streams) over a single fiber. Using different adapter cards, a mixture of ESCON, Fibre Channel, FDDI, and ATM links can be plugged into the device, which is protocol independent. The data is remodulated using distributed feedback laser diodes with wavelengths spaced 1 nm apart near the 1550 nm region, in C-band only. The optical signals are then combined using a diffraction grating with embedded fiber pigtails, and coupled into a single-mode fiber; another unit at the far end of the link demultiplexes the signals.

Fig. 5.9 A pair of IBM 9729 optical wavelength division multiplexers.

The maximum unrepeated distance for the 9729 is 50 km for data in the 200 Mbit/s range (such as ESCON or OC-3) with a 15 dB link budget, and 40 km for data in the 1 Gbit/s range (such as HiPerLinks) with a 12 dB link budget. Thus, the 9729 functions as a channel extender for most protocols, including ESCON, ETR, CLO, and HiPerLinks. Note that despite the low data rate of ETR links, they are limited to maximum distances of 40 km because of timing considerations. Proprietary signaling is used between a pair of 9729s to support OFC propagation beyond the 20-km limits imposed by the Fibre Channel Standard. For protection against a broken optical link between the units, an optional dual-fiber switch card is available, which consists of an optical switch and a second fiber link. The units automatically detect if the primary fiber link is broken, and switch operation to the secondary fiber within 2 seconds; this is not intended to provide continuous operation of the attached systems, only to restore the broken link capacity more quickly. The 9729s are managed through a serial data port, and can report their status or receive simple commands, such as switching to the second fiber link, from a personal computer or workstation running a software management package.

A typical GDPS installation can be implemented by routing all point-to-point links between site 1 and site 2 over the 9729; testing and performance of this system have been described previously [31]. Today, many large systems use point-to-point wavelength division multiplexing to reduce the total number of inter-site fiber optic links. However, there is considerable interest in extending this architecture into ring topologies, with additional features for data protection and management. In the following section, we describe an example of a next-generation DWDM system recently developed for GDPS and other datacom applications.

5.3.2. THIRD-GENERATION WDM: THE IBM 2029 FIBER SAVER (NORTEL OPTERA METRO 5200 MULTISERVICE PLATFORM)

In February 2000, IBM announced a third-generation DWDM solution, the 2029 Fiber Saver, as a follow-on to the first generation 9729 technology [32, 33]. This product is the result of a joint development relationship with Nortel Networks, and the 2029 is based on the same building blocks used in the Nortel Optera Metro 5200 MultiService Platform. The two products share a common set of hardware and software, although the 2029 supports only a pre-tested level of Optera hardware and code. While the Optera is provided as a standard telecom service provider package, the 2029 is repackaged by IBM with additional features, including turnkey installation with a pre-tested and configured PC, a class 1 laser eye safe cabinet for enterprise applications, integrated patch panels for native attachment of all datacom interfaces, and standard dual AC power supplies. For the sake of brevity, we will describe the design features of both products in this section, referring to the 2029 for our examples, and note those areas in which the two designs may be different.

The 2029 is a fundamentally different network architecture from the 9729 and offers many additional features that make it a more modular, scaleable approach to DWDM. Each 2029 model contains up to 2 shelves mounted on a 19-inch rack inside a standard-size datacom cabinet, as illustrated in Fig. 5.10. The cabinet also contains standard dual redundant power supplies, an optical patch panel, and (if required) an Ethernet hub for managing inter-shelf communications. We will discuss each of these functions in detail. We will discuss the link budgets and distances in more detail shortly; for now, we note that they allow a maximum distance of 50 km in a point-to-point configuration. Although the individual ITU grid lasers used

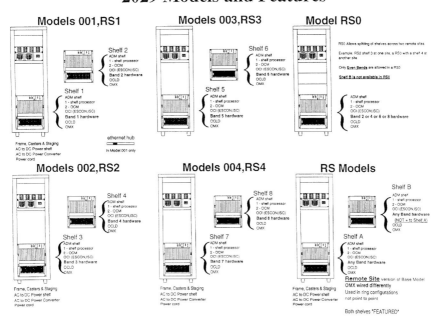

Fig. 5.10 The IBM 2029 Fiber Saver.

to achieve this distance meet international Class 1 laser safety requirements, the multiplexed link carrying more than 16 wavelengths is Class 1 in North America only; elsewhere it is a Class 3A device per IEC 825 standards. (For a discussion of laser safety standards, see Vol. 1, Chapter 7.) This is a standard requirement for all WDM systems supporting more than 16 wavelengths at this distance; the only way to avoid this and achieve worldwide class 1 operation of the entire link would be to either reduce the supported distance, reduce the number of wavelength channels, or implement some form of open fiber control interlock on all channels. (This use of OFC would not be industry standardized, and may introduce additional design tradeoffs related to loss of light propagation across the WDM network, especially for ring topologies.) Thus, an Optera system installed outside North America must be located in a locked room or similar environment with access restricted to individuals with appropriate laser safety training. However, the 2029 provides restricted access to this interface by means of a lockable cabinet, screw-down covers over the multiplexed fiber connections, appropriate safety labeling on the product, and supporting safety materials in the documentation; thus, the 2029 provides a self-contained laser safe environment and there are no restrictions on its installation outside North America. All of the interfaces for subtended equipment meet Class 1 laser safety requirements; a single client interface card supports many different protocols, so optical attenuators and adapters or hybrid fiber cables are required to attach datacom channels to the WDM device. The Optera provides plug-in optical attenuators for this purpose, which are configured to a separate rack mountable patch panel. The 2029 uses attenuated fiber optic cables as described in Chapter 4, with the appropriate connector types available on a patch panel integrated into the enterprise cabinet. There is a 1 to 1 mapping of interfaces on the patch panel to adapter cards in the shelves.

A detailed view of the shelf is given in Fig. 5.11; it consists of a card cage that holds different types of adapters, a maintenance panel with power supply breakers and connections for monitoring and telemetry, some fiber slack management, a dual redundant cooling unit, and a tray containing the optical multiplexing (OMX) modules. Each OMX module is a passive device that can multiplex up to 4 optical wavelengths into one fiber path; the OMX modules are wavelength specific and are identified by their band (1 to 8). As shown in Fig. 5.12, the shelf contains up to 20 active circuit cards of 4 different types:

- 1 Shelf Processor (SP) card; this is a programmable processor that does not handle data, but provides management information and IP

5. Optical Wavelength Division Multiplexing

2029 Shelf

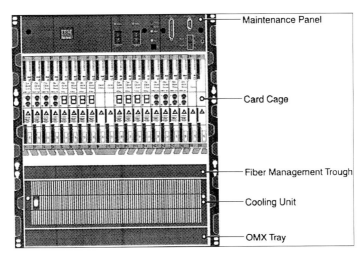

Fig. 5.11 Detail of a 2029 shelf.

Shelf Layout

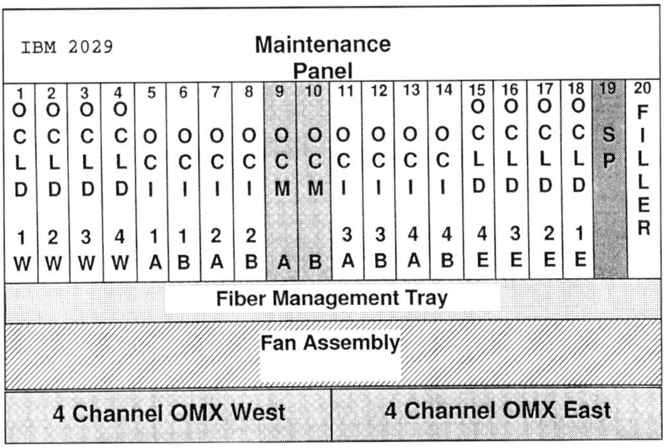

Fig. 5.12 Card types in a 2029 shelf.

addressing for the shelf. The SP card monitors all circuit cards in the shelf, using feedback from each card to provide performance monitoring, software and configuration management, and alarm reporting. It has only an electrical connection to the shelf backplane, and always occupies slot 19.

- 8 Optical Channel Interface (OCI) cards; these connect to the equipment to be multiplexed or demultiplexed through optical fiber attachments to the patch panel, and perform optical to electrical conversion of the data prior to multiplexing and electrical to optical conversion after demultiplexing. There are different types of OCI cards available; a low-speed card that supports single-mode protocols only up to a maximum data rate of 622 Mbit/s, a high-speed card that supports both multimode and single-mode protocols up to 1.25 Gbit/s, another high speed card that extends this range to 2.5 Gbit/s, a 4:1 TDM card that puts up to 4 signals at data rates of 270 Mbit/s or less over a single 1.25 Gbit/s wavelength channel, and a protocol-specific card for Parallel Sysplex coupling links. All of these operate at 1300-nm wavelength; a separate OCI card is available for short-wavelength (850-nm) operation. These cards are located in slots 5 to 8 and 11 to 14; they typically use SC duplex connectors, except for the 4TDM card, which uses MT-RJ interfaces.
- 8 Optical Channel Laser and Detector (OCLD) cards; these perform electrical to optical and optical to electrical conversion of the data onto the ITU grid long-wavelength lasers. There is a different OCLD for each wavelength; they are identified by wavelength band (1 to 8) and by channel (1 to 4). They must correspond to the wavelength of the OMX filters in the shelf. Specific channels have fixed locations in the shelf, occupying slots 1 to 4 and 18 to 15; they use a pair of FC optical connectors to attach to the OMX modules.
- 2 Optical Channel Manager (OCM) cards; these are dual redundant, and provide switching functions from the fully cross-connected backplane for backup protection switching of data. All of the data to be multiplexed flows through one or both OCM cards, which allow any OCI card to map to any OCLD in the same shelf. The OCM stores configuration and provisioning data, as well as copies of the IP address information for the shelf. The OCMs have fixed positions in slots 9 and 10, and have only an electrical connection to the shelf backplane.

The shelf can be divided roughly in half, with channels corresponding to either the "east" or "west" side (while this designation has been adopted by many telecom service providers, it does not carry any inherent meaning; the two paths can just as readily be referred to as North/South, Red/Blue, etc.). The two halves of the shelf act as dual redundant data paths in protected

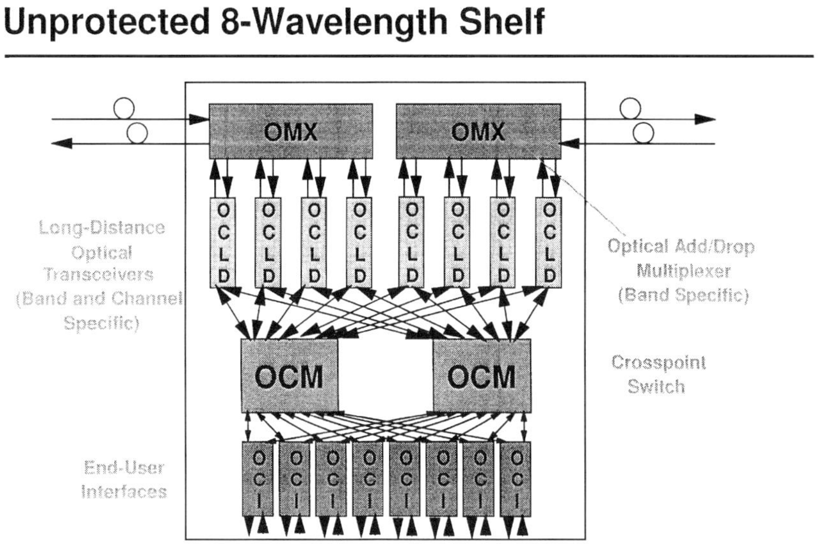

Unprotected 8-Wavelength Shelf

Fig. 5.13 2029 shelf traffic flow in base or unprotected mode.

or high availability mode. In unprotected or base mode, the data flow is as shown in Fig. 5.13; there are up to 8 duplex channels (OCI cards) per shelf, each with a corresponding OCLD. The output of each shelf is a single fiber link on the east and west sides, carrying a multiplex of up to 4 wavelengths. If there is a card failure, that channel is lost; if there is a fiber cut on either the east or west side, 4 channels are lost. By contrast, the protected or high availability configuration is shown in Fig. 5.14; in this case, only 4 channels are used, but the data is split after passing through the OCI card and travels over dual-redundant OCLD cards and fiber paths. There is no single point of failure in this configuration; an equipment failure or fiber break results in the data being switched to the redundant path within 50 ms. The system is a complete "3R" repeater (repeats, retimes, and regenerates the signal).

A single shelf thus supports up to 8 high availability or 4 base channels, without using TDM; with TDM, the capacity is increased by a factor of 4 (note that the 4TDM channels are treated as a single wavelength for protection purposes). Up to 4 models (8 shelves) can be daisy-chained together with optical fibers, which allows a maximum of 32 full duplex links to be multiplexed over a single pair of optical fibers (the 32 wavelengths are compliant with the ITU grid, half in C-band and half in L-band, and for convenience are grouped into 8 bands of 4 wavelengths each). In this way, a 2029 network contains up to 32 high availability or 64 base channels,

Protected 4-Wavelength Shelf

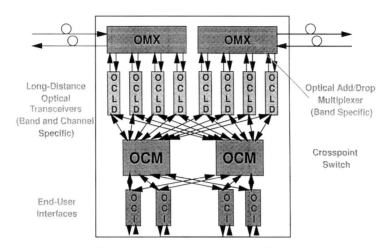

Fig. 5.14 2029 shelf traffic flow in high availability or protected mode.

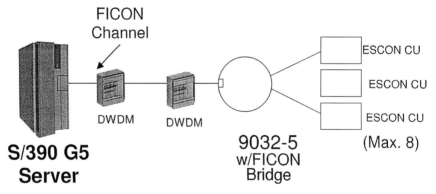

Fig. 5.15 DWDM used with the 9032-5 ESCON Director (FICON bridge feature) and S/390 model G5 enterprise server to achieve an effective 8 to 1 TDM of ESCON channels over WDM.

without TDM; using the 4TDM OCI card increases this capacity by a factor of 4. Note that the 2029 is compatible with external TDM devices as well; for example, as shown in Fig. 5.15, a 9032-5 ESCON Director with the FICON Bridge feature can be used, so that up to 8 ESCON channels occupy only a single WDM wavelength; in this manner, the total capacity of a 2029 network can be increased to 256 ESCON channels, or the full capacity of an S/390 mainframe computer.

Connectivity in a Point-to-Point Configuration

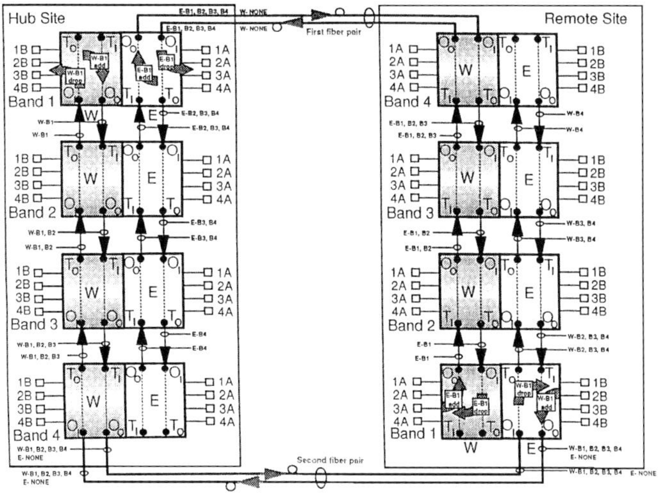

Fig. 5.16 2029 OMX connectivity, point-to-point configuration (4 shelves shown as an example).

Optical signal multiplexing is performed using passive thin film interference filters in the OMX cards. The daisy chain connections between OMX cards for both point-to-point and hubbed ring configurations are shown in Fig. 5.16 and Fig. 5.17 as an example. Other topologies, including ring mesh, are also available. The maximum distance for point-to-point links is 50 km/15 dB, while the maximum distance between any 2 nodes on a hubbed or dual hubbed ring is 35 km/10.5 dB. The difference is due to the inclusion of additional mux/demux stages in ring topologies to allow for passive add/drop of individual wavelength channels at any point on the ring, and optical pass-through for wavelengths destined for other nodes (shelves). For example, a point-to-point topology with 3 shelves is shown in Fig. 5.18, giving the resultant mapping of logical and physical connectivity; the flow of traffic in base and high availability modes is shown in Fig. 5.19 and Fig. 5.20. Note that the shelves are connected in reverse order at the hub and remote sites, so that all wavelengths experience an equal amount of delay propagating through the network; this principle is known as first add/last drop or last add/first drop. The corresponding diagram for a hubbed ring is given in Fig. 5.21; each 2029 shelf can act as a node, meaning that a hubbed ring will consist of up to 8 remote locations which logically communicate with a central hub site. High availability channels work in the

Fiber Connectivity in a 2029 hubbed-ring Configuration

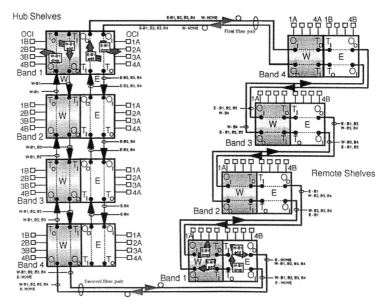

Fig. 5.17 2029 OMX connectivity, hubbed ring configuration (4 shelves shown as an example).

same manner between the hub and remote locations on a ring. A meshed ring supports re-use of wavelength channels, so that any 2 nodes may communicate with each other; the number of nodes in this case is limited only by the network configuration, not the DWDM technology. The multiplexers for each wavelength band are configured as either an optical seam or optical bypass, depending on their location in the ring, to facilitate the largest possible link budgets. For example, in Fig. 5.17 the hub site is configured as an optical seam, while the remote site is configured as an optical bypass. The dual counter-rotating rings are also self-healing (a break in the ring does not disrupt traffic to other points on the ring) and fault tolerant (an electrical power failure at any shelf does not impact other bands that pass through that shelf, since the OMX is passive), while the shelf supports concurrent maintenance (replacement of cards without affecting other channels). Because all data channels are 3R retimed, the 2029 also supports cascading of up to 4 networks with 32 channels each in series, which increases the total distance to 200 km in a cascaded point-to-point configuration.

5. Optical Wavelength Division Multiplexing

Point-to-Point Configuration

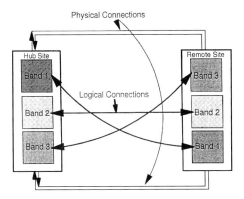

- Consists of only two sites
- Both sites have the same number of shelves (max. 8)
- The number of shelves between a shelf pair is the same

Fig. 5.18 Example of 3 shelf point-to-point configuration showing physical and logical connections.

Base Channels (unprotected mode)

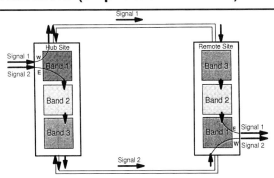

- Signal 1 coming from a channel on the West side of the shelf travels over the West fiber (max. 4 channels)
- Signal 2 coming from a channel on the East side of the shelf travels over the East fiber (max. 4 channels)
- If a fiber cut occurs on the West fiber, only West side traffic is affected
- The max. number of channels is 64 with 8 shelves at each site

Fig. 5.19 Example of 3 shelf point-to-point traffic flow in base mode.

High Availability Channels (protected mode)

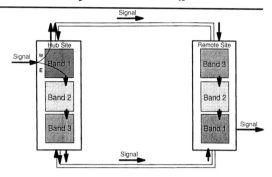

- A signal from a channel on either side of the shelf travels over both fibers
- The stronger signal will be chosen at the other end (this is determined at configuration time)
- The path can be switched manually or automatically (50 milliseconds)
- Max. number of channels is 32 with 8 shelves at each site

Fig. 5.20 Example of 3 shelf point-to-point traffic flow in high-availability mode.

Hubbed-ring Configuration

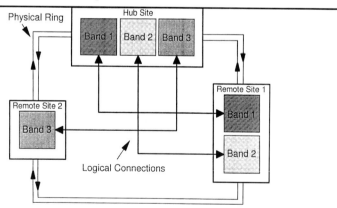

- Consists of up to nine sites
- Multiple shelves are at the Hub site and one or more at the remote sites
- The number of shelves between a shelf pair is the same

Fig. 5.21 Example of 3 shelf hubbed ring configuration showing physical and logical connections.

Both point-to-point and ring networks may be cascaded, which can result in many different topologies; a few of these were illustrated earlier. Optical amplifiers may also be used to increase the working distance to over 100 km point-to-point, or over 400 km in a 4 system cascade; the amplifiers occupy a separate shelf, and may be used either as pre-amps, post-amps, or in-line amps with various tradeoffs in the achievable distance and link budgets; appropriate equalization must also be used.

Unidirectional switching is employed in high-availability mode on the transmit and receive fiber separately to ensure that there are no single points of failure in a protected 2029 channel. With a worst case switching time of 50 ms, this is a significant improvement over the 9729 and allows uninterrupted operation of many protocols; some interfaces, such as HiPerLinks and sysplex timer, should still rely on link redundancy for continuous application availability. There is a third protection option available on the 2029, known as switched base mode; this uses a dual fiber optical switch to detect fiber breaks and switch all traffic to a redundant backup path. The switch is intended as a lower cost option than high availability for environments in which fiber breaks are more common than equipment failures, since the dual fiber switch protects only the fiber path and not the equipment cards. The backup fiber path is monitored with a heartbeat function to ensure that it remains available at all times. The switch also implements bidirectional switching, meaning that it can support some protocols, such as Sysplex Timer links, which may not function properly in high availability mode. There are 2 switches, one each for the east and west side of the network; this is shown schematically in Fig. 5.22.

This platform provides for protection and restoration of communication services within the optical (physical) layer of the network, without requiring protection at higher level network protocols. Conventional MANs have deployed SONET-based networking to support a variety of services including traditional asynchronous networks (OC-3, DS-1, etc.) as well as new services such as ATM, IP, compressed video, circuit switching, and others. In particular, ATM over SONET allows conventional "1 + 1" protection switching to be implemented over the network, providing redundancy in the event of a fiber break or equipment failure. Because of the proprietary nature of many new protocol signals, overlay networks must be implemented to support the full range of services; mapping all of the desired protocols to SONET is not always possible because many signal types are not compatible with SONET (for example, FICON and Parallel Sysplex links). New services offered in this manner cannot necessarily take advantage

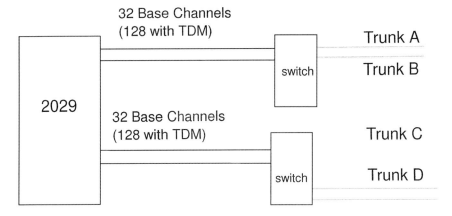

Fig. 5.22 Example of dual fiber switch implementation for east and west side of a 2029 system, switched base channels, point-to-point configurations only.

of protection switching and other SONET features, although these features are highly desirable in the design of fault-tolerant computing systems, which require high reliability and availability. Using the 2029, these multiple infrastructures are combined into a single, protocol-independent backbone implemented at the optical layer. The 2029 supports native attachment of all industry-standard protocols up to a maximum data rate of 1.25 Gbit/s, including ESCON, FICON, Fibre Channel, Parallel Sysplex links (HiPerLinks, ETR, and CLO), ATM 155 and 622 (OC-3 and OC-12), FDDI, Fast Ethernet, and Gigabit Ethernet (LX and SX). Propagation of OFC protocols on HiPerLinks is supported using proprietary signaling between 2029 devices, similar to the 9729; because GDPS protocols remain IBM proprietary as of this writing, the 2029 is the only currently available product tested and supported by IBM for Parallel Sysplex and GDPS applications. Many of these protocols use common hardware; the 2029 includes the necessary optical attenuators, patch cables, mode conditioning, and optical adapters as required for ESCON, ETR, CLO, and other links. Future work is planned to be compatible with other optical WAN products, including Nortel's recently announced 1.6 Tbit/s multiplexers, currently the largest in the world.

As the DWDM infrastructure scales, network management and security become increasingly important. Each data channel on the 2029 includes an

Cross-Band Communications

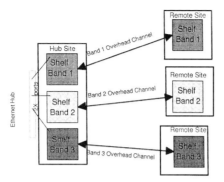

- With three or more shelves in the Hub site, an Ethernet hub is required
- All Shelves are connected to the Ethernet hub for the 2X ports with STP crossover cables

Fig. 5.23 Cross-band network management using per wavelength overhead channels and Ethernet hub interconnections at the hub site in a 2029 network.

overhead service channel multiplexed on the same path as the data at much lower speed; this allows individual channels to communicate in-band across the 2029 network. This approach avoids any single points of failure in the network monitoring systems, which is a potential concern with telecom standard systems that carry all management information over a separate outband wavelength channel. External management in a 2029 network is accomplished using Ethernet routing; all network shelves are connected to an Ethernet hub in the 2029 model 1 using a 2X crossover cable as shown in Fig. 5.23. The 2029 network management, commissioning, and provisioning is accomplished by Java-based applets that run System Manager software on an attached personal computer (PC); this provides a graphic user interface running under a Web browser. Either the Windows 95 or Windows NT operating system is currently supported for the attached PC. Using the System Manager software running on an attached PC, protection switching can be turned on or off for individual channels, and the user can always monitor which channels are being used to transport data. As shown in Fig. 5.24, the entire 2029 system can be viewed and managed from any point on the network; in practice, multiple PCs are used for redundancy and to provide information to users at different locations. The PC attaches to the 2029 via an Ethernet interface, and can be located anywhere on an

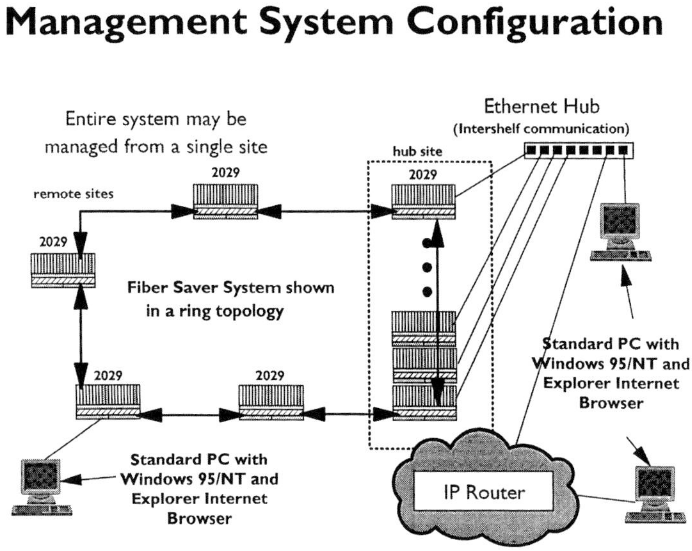

Management System Configuration

Fig. 5.24 Single point of control management of a 2029 network.

Ethernet LAN. The user provides a set of sub-netted IP addresses for the 2029 shelves; internally, the 2029 implements proxy ARP serving through one or more gateway interfaces, so that it can be treated in the same manner as any other network attached device. A typical IP network management system is illustrated in Fig. 5.25, which employs multiple points of control for both the end user and the fiber service provider. Security is provided by user-defined passwords, which permit logging into the 2029 as either an observer (able to view the configuration but not change it) or an administrator (able to both view and change the configuration). Ease of use features include a graphic display with user-defined names for each channel and system, and color-coded alarm banners (for example, green for a working channel, red for a failure). The management is based on Standard Network Management Protocol (SNMP) version 1.0, and the 2029 can be remotely managed under a variety of software applications such as NetView; alternately, TL-1 management is also supported. The 2029 is configured to send alarms and alerts to any other user-defined IP address on the Ethernet LAN; multiple gateway devices, such as routers, are supported for redundancy at different points on the 2029 network using OSPF protocols. The 2029 also provides integrated management from an IBM System/390 Enterprise Server Hardware Management Console (HMC), which implements an automatic call-home feature to proactively report equipment problems to a

5. Optical Wavelength Division Multiplexing 191

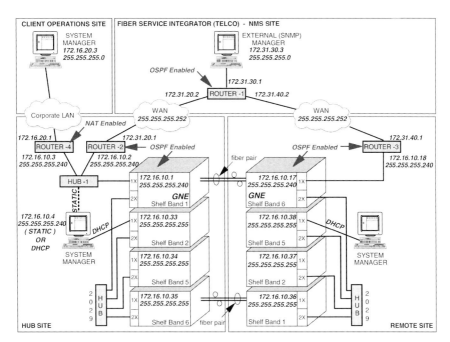

Fig. 5.25 2029 network management using service provider managed network service and external TCP/IP network to both the end user and service provider locations.

service center; this can be configured, for example, as part of the System Integrator software in a GDPS installation.

As part of product testing and qualification, a large data system using the 2029 network at the IBM TeraPlex Center is on the S/390 test floor in Poughkeepsie, New York, to evaluate its performance. Results of this work have been published elsewhere [32, 33]; we will provide a brief summary of the results here. A fully protected 32-channel 2029 system was used with a variety of input/output devices. Two IBM G5 Enterprise Servers (air-cooled CMOS) were configured in a 40-km GDPS using two sysplex timers. Four ESCON channels were routed through the 2029, optically looped back at the far end, and returned to the original processor to evaluate the effect of multiple passes through the 2029 network. Four FICON channels, each carrying a time division multiplex of 4 ESCON channels, were also run from the processor through the 2029s and into a 9032-5 ESCON Director where they were broken out into single channels and used to drive various storage devices, including an IBM 3945 automated tape library. One channel each of FDDI, ATM 155 over SONET, and Gigabit Ethernet LX was run from the G5 Open System Adapter (OSA) interface cards through the 2029 network to LAN connections; for example, the Gigabit Ethernet link was connected to a Cisco Catalyst 5000 router. Spools of fiber were used to simulate inter-site links, with the maximum 15 dB link budget and at least eight ST-type optical connectors in the link to verify that there was no effect from connector return loss or modal noise. The processors were logically partitioned into 15 processing zones, the maximum allowed for this machine type. Various applications were run on the sysplex to simulate stressful traffic under the OS/390 and MVS operating systems, including Lotus Domino and Notes servers, a UNIX-branded operating system partition (S/390 is officially branded as UNIX compatible) and transaction processing using secure encryption methods (IBM offers the only cryptography coprocessor certified as Level 4, the highest achievable level, by the National Institute of Standards and Technology). System code on the processors was used to log bit errors on all links as well as any other conditions indicating either failure or degraded performance; the 2029 system manager was also monitored during this testing. All fiber optic links operated error free (extrapolated to 10^{-15} bit error rate) over a 72-hour test run. By opening individual fiber sections throughout the system, pulling and reseating cards with the system powered on, and failing power supply components, we verified that there were no single points of failure in the system and no cross-talk effects between adjacent channels, either electrical or optical.

5.4. Intelligent Optical Internetworking

In recent years, the growing demand for bandwidth in MANs has driven the acceptance of DWDM technology. The nature of the metropolitan network is changing; many current networks are service specific, providing multiple overlay networks for different communication protocols and applications. This approach is well understood and has the advantage of a large installed base and simple network design rules. Such systems require adaptation at the network layer in order to accommodate both data and voice protocols; as the balance of traffic shifts from predominantly voice to data, the limited scalability of this approach makes it difficult to offer new services while maintaining support for legacy systems. For this reason, MANs are evolving into a more service-transparent structure based on DWDM. This offers the advantages of a highly scaleable, low-cost network infrastructure that is both bit rate and protocol transparent. Eventually, this roadmap may also lead to transparent all-optical networks with add/drop capability and over 10 Gigabit/second serial line rates per channel, although it appears that service transparency alone delivers most of the value in an optically transparent network with lower cost and less complexity. In the near term, a combination of circuit and packet switching is likely to prevail in the DWDM MAN or WAN. In this section, we will discuss some emerging trends and directions in technology and services that are expected to play an important role in the rapidly developing WDM market.

5.4.1. IP OVER WDM: DIGITAL WRAPPERS

The ability to transmit multiple data channels over a common physical media has helped to alleviate fiber exhaust in densely populated areas, where the cost and availability of optical fiber have become obstacles to the growth of new applications. However, it has become apparent that DWDM technology alone is not sufficient to address the requirements of growing MAN environments; some level of electronic signal processing is also required of the data in order to assure the necessary data integrity, quality of service, reliability, security, and manageability of the network infrastructure. Previously, these features have been provided over the telecommunications infrastructure using a combination of asynchronous transfer mode (ATM) over synchronous optical networks (SONET). Indeed, SONET-based traffic is already carried over a physical layer that makes extensive use of DWDM optical fiber interfaces. However, the growth of IP traffic has led to a complicated arrangement with up to 4 separate transport layers (IP over ATM over SONET over DWDM). In an effort to simplify this approach, there is a clearly emerging trend toward elimination of the ATM and SONET layers, and toward the direct transmission of IP over DWDM [34].

The combination of the most widely used networking protocol (IP) and the almost unlimited bandwidth of DWDM offers the potential for many new networking architectures. In practice, this term is a bit misleading: "IP over WDM" implies one of a multiplicity of mappings of IP onto fiber (or wavelengths) as illustrated in Fig. 5.26. To state that any one particular mapping represents IP over WDM is very disingenuous, and ignores the fact that every data network is unique in a marketplace governed by differentiation. For example, some of the mappings shown in this figure continue to exploit the advantages of the installed ATM infrastructure in the WAN, commonly known as "everything over SONET" (EOS). Many of these models can also be extended to encapsulation of other protocols such as ESCON or

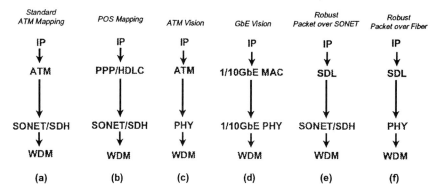

Fig. 5.26 Different types of IP over WDM (a) standard ATM mapping (b) point-of-service mapping (c) ATM future mapping (envisioned) (d) Gigabit Ethernet mapping (e) robust packet over SONET mapping (f) robust packet over fiber mapping.

Fibre Channel. Commensurate with the emergence of IP over WDM protocol mappings is the demand for transport networking at unprecedented levels of granularity — on the order of gigabits to tens of gigabits per second — and the evolution from exclusive use of SONET or SDH-based time division multiplexing to Optical Transport Networking (OTN) via WDM. This may be the next step in the evolution of the transport network, which will be predicated upon the support of operations, administration, maintenance, and provisioning (OAM&P) functions at the optical layer.

Current deployments of IP networks presuppose that data delivery will be on a "best effort" basis. However, many network operators want to provide new services that require traffic management of IP flows. One of the most significant of these is virtual private networking (VPN), where one customer's traffic is separated from another's in a single IP network with well-defined quality of service and security. While this is not possible with today's best effort IP network, it can be achieved in the future by means of a protocol that allows IP packets to be partitioned into traffic flows that can be manipulated in their own right. One example of such a protocol is MPLS. The partitioned flows must then be encapsulated to be carried over WDM (from the perspective of IP, it is of no consequence if IP is partitioned or not, only that it is encapsulated).

The management of OTNs using digital wrappers such as those in Fig. 5.26 will also undergo a profound evolution. Digital wrappers could enable protocol-independent OTNs with similar management features to those existing today only in SONET and SDH networks, while at the same time opening the optical layer to a wide variety of client traffic. This could allow service providers to accommodate different protocol stacks and network architectures for their next-generation IP backbones. This is known as service transparency; together with the required technologies for optical switching, protection, and restoration, this is critical for future WDM networks that will be governed by service differentiation and rapid service provisioning. This new model requires the DWDM layer to assume many of the traditional functions associated with ATM over SONET, such as protection switching. In general, the more intelligence that can be built into the product, the more value to the end users; but where the intelligence is placed in the network may ultimately determine what the final applications will be. We will discuss alternate approaches to design of next-generation DWDM devices and networks that use electronic signal processing to enhance the performance of optical interconnects.

5.4.2. WAN TRAFFIC ENGINEERING

Optical Internetworking is a concept introduced in 1998, when equipment was introduced that allowed the IP switching/routing layer to operate at the same data rates as the DWDM equipment that provides the physical layer interface. Typically this is realized by providing an OC-48c interface (2.4 Gigabit/second) on IP switches offered by vendors such as Cicso, Ascend, and others; in this model, the IP switch/router serves as the central location for all network intelligence. This data rate matching removes the need for SONET aggregation equipment in the core of a data network, which is seen as the first step in an evolutionary path toward full transmission of IP over DWDM. IP router interfaces may also be directly attached to optical ADM ports or OXCs. This forms an overlay network in which DWDM acts as the server layer and IP behaves like the client layer. Of course, there are many alternatives to forming a virtual topology between IP routers and DWDM equipment. Intelligence required to manage resources of both IP and WDM layers could reside in the IP router, or the router could simply act as a "black box" with optical interfaces.

There is vigorous ongoing discussion in the technical community concerning the best way to implement intelligence in the network topology, and thereby exploit the strengths of both optics and electronics. For example, design rules for WDM LAN and MAN routing have been proposed [35–41]. In electronics, operations such as buffering, adding and dropping packets, or merging packet streams, are done with ease. In optics, however, buffering is onerous. Operations such as retrieving a packet from a traffic stream affect the whole stream. Therefore buffering in the network, with the attendant issues of cost and possibility of overflow, is to be avoided, and consequently so is traffic merging. Traffic splitting, on the other hand, doesn't require buffering. While traffic to remote locations should be handled by switches, local traffic in the WAN need not be handled by these switches. In fact, it can be desirable to isolate local traffic from the vagaries of congestion associated with external traffic. Also, switches are expensive and make extendibility of the network more difficult; switch upgrades or outages may also disrupt the network. For these reasons, the network should allow traffic that does not need to be routed through switches to bypass them. Finally, routing should be well suited to packet-switched traffic but not rely on a specific protocol such as IP or ATM; instead, it should operate at the physical layer, below such protocols. This includes recovery of failed components at the optical layer, in a manner that is sufficiently simple, robust, and rapid that it does not trigger recovery attempts by the higher order protocols. Such an approach is well suited to overlay network engineering, where operations in one layer are pursued independently from those in other layers.

Traffic engineering solutions developed for either IP networks or WDM networks can be applied directly to their respective layers in this model, with little or no inter-layer coordination. In an IP over WDM approach, traffic engineering in the IP layer can theoretically be effected via IP routing algorithms that can adapt the IP packet routes and perform load balancing. Likewise, traffic engineering in the WDM layer can be effected through light path reconfiguration that adapts the IP network's virtual topology to the evolving traffic pattern. In the case of switched WDM, the optical layer can also adjust the number of wavelengths allocated to each virtual link, thereby affecting the bandwidth and contention statistics of the virtual link. In practice, because traditional IP routing algorithms are all oblivious to traffic loading, traffic engineering in current IP networks often relies entirely on link layer adjustments. For example, an IP over ATM network typically maintains a full IP mesh connectivity, and relies on the ATM layer to adjust the bandwidths provisioned for the ATM virtual circuits that support the IP links. These ATM layer actions are completely transparent to the IP layer routing algorithm. Overlay traffic engineering is most likely to be adopted in near-term deployment of overlay IP/WDM networks. Network engineering rules for IP over WDM will likely consist of a number of basic functional components, including the following [42, 43]:

- Traffic monitoring, analysis, and aggregation is responsible for collecting data traffic statistics from the network elements (IP routers and OXCs). These statistics are then analyzed and/or aggregated to prepare for the traffic engineering and network reconfiguration related changes; depending on the implementation, this may be either local or end-to-end performance monitoring. In addition, the optical channels may be monitored for other statistics, such as average optical power (to predict laser end of life and estimate bit error rates), optical signal-to-noise ratio (OSNR), or wavelength stability. Various types of traffic monitoring are possible in a WDM system, including schemes based on the power spectral density of the WDM signals.
- Bandwidth demand projection uses past and present measurement of the network and characteristics of the traffic arrival process to forecast bandwidth requirements in the near future. These projections are used for planning subsequent bandwidth allocations.
- Reconfiguration triggers consist of a set of network management policies that decide when a network level reconfiguration is to be performed. This can be based on traffic measurements, bandwidth predictions, and operational issues (such as allowing adequate time for the new network to converge and stabilize while minimizing transient effects).
- Topology design provides reconfiguration of the network topology based on traffic measurements and predictions. Conceptually, this can be viewed as optimizing some performance function (number of IP routers connected by optical paths in the WDM layer) to meet a specific objective (maximizing throughput) subject to certain constraints (interface capacity or nodal degree of the network) for a given set of load conditions applied to the network. This is generally a very complex problem. Because reconfiguration is regularly triggered by the continually changing traffic patterns, an optimized solution may not be stable. It may be more practical to develop a metric that emphasizes factors such as fast convergence and minimal impact to ongoing traffic rather than optimizing over the whole network.
- Topology migration refers to a set of algorithms or policies to coordinate the network migration from an old topology to a new one. As WDM reconfiguration deals with large capacity channels, changing allocation of channel resources with the resulting coarse

level of granularity has a significant impact on many end user communication links at the same time. Traffic flows need to adapt to the light path changes during and after each migration step; the effects can potentially spread over the routing pattern of the whole network and impact many user data channels.

In future networks, protocols such as MPLS, MPlambdaS, and related types of optical label switching (OLS) may provide a unified control plane across both IP and WDM layers. MPLS would effectively serve as the intermediate layer between IP and WDM. This could make it possible to optimize performance across both layers together with some form of integrated traffic engineering. All of the IP and WDM network traffic management discussed above still exists, but could now be coordinated together. Specific traffic management methods can be applied to different layers, subject to the prevailing network traffic load, granularity, and time scale considerations. Note that another consequence of the emerging converged IP/WDM environment in the WAN is a change in the network traffic patterns, the provisioning of transport resources, and the sources of network traffic. Traditionally, nationwide transport networks were operated by a single telecommunications provider to meet the needs of their subscribers. A new model of dynamically reconfigurable networks is emerging, in which the OTN operator provides capacity on demand in the form of lightpath connections (wavelengths) to independent users such as independent service providers (ISPs), storage solution providers (SSPs), or even telephone companies. Because the light paths are service transparent, a leased wavelength could be managed according to the needs of a specific application. Today these are primarily narrowcast (NC) services such as telephony or video on demand, but they will expand to include many other features as well. This also implies that traditional methods of forecasting average traffic demand in the WAN based on past history (so-called static traffic capacity planning [44]) becomes very difficult to apply. Not only do emerging networks offer new services for which no prior history exists, they also ignore both the number of clients in use at any given time in the future and what kind of service will be required for these clients. This has led to ongoing research into dynamic traffic modeling, which means that lightpath requests can arrive at random intervals while other connections are already active. If the new request can be accommodated without disrupting existing connections, a new lightpath is created and maintained for the duration of the service; otherwise the request is blocked. Dynamic network performance is usually measured by the blocking probability. As one might expect, this is a multivariable optimization problem; for example, in some cases blocking probability decreases by increasing the nodal degree (number of add/drop points in the network) which is a network topology consideration.

Another measure of network performance compares the blocking probability of connections between adjacent nodes and widely separated nodes; this is called the fairness of the network. For example, it has been suggested [45] that voice and video traffic follow a Poisson distribution, while IP data is a self-similar profile. Various models incorporating link utilization metrics, packet loss rates, and different loading considerations have been proposed [45], including migration plans from existing legacy networks. For example, it is now possible to construct separate ATM and IP overlay networks that attach directly to DWDM equipment; then, as the mix of traffic in the MAN tends toward more data than voice, especially IP data, the two networks can gradually be converged into a single backbone. Previously, network service providers have deployed dedicated networks for each type of traffic; these overlay networks do not scale well, due to unplanned and rapid growth in many different types of data protocols (Fiber channel, Gigabit Ethernet, ATM, etc.). Basic networks have been optimized for time division multiplexed circuit-switched voice traffic; the new push is toward packet switched, data centric networks.

5.4.3. EMERGING STANDARDS

Mapping all of the desired protocols into one network is not always possible because many signal types are not compatible with SONET (for example, Fibre Channel and Parallel Sysplex links). New services offered in this manner cannot necessarily take advantage of protection switching and other SONET features, although these features are highly desirable in the design of fault-tolerant computing systems, which require high reliability and availability. For these reasons, there has been a need for protocol-independent DWDM solutions that combine these multiple infrastructures into a single backbone at the optical layer. Some more aggressive regional telecom service providers have chosen to skip the intermediate step altogether, and jump directly to IP over DWDM. The IP switch must also provide protection switching, quality of service, and related functions.

An alternate viewpoint is known as the "intelligent optical network"; this maintains that IP switches/routers can't provide the wavelength switching, service provisioning, and rapid path restoration functions required by future networks, and these functions best reside in the optical layer. Such devices would either be integrated with or interface with existing DWDM equipment, and would provide the above functions in addition to handling the details of signal transport. Generally, these two are viewed as complimentary technologies; relatively small, purely IP networks will be built following the optical internetworking vision, but larger more complex networks will require intelligent optical networking to either supplement or replace these functions. Optical internetworking is better suited to pure data environments, especially IP data; multiservice provisioning is better suited to an intelligent optical network. Intelligent networking offers the advantage of bandwidth management, including integrated TDM functions at all data rates. Also, it can respond to protection switch events within the maximum SONET switch time of 50 ms, with typical switch times of 25 ms or better; this helps prevent conflict between the two versions, as intelligent optical networks will respond faster to remedy service problems before switch/routers can respond. However, standards are not yet defined to negotiate the division of responsibility in networks that contain both elements.

Generally, DWDM can best respond to physical path problems, while switch/router can better respond to application and quality-of-service issues. Equipment must also be able to interface with many different kinds of management systems, both legacy telecom carrier (TL-1) and modern datacom (simple network management protocol or SNMP-based) management, and there are distinct advantages to integrating both element and network management systems without requiring outband management. Future network devices such as switches and routers based on Fiber Channel or similar standards will likely be able to automatically recognize attached devices and perform a configuration map of the network, as well as respond to in-band management commands. Currently, DWDM devices typically employ their own network management interface which is transparent to the attached devices; this can create conflicts, for example DWDM implementation of ring topologies that are not necessarily compliant with Fiber Channel arbitrated loop or other topologies. Existing industry standards do not provide for the management of transparent, out-of-band devices such as repeaters, protocol converters, or multiplexers; thus, some level of intelligent DWDM is needed to assist with network management of these devices. Wavelength switching and routing are also key advantages of intelligent networks; this enables the creation of virtual optical data pipes with bandwidth management and simultaneously provisioning IP services. It is an advantage if current DWDM designs offer fully cross-connected backplanes, as this should allow the architecture to scale into wavelength switching and routing in the future or integrate with WAN devices using high-speed parallel optical backplane bus extensions. An advantage of integrating the intelligence with the multiplexer is that DWDM is a modular platform, providing building blocks with individual channel control; this leads to improved reliability, availability, serviceability (concurrent maintenance), and scalability between MAN and WAN networks. There is ongoing debate whether this intelligence should reside at the edge of the network (into the end-user premises or private networks) vs. in the central office of telcom or service providers who may wish to offer line-drop services.

As SONET-based termination of circuit-switched, TDM environments is being embedded in other packet-based network layers, future network designs will require new ways of allocating network bandwidth to IP traffic. With the advent of "always-on" network services and the growing need to link high-speed optical backbones with slower legacy equipment, a coalition of optical networking experts and IP specialists would seem to be a natural step for the next generation of internetworking standards. Indeed, there are a number of promising efforts underway in the telecommunications industry to more closely integrate the electrical and optical signaling domains. In particular, two standards efforts are currently underway to link IP data packets directly to DWDM optical wavelengths, so that the optical network can take some advantage of the intelligence embedded in IP traffic. This so-called "optical IP" effort could eventually allow customers to dynamically request portions of a fiber cable's bandwidth for a particular time or service. It also represents a radical change for service providers, who would be able to build intelligent optical networks with a seamless interface between switched packet and optical services. One standards effort that held its first meeting in Boston this past January is the Optical Domain Service Interconnect (ODSI) coalition [46], a loose connection of vendors providing optical transmission equipment, access services, terabit routers, switches, and network provisioning software. ODSI seeks to define common control interfaces between optical or electrical physical layers and IP media access layers of the Open Systems Interconnect (OSI) model. This work may ultimately rely on derivatives of the Internet Engineering Task Force (IETF) MPLS standard, which basically provides a means of defining IP flows and is already being used in the electrical signaling domain.

ODSI will propose low-level (below Layer 3) control plane standards that must be met by vendors of both optical transmission equipment and broadband IP switches/routers; these standards would also be offered to the IETF, the OIF, and other standardization groups.

A separate but complementary effort, also based on MPLS, is being drafted within the IETF itself. Although this effort is somewhat smaller at present than ODSI, it has proposed methods to link optical cross-connects and gigabit routers through a Layer 3 switching methodology called Multi-Protocol Lambda Switching (MPLmS; lambda refers to switching by native wavelengths). Operating at a higher level than ODSI, this proposal has the advantage of not requiring a new set of protocols for quality-of-service and bandwidth control. This work may have implications to many other OSI layers as well; for example, it has been suggested that MPLmS could be used as a control mechanism for optoelectronic interfaces. Other proposals for the control plane of OTNs are also under consideration by the IETF, ODSI, ANSI T1, and ITU standards bodies [47].

These efforts are still in the early stages of development, providing opportunities for technical input and new services to evolve. First interoperability demonstrations of ODSI are planned for the latter half of 2001, but there have already been product demonstrations of an MPLS-based router as a conduit between IP and optical networks, which associates optical wavelengths directly with IP services. It has also been suggested that these efforts could lead to a "collapsed central office" architecture, in which optical assignment switching is performed at locations far from a telecom carrier central office and the typical distinctions between the MAN and WAN environments begin to blur together.

5.4.4. CONVERGED AND HIERARCHICAL NETWORKS

Within the datacom industry, there are a number of ad hoc efforts to standardize on the next-generation input/output (NGIO) interfaces [48] for computer equipment. The recent emergence of SANs, or switch-based private networks interconnecting computer processors with remote storage devices, has helped accelerate this trend. One promising effort is centered around the Infiniband consortium [49], which seeks to define the next generation of parallel copper and parallel optical interfaces for I/O subsystems. Although most efforts to date have concentrated on small form factor parallel copper interfaces, there has also been a recent draft proposal for multi-channel parallel optical interconnects. This standards effort is expected to drive a need for higher density optoelectronic packaging and closer integration between the parallel optical transceivers and associated electronics for data manipulation.

There is also a strong push toward terabit networking in both the datacom and telecom industries. As discussed earlier, DWDM has emerged as a solution to bandwidth constraints at the physical layer; survivability has been integrated into the physical, or transport, layer (extending traditional 1 + 1 SONET protection to all protocols) and bandwidth optimization has been provided by sub-rate TDM over DWDM; future systems may also take advantage of code division multiple access or other bandwidth provisioning techniques. However, this approach will require the service layer (which provides IP, ATM, and other protocols) to upgrade more than twice as fast as the transport layer, or roughly double capacity every six months, in order to keep pace with the demand for bandwidth. Some estimates have shown the service layer growing over 70 times by 2003. A more realistic approach is to have the transport and service layers evolve together, although this still requires the service layer capacity to double on a yearly basis.

This trend is likely to be accompanied by a blurring of the conventional boundaries between the LAN (traditionally 0–10 km), MAN (10–100 km), and long haul or WAN (100 km +). Convergence of the datacom and telecom environments may drive traditional carriers and service providers to deploy networks with hierarchical logical structures [50]. In a hierarchical network, each layer is designed to hide the details of its operations from the layers above, thus enabling cost-effective and scalable network growth. At each layer, the appropriate traffic granularity and switching are selected, as well as suitable traffic "grooming" (sorting services by type and destination to wide bandwidth optical paths); both physical and logical paths are then selected. Significant cost benefits can be realized with this approach, including both equipment capital cost and operational expenses. Multi-tiered architectures can take advantage of cost-effective banded switching technologies, enable a large amount of optical bypass (which reduces the amount of required electronic termination equipment), and reduce the number of elements that need to be managed. A hierarchical scheme also provides efficient use of restoration capacity, reduces the amount of electronic access equipment required in order to effectively share this capacity, and simplifies the restoration policy algorithms [51].

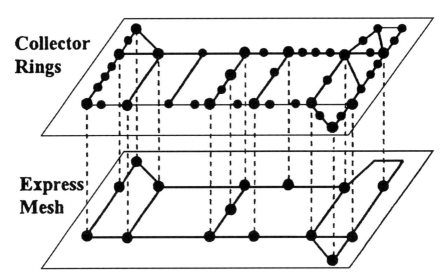

Fig. 5.27 Example of a two-level hierarchical network in the WAN, where a group of collector rings aggregates and grooms fine granularity traffic before delivering it to a streamlined, coarse granularity express mesh.

One example of a hierarchical architecture is that of the two-tiered scheme shown in Fig. 5.27 and 5.28 [50]. In this architecture, the upper layer is comprised of localized collector rings, operating at granularities of DS-3 or OC-3, whereas the lower layer consists of a nationwide "express mesh" operating at granularities of OC-48, OC-192, or higher. Each collector ring is typically populated with ADMs, with all nodes in the ring being logically connected (although not necessarily at all wavelengths). The functions of the collector ring are to route traffic within the ring and to collect inter-ring traffic for delivery to the large mesh nodes located on that ring. The ring topology is suitable for this layer due to the limited geographic extent of the collector rings and the simple protection properties of rings. Furthermore, many carriers and ISPs already have a large legacy investment in SONET ring-based networks, whose capacity is presently being exhausted by the tremendous growth of traffic, exacerbated by the inefficiencies of inter-ring routings. These legacy rings can be more effectively used if their role is restricted to serving as the collector network; alternately, this network could be implemented as a mesh architecture with grooming cross-connects.

The collector network is designed with fine granularity and limited distances; by contrast, the lower-layer express mesh of Fig. 5.27 must carry

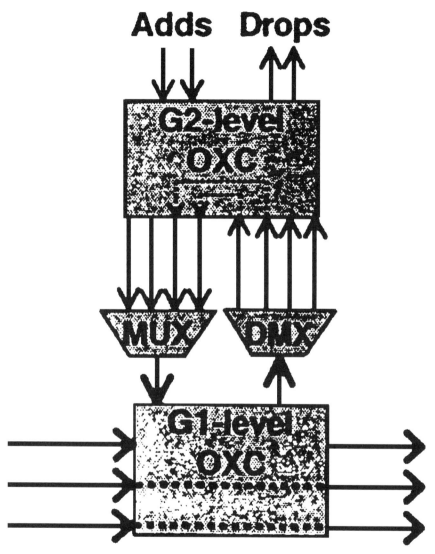

Fig. 5.28 Advantages of hierarchical switching; at the lowest level, an optical cross-connect (OXC) switches traffic at a coarse granularity 61, and only a small percentage of the traffic needs to be switched at a finer granularity (62).

large amounts of traffic for longer distances. These requirements could be met with emerging high-capacity, ultra-long-reach transport networks, possibly using all-optical switches. Because the collector rings perform an aggregation function, the express mesh could carry traffic at coarser granularities (OC-38 and above). The logical mesh does not include all network nodes, only those that generate traffic at levels comparable to the granularity of this level, or those that are strategically located. As a result, the mesh may completely bypass many of the nodes, depending on network traffic patterns; the express mesh may consist of only about 20 nodes nationwide. To take full advantage of this approach, traffic needs to be routed between and through lower-layer nodes without any intermediary regeneration; this requires the development of very long reach amplified optical networks. The combination of a sparsely populated express mesh with nationwide transport distances eliminates a large part of the electronic terminations in this proposed network [51]. In addition to limiting the nodes that are included in the lower layer, it is also important to create a streamlined physical topology; thus, fiber links that do not pass by major nodes may not be included in this layer. A mesh topology is most appropriate for the lower level since it can use switches to create virtual topologies that can be optimized for traffic routing and sharing of protection bandwidth. A significant benefit of the hierarchical approach is that it enables deployment of a hierarchical switch architecture, which has advantages in terms of cost and scalability. The streamlined design of an express mesh creates large bundles of traffic that can be switched as a single unit. For example, increasing the switching granularity by a factor of B decreases the switch fabric size by a factor between $B \log B$ and B^2, depending on the switch topology [50]. This translates into a savings in switch cost, size, power, number of physical interfaces, and states that must be managed.

The preceding example has discussed a two-tiered approach to network design. Ultimately, there may be a multi-layered hierarchal approach, where the number and granularity of layers is determined by the available technology and relative cost. For example, assume that the cross-sectional traffic on any link in a particular layer of the network is T, and that the finest level of granularity required at that layer is G. Then the number of elements per link that are potentially switched is T/G. At a node of degree N, this translates to a switch size of about $(NT/G)^2$ [50]. Due to limits of technology and physical space, the switch size may be limited, resulting in multiple tiers, where the factor *NT/G* is chosen for each tier to enable the network to scale easily. This approach may also lead to a compromise between the two competing views of the WAN as an opaque network (regeneration provided on every channel at every configuration point) vs. the construction of smaller, transparent sub-networks interconnected by regeneration sites (the so-called "islands of transparency" model) [51].

5.5. Future Directions and Conclusions

DWDM is being adopted in large data communication systems as a practical solution to problems of fiber exhaust, and a way to offer new high-availability services with fault recovery at the physical layer that do not need to be based on a SONET infrastructure. DWDM has enabled new computer architectures such as Parallel Sysplex, and made it practical to implement real-time disaster recovery and data mirroring at extended distances. New datacom network topologies are also enabled by DWDM coupled with flexible network management and security. Efficient bandwidth utilization is possible by using a TDM front end in combination with DWDM. Future systems are expected to continue this trend toward higher data rates, extended distances, more complex topologies, and new applications or service offerings; emerging technologies including multi-terabit routing and space-time conversion are expected to play an increasingly important role in these systems, which may be the first step toward all-optical networks.

There has been an explosion of research and development activity in WDM recently, including many new product announcements and developing industry standards. Given the rapid pace of growth and change, it is difficult to predict future trends in this area. Generally speaking, we can expect WDM systems to increase their maximum data rate per channel from OC-48 to 10 Gbit/s or 40 Gbit/s and beyond, with some form of digital wrapper or sub-rate TDM scheme to provide efficient bandwidth management. The number of wavelengths is expected to grow as well, from the current 32 up to 128 or more. With the combined aggregate bandwidth of WDM networks approaching multi-terabit rates, even in the MAN environment, significant challenges will emerge for network management, restoration, monitoring, topology configuration, and intelligence. The convergence of MAN and WAN applications will continue to drive use of optical amplifiers for extended distance, as part of either 2R or 3R repeater designs. Use of new topologies including ring mesh in the MAN or hierarchical networks at the nationwide level, combined with optical switching and cross-connects, promises to open up even more applications.

Because there are no ultra-dense WDM systems commercially available yet, a combination of TDM and DWDM is currently used to maximize use of existing installed fibers, enable new applications such as Parallel Sysplex, and drive down the per channel cost of datacom systems. This approach offers some advantages; because TDM is an established technology, it can be implemented immediately with good reliability and reasonable cost; it also does not require changing the DWDM wavelengths currently in use, so that it can be made backward compatible with existing systems. There is also a well-established roadmap for incorporating TDM into datacom networks; care must be taken in extending the SONET TDM approach to protocol-independent TDM systems, because the timing jitter and other factors may vary significantly as a function of data rate. Traditionally, optical communication solutions have been successful in those areas that require higher data rates and longer unrepeated distances than can be provided by copper solutions. In the 1960s most computer interconnections were performed by using parallel copper cables with 8 to 16 wires in a duplex link; these attachments were limited in distance to about 100 meters, and to data rates of about 3.5 Mbyte/s. As the demand for higher data rates became apparent and fiber optics became affordable, many low-speed parallel copper links were combined into a single high-speed serial data link. The classic example of this approach is the ESCON protocol, which provides a 17 Mbyte/s serial data stream and performs serial-to-parallel conversion at either end of the link to recover the low-speed data links. Over time, this approach has been applied again and again to increase the effective data rate. Currently, FICON links are capable of time division multiplexing up to 8 ESCON links over one gigabyte link using the FICON Bridge feature in the 9032-5 ESCON Director. In the near future, we can expect this trend to continue as the most cost-effective way to implement higher bandwidth connectivity.

Recently, TDM has been supplemented by DWDM techniques; for example, combining the FICON Bridge with the 2029 allows up to 128 ESCON channels to be multiplexed over a single fiber pair. Currently, the 2029 provides 40 Gbytes total bandwidth over a fiber pair; as the per channel data rate of DWDM is increased, future plans include a protocol-independent TDM sub-rate multiplexing (SRM) function integrated with the DWDM products. For example, a 4:1 TDM of ESCON channels (at 100% utilization) could be accommodated in a single 1.25 Gbit/s DWDM channel; if the 4:1 TDM card occupies only a single slot in the DWDM card cage, then the total capacity of the 2029 can be increased from 32 to 128 protected channels or 256 unprotected channels (the total capacity of an S/390 Enterprise Server). The use of TDM as a front-end for DWDM in this manner offers significantly improved bandwidth utilization and lower cost per channel.

Next, we will discuss other approaches that have the potential to be used in future applications. Before we consider these future systems, let us consider the fundamental building blocks of time-bandwidth allocation or space-spatial bandwidth allocation in a communications network. If we consider time-based systems only we can implement many well-known operations such as TDM, which is also known in wireless communications as time division multiple access (TDMA), frequency multiplexing or frequency division multiple access (FDMA), and code division multiple access (CDMA) networks. The consideration of space-based systems leads us to examples such as parallel fiber ribbons and linear arrays of optical transmitters and receivers. Among the time-based systems, TDMA and FDMA have been the most widely available technologies; however, CDMA has received increasing attention lately as the technology has become more advanced. Wireless systems using CDMA with multiple antenna arrays are an example of combining time-based and space-based bandwidth considerations into a single system. By contrast, although fiber optic networks are quite advanced they still do not possess the necessary building blocks to combine time-based and space-based approaches. As an example, consider that the bandwidth of a single-mode fiber is on the order of 20 THz; thus, we should be able to work with optical pulses as small as 50 femtoseconds (actually, in the future we can envision a parallel array of semiconductor femtosecond pulses lasers, just as we currently use nanosecond radio frequency (R.F.) pulses for electronic communication). We can then consider a simple TDMA system with M channels; as a numerical example, we will consider $M = 32$. Assume further that each channel carries a signal of 10 Gbit/s; this could represent advanced channels in a future Parallel Sysplex, or perhaps the emerging 10 Gigabit Ethernet standard. The total system throughput is thus 320 Gbit/s, with each input channel modulating a series of 50 femtosecond pulses, and the TDMA circuitry running at 320 GHz. In this example, channel one modulates pulses $0, M, 2M$, etc.

while channel two modulates pulses 1, $M+1$, $2M+1$, and so on. The modulated pulses are combined and transmitted along a single fiber; the combiner is really acting as a parallel-to-serial converter or a multiplexer. At the receiver we need to perform demultiplexing or serial-to-parallel conversion capable of handling 50-femtosecond pulses. Although elements of this system have already been demonstrated, including the 50-femtosecond lasers, the technology required to produce the serial/parallel and parallel/serial conversions is very difficult to realize and has been the bottleneck to producing systems like the one in our example. Before we discuss a possible solution to this problem, we point out that demultiplexing from a series of fast pulses to a parallel group of slower pulses is a well-known, fundamental problem that arises whenever high bandwidth optical fiber is coupled to comparatively low bandwidth electronics. Sometimes this serial-to-parallel conversion is called a time-space convolver, N-path convolver, or channelizer; its equivalent model uses z-transforms and multirate filters [34]. In the history of networking, such devices are often the bottleneck whenever a new, faster technology is introduced and it is desirable to interoperate with legacy, slower technology. There are many examples of this, including the replacement of parallel copper cables with ESCON fiber optic channels in many IBM enterprise servers.

Although there are many design proposals and laboratory demonstrations of methods to deal with this problem, we will note two possible variations of the so-called time-space multiplexer that are promising for long-term implementation. Of these, the option that uses DWDM as part of its functionality is probably more viable. In the optics community, the serial-to-parallel converter is generally referred to as a time-space converter because the input time signal is spread into M channels, which are distributed in space. One example of a time-space converter based on a nonlinear crystal or holographic element has been demonstrated [52]. This device has the ability to manipulate the amplitude and phase of each channel. A second approach uses DWDM filters to replace the nonlinear optical crystal in this implementation. As in the previous case, input signal pulses are multiplied with a time chirp signal, which basically maps the input pulses into different wavelengths (or frequencies). This can be done using a chirped fiber grating or dispersive filter, for example. The output is then applied to an M-element DWDM filter, which separates the pulses. Note that in this case, the phase information is lost because we are using a chirp signal. There are many technology and engineering issues to be resolved for these two serial-to-parallel converter designs. However, it appears that there is no fundamental limitation that prevents their realization; in particular, the WDM filters required for the second approach are already commercially available.

This approach leads naturally to the capability of implementing CDMA using femtosecond pulses, similar to the proliferation of this technology in wireless communications. The same time-space convolver discussed earlier can be easily modified to implement spread spectrum coding, by placing a transmission mask at the input spatial signal plane [52]. As the signal is multiplied by a chirp, it can be shown that the spatial distribution of light in this plane of the figure is the equivalent of a spatial Fourier transform of the input signal. By placing a mask in this plane whose transmittance function contains the Fourier transform of a spread spectrum code, the desired encoded pulse is obtained at the output plane. This is possible because the last half of the optical system shown is equivalent to performing the inverse Fourier transform. Although there are still many technology challenges involved in adopting R.F. solutions to the optical domain, once devices such as this come out of the laboratory and into commercial use, it is possible that optical networking will follow the history of wireless communications in the adoption of CDMA.

ACKNOWLEDGMENTS

The author would like to thank the IBM TeraPlex Center support staff (Rich Hamilton, Mario Borelli, and Jack Myers) for support during the 2029 testing, the IBM 2029 product development team and GDPS support team (Ray Swift, Gary Vitullo, Simon Yee, JoAnn Transue, John Matcham, Ernie Swanson, John Torok, Jim Keller, John Deatcher, Steve Rohersen, Ken Knipple, Juan Parrilla, Dave Petersen, and Noshir Dhondy) for their months of dedicated work on this program.

The terms System/390, S/390, ESCON, FICON, Generation 5, G5, Generation 6, G6, Parallel Sysplex, GDPS, and 2029 Fiber Saver are trademarks of IBM Corporation. The terms Windows 95, Windows NT, and Internet Explorer are registered trademarks of Microsoft Corporation. The term Netscape Navigator is a registered trademark of Netscape Corporation. The terms Domino and Notes are trademarks of Lotus Corporation.

References

1. DeCusatis, C., E. Maass, D. Clement, and R. Lasky, eds. 2001. *Handbook of Fiber Optic Data Communication.* New York: Academic Press.
2. *IBM Journal of Research & Development.* 1992. Special issue on "IBM System/390: Architecture and Design," vol. 36, no. 4.
3. DeCusatis, C. December 1998. "Optical data communication: fundamentals and future directions." *Opt. Eng.* **37**(12): 3082–3099.

4. "Market Report on Future Fiber Optic Technologies." May 1999. Available from Frost and Sullivan, New York, N.Y. For details on 10 Gbit/s Ethernet, see also the homepage of the Optical Internetworking Forum (OIF) at http://www.oiforum.com.
5. "Optical interface for multichannel systems with optical amplifiers." 1999. Draft standard G.MCS, annex A4 of standard COM15-R-67-E, available from the International Telecommunication Union.
6. Ferries, M. 1999. "Recent developments in passive components and modules for future optical communication systems." Paper MI1, *Proc.* OSA Annual Meeting, Santa Clara, Calif., p. 61.
7. Gyaneshwar, C. G., et al. 2000. Paper TuJ7-2, *Proc. OFC 2000*, Baltimore, Md.
8. Simon, J. C., L. Billes, and L. Bramerie. 2000. "All optical regeneration." Paper FC4.1, *Proc. IEEE Summer Topical Meeting*, Boca Raton, Fla.
9. Meagher, B., X. Yang, J. Perreault, and R. MacFarland. 2000. "Burst mode optical data switching in WDM networks." Paper FC3.2, *Proc. IEEE Summer Topical Meeting*, Boca Raton, Fla.
10. Wei, J. 2000. "The role of DCN in Optical WDM Networks." *Proc. OFC 2000*, Baltimore, Md.
11. Turnet, J. S., 2000. "WDM burst switching for petabit data networks." *Proc. OFC 2000*, Baltimore, Md.
12. Chang, G. K., et al. 2000. "A proof-of-concept, ultra-low latency optical label switching testbed demonstration for next generation internet networks." *Proc. OFC 2000*, Baltimore, Md.
13. Henmi, N., et al. 1998. "OADM workshop." *EURESCOM P615*, p. 36.
14. Henmi, N. 2000. "Beyond terabit per second capacity optical core networks." Paper FC2.5, *Proc. IEEE Summer Topical Meeting*, Boca Raton, Fla.
15. Telcordia Technologies standard GR-485-CORE. May 1999. "Common language equipment coding procedures and guidelines: generic requirements," issue 3.
16. Dutton, H. 1999. *Optical Communications.* San Diego, Calif.: Academic Press.
17. DeCusatis, C., and P. Das. 1990. "Spread spectrum techniques in optical communication using transform domain processing." *IEEE Trans. Selected Areas in Communications* **8**(8): 1608–1616.
18. Brown, T. 2000. "Optical fibers and fiber optic communications." Chapter 1 in *Handbook of Optics* vol. IV, OSA Press.
19. Hill, K. 2000. "Fiber Bragg Gratings." Chapter 9 in *Handbook of Optics* vol. IV, OSA Press.
20. Poumellec, B., P. Niay, M. Douay et al. 1996. "The UV induced refractive index grating in $Ge:SiO_2$ Preforms: Additional CW experiments and the

macroscopic origins of the change in index." *Journ. Of Physics D, App. Phys.* 29: 1842–1856.
21. Hill, K. O., B. Malo, F. Bilodeau, et al. 1993. "Photosensitivity in optical fibers." *Ann. Review of Material Science* 23: 125–157.
22. Kawasaki, B. S., K. O. Hill, D. C. Johnson, et al. 1978. "Narrowband Bragg reflectors in optical fibers." *Opt. Lett.* 3: 66–68.
23. Hill, K. and G. Meltz. 1997. "Fiber Bragg grating technology: fundamentals and overview." *Journ. of Lightwave Tech.* 15: 1263–1276.
24. Stolen, R. 1979. "Nonlinear properties of optical fiber." Chapter 5 in *Optical Fiber Communications.* S.E. Miller and A.G. Chynoweth, eds. New York: Academic Press.
25. Li, C. S., and F. Tong. 1998. "Emerging technology for fiber optic data communication." Chapter 21 in *Handbook of Fiber Optic Data Communication*, pp. 759–783.
26. Xin, W., G. K. Chang, and T. T. Gibbons. 2000. "Transport of gigabit Ethernet directly over WDM for 1062 km in the MONET Washington, D.C. network." Paper WC1.4, *Proc. IEEE Summer Topical Meeting*, Boca Raton, Fla.
27. Bouchat, C., C. Martin, E. Ringoot, et al. 2000. "Evaluation of Super PON demonstrator." Paper ThC2.3, *Proc. IEEE Summer Topical Meeting*, Boca Raton, Fla.
28. Manzalini, A., A. Gladisch, and G. Lehr. 2000. "Management optical networks: view of the EURESCOM project P918." Paper ThC1.2, *Proc. IEEE Summer Topical Meeting*, Boca Raton, Fla.
29. DeCusatis, C., D. Petersen, E. Hall, F. Janniello. December 1998. "Geographically distributed parallel sysplex architecture using optical wavelength division multiplexing." *Optical Engineering*, special issue on Optical Data Communication **37**(12): 3229–3236.
30. DeCusatis, C. 1995. "Wavelength and channel multiplexing for data communications." *Proc. OSA Annual Meeting*, Portland, Ore., p. 115.
31. DeCusatis, C., D. Stigliani Jr., W. Mostowy, M. Lewis, D. Petersen, and N. Dhondy. Sept./Nov. 1999. "Fiber optic interconnects for the IBM S/390 Parallel Enterprise Server G5." *IBM Journal of Research and Development* **43**(5/6): 807–828.
32. DeCusatis, C., and P. Das. 2000. "Subrate multiplexing using time and code division multiple access in dense wavelength division multiplexing networks." *Proc. SPIE Workshop on Optical Networks*, Dallas, Tex., pp. 1–8.
33. DeCusatis, C., and D. Priest. June 2000. "Dense wavelength division multiplexing devices for metropolitan area datacom and telecom networks." *International Conference on Applications of Photonics Technology (Photonics North)*, Quebec, Canada, pp. 8–9.

34. Wei, J., C. Liu, and K. Liu. 2000. "IP over WDM traffic engineering." Paper WC1.3, *Proc. IEEE Summer Topical Meeting*, Boca Raton, Fla.
35. Awduce, D. O., Y. Rckhter, J. Drake, et al. "Multiprotocol lambda switching: combining MPLS traffic engineering control with optical crossconnects," IETF draft proposal at draft-awduche-mpls-te-optical-01.txt
36. Chaudhuri, S., G. Hjalmtysson, and J. Yates. "Control of lightpaths in an optical network." IETF draft proposal: draft-chaudhuri-ip-olx-control-00.txt
37. Basak, D., D. O. Awduche, J. Drake, et al. "Multiprotocol lambda switching: combining MPLS traffic engineering control with optical crossconnects." IETF draft proposal: draft-basak-mpls-oxc-issues-01.txt
38. Krishnaswamy, M., G. Newsome, J. Gajewski, et al. "MPLS control plane for switched optical networks." IETF draft proposal: draft-krishnaswamy-mpls-son-00.txt
39. Rajagopalan, B., D. Saha, B. Tang, et al. "Signalling framework for automated provisioning and restoration of paths in optical mesh networks." IETF draft proposal: draft-rstb-optical-signaling-framework-00.txt
40. Wang, G., D. Fedyk, V. Sharma, et al. "Extensions to OSPF/IS-IS for optical routing." IETF draft proposal: draft-wang-ospf-isis-lambda-te-routing-00.txt
41. Fan, Y., P. A. Smith, V. Sharma, et al. "Extensions to CR-LDP and RSVP-TE for optical path set-up." IETF draft: draft-fan-mpls-lambda-signaling-00.txt
42. Wei, J., C. Liu, and K. Liu. 2000. "IP over WDM traffic engineering." Paper WC1.3, *Proc. IEEE Summer Topical Meeting*, Boca Raton, Fla.
43. Bonenfant, P., A. Moral. 2000. "Digital wrappers for IP over WDM systems." Paper WC1.2, *Proc. IEEE Summer Topical Meeting*, Boca Raton, Fla.
44. Medard, M., S. Lumetta. 2000. "Robust routing for local area optical access nctworks." Paper FC1.4, *Proc. IEEE Summer Topical Meeting*, Boca Raton, Fla.
45. Chung, Y. C. 2000. "Optical monitoring techniques for WDM networks." Paper FC2.2, *Proc. IEEE Summer Topical Meeting*, Boca Raton, Fla.
46. Information on ODSI is available in Electrical Engineering Times issue 1096 p. 1 available at http://www.eet.com.
47. See, for example, ANSI X3.230–1994 rev. 4.3, Fibre channel — physical and signaling interface (FC-PH) May 30, 1995; ANSI X3.272-199x, rev. 4.5, Fibre channel — arbitrated loop (FC-AL), June 1995; ANSI X3.269-199x, rev. 012, Fiber channel protocol for SCSI (FCP), May 30, 1995.
48. For the most current information on NGIO, see http://www.intercast.de/design/servers/future_server_io/documents/071cameron/slide014.htm.
49. For the most current information on Infiniband, see Infiniband (SM) trade association at http://www.connectedpc.com/design/servers/future_server_io/link_spec.htm.

50. Simmons, J. 2000. "Economic and architectural benefits of hierarchical backbone networks." Paper WC1.1, *Proc. IEEE Summer Topical Meeting*, Boca Raton, Fla.
51. Simmons, J. 1999. "Hierarchical restoration in a backbone network," *Proc. OFC*, San Diego, Calif.
52. Sun, P. C., Y. T. Mazurenko, W. S. C. Chang, P. K. L. Yu, and Y. Fainman. 1995. "All optical parallel-to-serial conversion by holographic spatial-to-temporal frequency encoding." *Optics Letters* 20: 1728–1730.

Chapter 6 | Optical Backplanes, Board and Chip Interconnects

Rainer Michalzik
University of Ulm, Optoelectronics Dept., Ulm, Germany

6.1. Introduction

The present chapter is intended to give some insight into the rapidly advancing field of research into optical interconnects within data processing systems. This topic is dealt with in a huge number of publications at conferences, in scientific journals and their special issues, as well as book chapters, good access to which is obtained through [1–14]. Given the extremely dynamic environment, it becomes clear that we cannot attempt to provide a complete overview but rather have to restrict ourselves to several examples from current research, which the readers hopefully will find representative. We start off by looking into recent developments in the 10 Gbit/s data rate regime at the intersystem level. These are high-speed 850-nm vertical-cavity surface-emitting lasers, parallel optical links, and new-generation silica as well as plastic optical multimode fibers. Within a single box environment we follow down the usual hierarchy from optical backplanes to intraboard to inter- and perhaps even intrachip levels. For all these categories we attempt to list available technology options and give practical examples from ongoing project work.

The overwhelming majority of digital data inside and between even peak performance computer systems is nowadays carried by electrical signals traveling on metallic lines. Besides the fact that high-speed electronic rather than photonic data transmission has historically been the first available technique, clearly there are a number of advantages that speak in its favor. In the computer environment the most important ones are ease of implementation and handling, low cost, and high reliability. However, with the exponentially increasing performance of electronic processing systems, more and more drawbacks of the conventional approach become apparent. The pronounced waveguide dispersion characteristics of electrical lines correspond to a relatively small bandwidth–distance product so that at higher clock rates it is an increasingly difficult task to bridge the required distances even within a single box. Electromagnetic cross-talk is forcing circuit designers to increase the width of data buses rather than the bus frequency. Susceptibility to electromagnetic interference requires proper shielding and grounding, which can turn thin wires into bulky copper strands, thus limiting the overall interconnect density. Finally signaling rates in the hundreds of megahertz regime make impedance matching a necessity, which is not easy to achieve in practice and only works over a very limited frequency range. Apart from the clock rate's driving force, novel distributed or parallel computing approaches fuel the need for high bandwidth linking of individual data processing subsystems. Bearing these challenges in mind it is widely recognized that the consequences of an apparent electrical signaling bottleneck are experienced at shorter and shorter link lengths and that optical interconnects hybridly integrated with electronics have attractive solutions at hand or are at least potentially able to offer them.

It has become customary to classify optical interconnects into distinct, albeit overlapping, categories. Some of those are illustrated in Fig. 6.1, ranging from the longest to the shortest transmission distance:

- Rack-to-rack, also called frame-to-frame;
- Board-to-board;
- Multi-chip module (MCM)-to-MCM or intraboard;
- Chip-to-chip on a single MCM;
- Intra- or on-chip.

On the frame-to-frame level, it is a relatively easy task to replace space-consuming and performance-limited copper cables with lightweight fibers. Single-channel or space-parallel optical transceiver modules are already commercially available at reasonable cost. Board-to-board interconnection within a rack can be accomplished via edge connections to optical waveguides placed on a hybrid electrical/optical backplane or via free-space transmission. Within a printed circuit board, routed fiber circuits or integrated

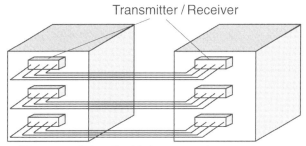

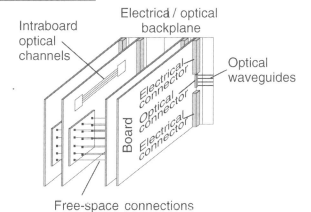

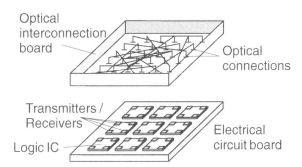

Fig. 6.1 Major optical interconnect categories.

channel waveguides can be applied for inter-MCM communication. Data transfer between or even within chips finally might be achieved by optical fiber bundles or an optical overlay providing guided-wave or free-space data transport. In most of these scenarios the demand for high-throughput interconnection can be satisfied only by one- or two-dimensional high-density arrays of optoelectronic components, the surface-normal operation of which usually proves to be extremely beneficial.

6.2. Frame-to-Frame Interconnections

Owing to the benefits addressed before, it is an increasingly common practice to take advantage of optical fibers to implement the data links between a high-performance computer box and its outside world, which might consist of other processing units, a storage farm, or a network server. In this section we thus review some recent achievements in graded-index fiber-based optical interconnects approaching the 10 Gbit/s data rate regime, which in the future are likely to be employed at the frame-to-frame level, replacing 1 Gbit/s or lower speed modules. In particular these advances rely on progress in the fabrication of high-speed vertical-cavity surface-emitting laser arrays and new-generation silica as well as polymer optical multimode fibers.

6.2.1. ONE-DIMENSIONAL VCSEL ARRAY DEVELOPMENT FOR NEXT-GENERATION PARALLEL OPTICAL DATA LINKS

The vertical-cavity surface-emitting laser (VCSEL) [15–15b] is a fine specimen of a novel compound semiconductor device that has been successfully commercialized in the last few years. Operation principles and laser technology are treated in some detail in Vol. 1 Chapters 2 and 16 of this handbook. Among the various VCSEL applications, optical datacom is the primary driving field. Especially Gigabit Ethernet (GbE) and related transceivers for graded-index (GI) multimode fiber (MMF) data transmission have become inexpensive mass products by relying on 850 nm short-wavelength VCSEL technology. Generally speaking, the most attractive features of datacom VCSELs include on-wafer testing capability, mounting technology familiar from the low-cost light-emitting diode market, circularly symmetric beam profiles for ease of light focusing and fiber coupling, high-speed modulation with low bias currents, driving voltages well compatible to silicon VLSI electronics, temperature insensitive operation characteristics, and obvious forming of one- or two-dimensional arrays. Transceiver modules are employed for a variety of tasks such as in-building backbone links, interconnection of computer clusters or of telecom gear in central offices. The aggregate data throughput of single-channel modules can easily be increased by using the space-division multiplexing technique described, e.g., in Vol. 1 Chapter 11, where optical signals are transmitted in parallel through a MMF ribbon cable. A useful overview with many references on early fiber ribbon data links has been compiled in [16]. Although significant cost breakthroughs have been achieved with high yield one-dimensional laser [17] and photodetector arrays as well as alignment tolerant packaging approaches [18], a GI MMF ribbon cable is still a relatively expensive component, especially if interchannel skew is to be minimized. Therefore parallel links are currently competitive only for several tens of meter transmission length. State-of-the-art modules operate at 2.5 Gbit/s channel data rate and thus achieve 30 Gbit/s throughput for a 12-channel system [19–21]. Obviously, intensive work toward modules with 10 Gbit/s individual channel data rate has commenced.

Figure 6.2 shows a photograph, bit error rate (BER) characteristics, and an eye diagram of a 1 ∗ 10 elements VCSEL array that is being developed for these next-generation parallel optical transceivers. The oxide-confined VCSELs are arranged on a 250-μm pitch that is compatible to MT (Mechanical Transfer) -type multifiber connectors [22] and each unit cell of the array contains p- and n-contacts separated by 125 μm. The etched VCSEL mesas are planarized by polyimide to obtain a low parasitic capacitance, coplanar contact layout. With this design, modulation corner frequencies in excess of 12 GHz are achieved [23] which are well-suited for data transmission in the 10 Gbit/s regime. In the given experiment, about 3 μm active diameter lasers with an average threshold current $I_{th} = 340\ \mu$A, emitting in a single transverse and longitudinal mode at 850-nm wavelength, have been driven at identical 1.65 mA bias current and 0.65 V_{pp} modulation voltage [24], yielding a dynamic on-off ratio of 6 dB. Figure 6.2 reveals that the BER curves thus obtained for back-to-back operation almost coincide and that error rates of 10^{-9} are reached with less than -15 dBm optical power incident onto a pin photodiode and transimpedance-amplifier-based fiber pigtailed receiver. Although 10 Gbit/s-compatible prototype VCSEL arrays are available today, the manufacturing of complete interconnect modules still requires some challenges to be addressed. Among those are the realization of high-sensitivity MMF-compatible photoreceiver arrays and the dense hybrid integration of optoelectronic chips with high-speed, probably silicon germanium (SiGe) based electronics [24a].

6. Optical Backplanes, Board and Chip Interconnects

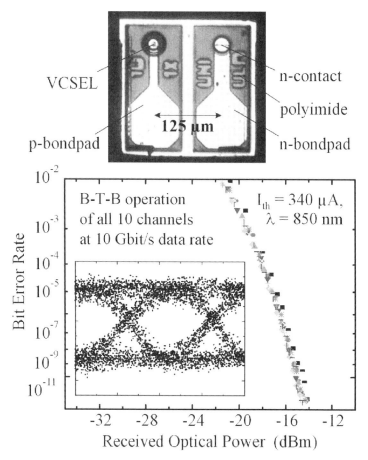

Fig. 6.2 Cleaved unit cell of a linear 1 ∗ 10, 250-μm spacing VCSEL array with coplanar, 125-μm pitch contact arrangement (top) and bit error rate characteristics for back-to-back (B-T-B) operation of all 10 channels at 10 Gbit/s modulation with a representative eye diagram (on a 200 ps horizontal and 900 mV vertical scale) in the inset (bottom).

Figure 6.3 shows a typical example of a complete parallel optical link module. Generally, effective heat sinking is required to remove the excess power generated by the driver and receiver electronics as well as the multiplexing and demultiplexing circuits.

6.2.2. LONG-DISTANCE 10 GBIT/S MMF DATA TRANSMISSION

Even before the GbE standard had been finalized it had already become evident that backbones operating at 1 Gbit/s speed would only for a short

Fig. 6.3 Infineon Technologies' parallel optical link (PAROLI) module with and without an attached 12-fiber ribbon cable [19] (left, © 2000 IEEE) and bottom view of an open module [19a] (right).

time be able to satisfy the ever-increasing bandwidth demand in local area network environments. Thus, in March 1999 the Institute of Electrical and Electronics Engineers, Inc. (IEEE) established a Higher Speed Study Group to explore available technology options, and the current 10-Gigabit Ethernet proposal is expected to be adopted as a standard termed IEEE 802.3ae [25] during 2002. With regard to a multimode fiber medium that is attractive for low apparatus cost in-building cabling, several options exist to reach the target data rate on the order of 10 Gbit/s. Coarse wavelength-division multiplexing (CWDM, sometimes also called wide-WDM), usually with more than 10 nm channel spacing, is an attractive option for increasing the data throughput while utilizing the often low-bandwidth installed MMF base. CWDM modules based on 780 to 860 nm wavelength VCSELs for $4*2.5$ Gbit/s data transmission over 100 m of 62.5 μm core diameter MMF [26] or for $8*155$ Mbit/s operation [27] have been demonstrated. For the 1300 nm long-wavelength regime, the use of edge-emitting DFB lasers has been proposed for $4*2.5$ Gbit/s transmission over either multimode or single-mode fiber [28]. However, several demonstrations of 10 Gbit/s VCSEL transmitters and MMF receivers together with the progress of CMOS and SiGe electronics clearly show the feasibility of a straight forward serial high-speed solution without the added complexity of optical multi- and demultiplexing. Indeed, the short-wavelength CWDM approach is presently not considered for the IEEE 802.3ae physical layer [28a]. Subcarrier multiplexing and multilevel coding [29] or adaptive electronic equalization [30] as alternative upgrading methods have not yet been sufficiently evaluated for practical use but might become competitive in the future. On the other hand, an adoption of the 850 nm serial solution into the 10-GbE standard required the availability of a MMF with improved modal bandwidth.

6. Optical Backplanes, Board and Chip Interconnects

The maximum transmission distances achievable with a MMF are limited by intermodal dispersion resulting in pulse broadening and thus intersymbol interference. As an example, for the 850-nm short-wavelength regime the GbE standard defines an operating range of 275 m for a 62.5-μm core diameter GI MMF with a bandwidth–distance product of 200 MHz $*$ km. Due to lower numerical aperture and smaller core area and thus better manufacturing control over the refractive index profile, a 50-μm diameter fiber featuring 500 MHz $*$ km normalized bandwidth will suffice up to distances of 550 m [31, 32]. Although transmission over longer lengths has been demonstrated by taking advantage of restricted fiber launch from a laser source [33], average quality MMFs are not able to support 10 Gbit/s signals over sufficiently long distances.

Minimum cabling length requirements can be extracted from the results of a survey of 107 U.S. companies of diverse size, location, and industry conducted by the IEEE in July 1996 during the development of GbE [34]. According to Fig. 6.4, more than 90% of the in-building fiber backbones are less than 300 m in length, which is thus defined as a target value. In early 2000, Lucent Technologies was the first company to introduce a 50-μm core diameter MMF with an increased bandwidth–distance product of 2.2 GHz $*$ km at 850-nm wavelength [34]. Besides the adoption of GbE and related standards, commercialization of VCSEL technology, and the realization of low-cost connector and other hardware for fiber local area networks (LANs), this development is regarded as a key opportunity for further

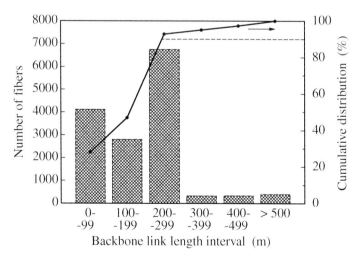

Fig. 6.4 In-building fiber backbone distance distribution from a 7/1996 IEEE survey [34].

penetration of optoelectronics into data networks [35]. Fiber optimization consists of a reproducible and improved realization of a close to parabolic refractive index grading that minimizes the group delay differences between the multiple guided modes over a certain target wavelength regime. Of special importance is the suppression of the center dip in the index profile, which is easily established during fiber preform preparation [36], leading to propagation delay particularly for the fundamental mode. Despite being designed for the short-wavelength region, the bandwidth requirements of 1.3 μm wavelength signals as defined in the IEEE 802.3z GbE standard 1000BASE-SX physical layer are still satisfied by the new fiber.

Figure 6.5 displays the results of time-domain differential mode delay (DMD) measurements carried out on a conventional as well as a new generation MMF [37]. In this technique, an optical pulse is launched from a single-mode fiber into the MMF under test at different radial offset positions and the resulting pulse shapes at the MMF end are recorded. Due to the excitation of different mode groups with different group velocities, the conventional fiber shows significant pulse splitting and wandering of the peak

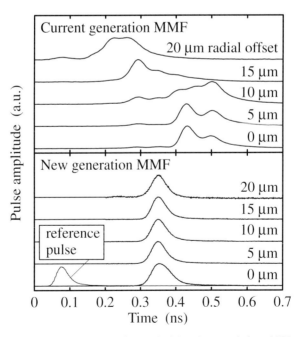

Fig. 6.5 850-nm wavelength differential mode delay characteristics of 500-m spools of a typical deployed MMF (top) and of a new-generation optimized MMF (bottom), both with 50-μm core diameter [37].

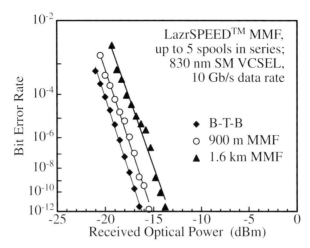

Fig. 6.6 Bit error rate characteristics for 10 Gbit/s data transmission with an 830-nm wavelength transverse single-mode (SM) VCSEL over up to 1.6 km of a high-performance multimode fiber as well as for back-to-back (B-T-B) operation [38] (© 1999 IEEE).

position with varying launch offset. In contrast, the bandwidth-optimized fiber features nearly ideal pulse propagation behavior over almost the entire core area with only slight broadening versus the directly detected reference pulse. Thus, alignment-tolerant data transmission properties can be expected, which retains the cost benefits of a MMF-based system.

Figure 6.6 shows bit error rate measurements for 10 Gbit/s data transmission with an 830-nm VCSEL source over this fiber medium [38]. Error-free operation over up to 1.6 km distance is achieved with only 3 dB power penalty. Five connectorized fiber spools of 500, 400, 300, 300, and 100 m length are used in the experiment. Further including a 2-km-long fiber piece, transmission over even 2.8 km was shown to be possible [39]. Although in these two demonstrations single-mode emission turned out to be an essential feature due to chromatic dispersion limitations, highly multi-moded VCSELs can be employed as well for transmission over distances that will be required for 10-GbE [40].

With a further increase of bandwidth demand beyond the direct current modulation capabilities of laser transmitters, CWDM approaches as mentioned before will certainly gain importance for premises backbone links. With a channel spacing of several nm, the requirements for wavelength stabilization of optical sources as well as on the demultiplexing process can be considerably relaxed compared to dense WDM in telecommunications. For a first laboratory realization of a 40 Gbit/s MMF

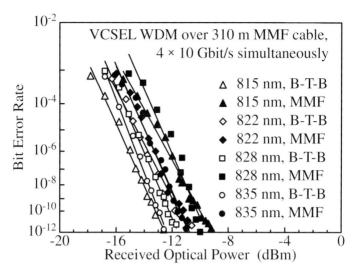

Fig. 6.7 Bit error rates in a 4-channel CWDM experiment showing $4 \ast 10$ Gbit/s data transmission over 310 m of a 50-μm core diameter MMF cable [41].

system [41], oxide-confined high-speed transverse single-mode VCSELs in the 3 to 4 μm active diameter range with emission wavelengths of 815, 822, 828, and 835 nm have been selected. VCSEL beams are combined through polarizing beam splitters and a MMF 3-dB coupler and a grating is used for free-space demultiplexing. The transmission experiment is performed by driving all VCSELs simultaneously with decorrelated 10 Gbit/s pseudo-random bit sequences from a pattern generator, whereas only one of the channels is detected. Figure 6.7 shows the measured BERs before and after the insertion of a 310-m-long high-bandwidth MMF cable, where error-free operation is achieved for all channels. Power penalties between 1 and 3 dB are mainly caused by polarization noise generated during beam multiplexing and a below 15 dB adjacent wavelength suppression in the present demultiplexing setup. Integration of devices as demonstrated at lower speeds in [27] or [28] is likely to yield very compact affordable modules for 40 Gbit/s or higher data rates in the future.

6.2.3. DATA TRANSMISSION OVER PLASTIC OPTICAL FIBERS

Traditionally plastic (sometimes also denoted polymer) optical fibers (POFs) are mainly applied in lighting, display, image transmission, or simple optical interconnection of consumer electronics equipment [42]. For several years, however, POFs have produced much interest for use in high-speed data communications. POFs can be fabricated with larger core diameter than any silica glass fiber while being much more flexible and resistant to breaking. As a consequence, large alignment tolerances allow the use of injection-molded parts in connectors and transceivers, which reduces the apparatus cost to a minimum. At the same time, higher bandwidth, lighter weight, and inherent immunity to electromagnetic interference can give plastic fibers an edge over conventional copper cable technology. Progress in POF fabrication, system applications, and corresponding passive and active components is well documented in the Proceedings of the International Plastic Optical Fiber Conferences [43–45a]. Fiber manufacturing is concentrated mainly in Japan with Mitsubishi Rayon Co., Ltd., Asahi Kasei Co. (previously Asahi Chemical Industry Co., Ltd.), Toray Industries, Inc., and Asahi Glass Company as the main players.

One of the most important benchmarks for an optical fiber is its spectral attenuation characteristics. Figure 6.8 shows such spectra for the well-known silica glass fiber and the very low-cost PMMA step-index (SI) POF [46]. For data communication wavelengths of around 900 nm, these fiber types have hugely different attenuation coefficients of about 1 dB/km and 10^4 dB/km, respectively. The various attenuation peaks in the PMMA

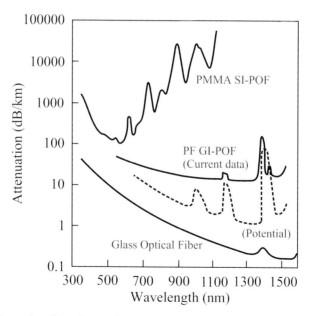

Fig. 6.8 Currently achieved as well as potential attenuation characteristics of a perfluorinated polymer (CYTOP) fiber compared with standard silica glass and polymethyl methacrylate (PMMA) material [46].

spectrum are mainly caused by resonances of carbon-hydrogen bonds. In the more recent perfluorinated (PF) POF, basically hydrogen is replaced by heavier fluorine atoms, thus effectively suppressing most of the resonances and allowing achievement of intermediate (on a logarithmic dB-scale) attenuation coefficients of about 100 dB/km. The potential of the promising perfluorinated material is exploited in graded-index fibers containing a parabolic-like refractive index profile in the core region. The so-called Lucina^{TM} fiber with 120-μm core diameter, fabricated from a glass-state PF polymer by Asahi Glass, offers an over-filled launch bandwidth in the 200 MHz $*$ km range (see [47] for a discussion of maximum theoretical bit rates) and losses below 50 dB/km throughout the entire data communications window from 700 to 1300 nm. Minimum losses down to 10 dB/km seem to be within reach [46].

Table 6.1 provides an overview of recent system experiments performed with various kinds of POF and either Fabry-Pérot-type, DFB, or VCSEL sources in different wavelength regimes. Due to high attenuation of 1 to 3 dB/m in the usual 780- to 860-nm short-wavelength datacom operation window (and even considerably higher losses for longer wavelengths), data transmission over PMMA-based fibers is feasible only for meter-long links within a cabinet [57] or on the circuit board level, as mentioned in

Table 6.1 **POF-Based Data Transmission Experiments Fiber Types are PMMA SI-POF (PMSI), PMMA GI-POF (PMGI), or Perfluorinated GI-POF (PFGI)**

Fiber type	Core diameter	Bit rate	Distance	Laser	Wavelength	Ref.
PMSI	1 mm	0.5 Gbit/s	100 m	FP	650 nm	[48]
PMSI	1 mm	1 Gbit/s	15 m	VCSEL	780 nm	[49]
PMSI	120 μm	2.5 Gbit/s	2.5 m	VCSEL	835 nm	[50]
PMGI	500 μm	2.5 Gbit/s	200 m	FP	645 nm	[51]
PFGI	170 μm	2.5 Gbit/s	300 m	FP	645 nm	[52]
PFGI	130 μm	2.5 Gbit/s	550 m	VCSEL	840 nm	[53]
PFGI	130 μm	2.5 Gbit/s	550 m	DFB	1310 nm	[53]
PFGI	130 μm	9 Gbit/s	100 m	VCSEL	830 nm	[54]
PFGI	155 μm	7 Gbit/s	80 m	VCSEL	935 nm	[55]
PFGI	130 μm	11 Gbit/s	100 m	FP	1300 nm	[56]

Sect. 6.5. Losses of about 0.2 dB/m allow for somewhat longer distances at around 650 nm, however, here the fabrication of array type laser sources with sufficient temperature stability imposes great difficulties. Although remarkable progress has been reported on red-emitting VCSELs in the last few years [58–60], high-speed operation at elevated temperatures still needs to be demonstrated. In addition, the PMMA attenuation spectrum in Fig. 6.8 exhibits a rather narrow dip at around 650 nm and unfortunately, laser performance degrades with almost every nanometer below 670 nm wavelength [61].

Due to comparatively low attenuation over a several 100-nm-wide spectral range, the PF GI-POF appears to be a candidate for data transmission even up to 10 Gbit/s over 100 m distance. Graded-index POFs indeed generally show higher bandwidth–distance products than would be expected from their refractive index profile. Both strong mode coupling between low order mode groups [62] and differential mode attenuation [63] are discussed as underlying mechanisms. Furthermore, the bandwidth can be enhanced through under-filled launch from a small numerical aperture source such as a VCSEL. Apart from the special modal dispersion properties it is also interesting to note that the material dispersion of perfluorinated material is lower than that of silica glass or PMMA [64]. However, as can be concluded from Table 6.1 there is no standardized fiber diameter as yet and long-term stability is still a much-debated issue. These factors all contribute to limited commercial fiber availability to date.

With the aforementioned fiber properties, the application fields included in Fig. 6.9 are envisioned for POFs [64]. Step-index POFs with conventional or lower (0.2 to 0.3 rather than 0.4 to 0.5) numerical aperture (NA) could compete with twisted pair and coax electrical cables and be employed in much discussed [43–45a] digital home area networks (HANs) based on the IEEE 1394 (FireWire) standard. Also, data buses for entertainment, control, or communication in mobile systems such as cars, trains, and airplanes have moderate bandwidth and length requirements but do benefit from electromagnetic interference immunity and small size and weight of POFs [65, 66]. PMMA GI-POFs could be used for data rates exceeding 400 Mbit/s but required 650-nm transceiver components. Fiber cables for horizontal as well as vertical connections with below 300 m length in premises networks are potential applications of PF POFs. With further expected increases in bandwidth and reductions in loss, these fibers might even be used in campus LANs or penetrate into access networks for HANs, business districts, or residential areas. Currently however, the performance of PF POFs is by far

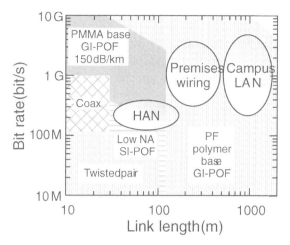

Fig. 6.9 Bit rate vs length limits and future application areas of POFs [64].

inferior to that of silica fibers, especially when considering the progress in high-bandwidth MMFs capable of 10 Gbit/s transport over more than 300 m, as reported in the previous section. Also it has been shown [67] that non-standard, extended core diameter (148 or 185 μm) and thus more alignment tolerant GI silica MMFs allow data transmission over much larger distances (2.5 Gbit/s over 4 km at 1300 nm wavelength) than PF POFs owing to significantly lower losses.

In any case, whereas it is relatively easy to launch the light from a laser source into a POF, the receiving end remains the most demanding part of a high-speed Gbit/s plastic-based data link. First of all, the diameter of the photodiode has to be chosen large enough in order not to waste signal power and to minimize modal noise penalties, which incurs bandwidth limitations through the increased diode capacitance. Mostly, additional optics have to be used. On the other hand, the higher fiber loss compared to silica decreases the available link power margin so that high-sensitivity receivers have to be employed, especially if the links are intended to be operated within the eye safety margin of the infrared short-wavelength regime.

6.3. Optical Backplanes

The first penetration step of optical data links into single-box, high-performance data processing systems most likely will be the introduction of optical backplanes to interconnect the usual multitude of printed circuit boards (PCBs). For clock frequencies exceeding values of roughly 100 MHz, parallel electrical lines already suffer from well-known drawbacks such as RC time constant limitations, ohmic losses, cross-talk, reflections due to impedance mismatch, or ground bounce. The usual design strategy consists of increasing the data bus width rather than the data rate. Operating a bus at higher frequencies is usually not possible, requiring a complete redesign. Apart from pure intra-backplane transmission, the interfaces between boards and backplane represent a performance bottleneck themselves. Although array connectors capable of handling differential 1 Gbit/s signals (at the expense of a higher number of pins) in a below 1 mm pitch are available on the market, high pin count electro-mechanical connections constitute a particularly delicate part of such a system. Optical backplane interconnects on the other hand are free from transmission line effects and are thus able to provide wider bandwidth at reduced and, most important, frequency independent cross-talk, greatly improved scalability of bus frequencies, voltage isolation, higher channel density, and reduced mechanical insertion force [68] contributing to higher reliability.

According to Fig. 6.10, approaches to optical backplane interconnects are subdivided here into five categories which are discussed in the following sections.

6.3.1. FIBER CABLED BACKPLANE

In the most straightforward case, depicted in Fig. 6.10(a), the optical channels of a hybrid electrical/optical backplane are composed of well-standardized duplex optical fiber jumper cables, which constitutes a readily available solution, especially if transceiver modules (like for GbE) are placed on the boards [69]. For higher throughput fiber ribbon cables special optical multi-fiber backplane connectors are being developed that lend themselves to replacing existing electrical connection schemes. In [70] a backplane connector incorporating 48 channels in a total area of $25 * 21$ mm^2 is proposed that is fully compatible with the electrical SIPAC (Siemens Packaging System) system. Small backplane openings are certainly required to minimize possible electromagnetic interference from outside sources. A current connector solution for rack-to-rack linking is shown in the top part of Fig. 6.11, where up to six standard MT ferrules holding a total of up to 72 fibers fit into an adapter that requires about $18 * 46$ mm^2 backpanel opening. Alternatively, parallel transceiver modules such as mentioned in Sect. 6.2.1 can be placed at the board edge, as

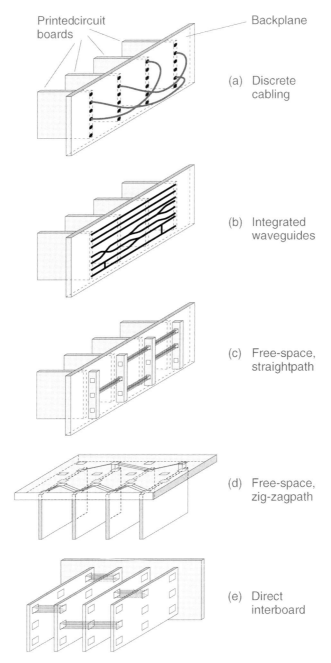

Fig. 6.10 Optical backplane categories.

6. Optical Backplanes, Board and Chip Interconnects   233

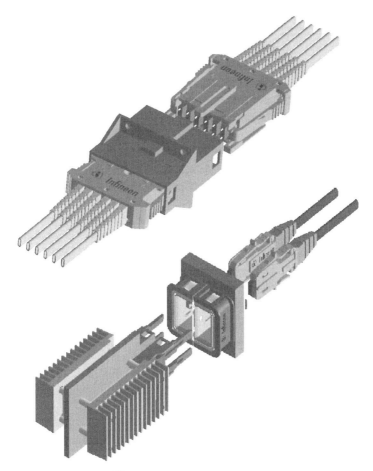

Fig. 6.11 Infineon's MMC™ (Multi MC) connector and adapter system for up to 6 MT connectors with 12 fibers each at 250-μm pitch (top) and PAROLI transmitter and receiver arrangement for direct coupling into a duplex-type backplane connection module [70a] (bottom).

illustrated in the bottom part of Fig. 6.11. High input/output (I/O) density is reached in a "belly-to-belly" configuration of transmitter and receiver modules and the use of an SMC- (Single MT Connector) Duplex connector. A minimum module pitch of 30 mm would then translate into an average port density of 8 per cm or a bit rate density of 20 Gbit/s/cm assuming 12-channel transceivers operating at individual 2.5 Gbit/s. The backplane socket incorporates a floating center part that assists the mating process so that no tighter mechanical tolerances are required compared to electrical solutions.

A fine example for the reduction of volume and weight of connectors and cables and the additional benefit of increased data throughput that can be achieved by replacing electrical with optical board edge interconnections is documented in [71].

6.3.2. INTEGRATED WAVEGUIDES

Certainly the wiring harness that can result from discrete fiber cabling [69] is not overwhelmingly elegant since it consumes extra space, is prone to assembling errors, and is not really cost-effective. While backplane connectors of the type in Fig. 6.11 are best suited as an interface for frame-to-frame interconnection, optical waveguides integrated into the backplane according to Fig. 6.10(b) appear more appropriate for high interboard connectivity. For this scenario, three important topics immediately need to be addressed, namely ease of waveguide integration, amount of power attenuation, and possible ways of bending the photon path by $90°$.

First of all it is clear that multimode rather than single-mode waveguides are preferred since they offer relaxed alignment tolerances but still provide sufficient bandwidth for less than 1 m transmission length. Second, at first glance, off-the-shelf optical fibers seem to be an attractive choice because of their low transmission loss. In [72] 500-μm-diameter POFs are laid into grooves and $45°$ notches incorporating hemispherical lenses are provided in the backplane for vertical in- and outcoupling. Although glass fibers have almost negligible losses, fiber termination at exact predetermined lengths is even less feasible. Within the European Commission (EC)-funded HIBITS project (High Bitrate ATM Termination and Switching, project duration Jan. 1992–Dec. 1995, part of the RACE II program), 62.5-μm core diameter silica MMFs have been encapsulated at 250 μm pitch in an epoxy layer on a printed circuit board [73], where fibers terminate in custom-made surface-mounted connector parts involving LIGA[1] fabrication technology. Fiber flexfoils for parallel board or backplane links have been fabricated in the EC SPIBOC project (Standardised Packaging and Interconnect for Inter- and Intra-Board Optical Interconnections, Nov. 1993–Nov. 1996, ESPRIT III program[2]) [76, 77].

---

[1] The German acronym stands for the terms lithography, electroforming, and plastic molding [74].
[2] Work on optical backplanes had already commenced in the precursor ESPRIT II project OLIVES (Optical Interconnections for VLSI and Electronic Systems), Jan. 1989–Dec. 1991 [75].

More recently, 62.5-μm core diameter glass-glass-polymer graded-index fiber arrays are laminated onto either PCB or backplane within the U.S. DARPA and Air Force funded ChEEtah/OMNET (Optical Micro-Networks) program launched in June 1997 with 36+ months duration [78]. In both latter examples, multifiber connectors are used at the parallel optoelectronic transceiver and board-to-backplane interfaces. Low loss and very high bandwidth clearly speak in favor of the use of fibers within a computer box, however, several shortcomings are quite obvious [68]. Among those, labor-intensive assembly, difficult photon bending, and the lack of a cost-reducing planar fabrication and packaging technology are most noteworthy.

Polymer waveguides deposited onto a backplane are the second and rather more viable choice for guided-wave propagation. At the end of the 1970s such waveguides had already been reported for use in multimode optical fiber systems [79]. Recently much work on board integration has been invested in the U.S. within the Polymer Optical Interconnect Technology (POINT) project. Several polymers with attenuation coefficients in the 0.1 dB/cm range at 850 nm wavelength are reported in [80–82]. A 288-mm-long 144-channel backplane with 50-μm-wide rib waveguides arranged in a 100-μm pitch has been demonstrated for about 1 Gbit/s speed data transmission. In addition a novel packaging solution for VCSELs employing a planar fabrication process has been devised in this project. Components are picked and placed on a common ceramic or plastic substrate and electrical lines are fabricated on a polyimide layer laminated over the devices. Passive alignment structures for planar Polyguide (a trademark of DuPont) [83] waveguide arrays are produced by laser micromachining [81].

In [84] transmission losses of $200 * 200$ μm^2 cross-section polymer waveguides as low as 0.03 dB/cm have been achieved at 780-nm wavelength. A practical system based on this technology is discussed in Sect. 6.3.6 in greater detail. Apart from rigid substrates, multichannel ribbon waveguides can also be realized on flexible foils like PVC [85] or Kapton [81].

Whereas fiber circuits on a backplane would mostly serve to establish straight or perhaps slightly bent paths without crossings (which would require successive fiber layers) as in the upper part of the schematic backplane in Fig. 6.10(b), it is quite possible to implement polymer waveguide splitters, combiners, and intersections that give additional degrees of freedom for signal routing. In [86] excess losses as low as 2.2 dB for 100 rectangular crossings have been achieved with deuterated PMMA waveguides of about $40 * 40$ μm^2 core size. Propagation loss of straight waveguides was less than 0.02 dB/cm at 850-nm wavelength. Because minimum bending radii for tolerable attenuation usually are in the millimeter range and the incorporation of low-loss 90° deflections often proves difficult, planar waveguide circuits are rather space consuming. Some researchers raise a lot of hope by predicting that these limitations will be overcome in the future by exploiting the unique properties of photonic bandgap structures. Given the extremely stringent requirements on precision manufacturing it seems clear, however, that the scattering losses of such waveguides cannot be reduced to acceptable levels with any presently known microstructuring technique [87].

6.3.3. FREE-SPACE BACKPLANE WITH STRAIGHT PHOTON PATH

Very high density optical interconnection within a backplane can in principle be achieved with two-dimensional arrays of free-space beams, as schematically illustrated in Fig. 6.10(c). Vol. 2, Chapter 10 of this handbook is devoted to the smart pixel technology [4, 5, 88] to be employed for its realization. Although such a configuration is projected to be able to transport and process Tbit/s amounts of data [89, 90], the interboard communication bottleneck still remains at the board-to-backplane interfaces.

According to Table 6.2, free-space (FS) interconnections potentially offer numerous advantages over guided-wave (GW) approaches. However, if it comes to practical implementations the last topic on the list often turns out to be most restricting. Concepts like stacked planar optics [91] have been proposed rather early to overcome the alignment and stability issues but are usually not compatible with the conventional design of electronic data processing systems. With respect to volume consumption it has been shown in [92] that a space-invariant fiber image guide (see Sect. 6.5.2.2) based interconnect can be superior to a free-space solution. For a space-variant interconnect case, the guided-wave approach features a more favorable scaling with the number of channels to be imaged.

6.3.4. FREE-SPACE BACKPLANE WITH ZIG-ZAG PHOTON PATH

Zig-zag-type beam propagation (see Fig. 6.10(d)) within a millimeter-range thick optically transparent substrate can be regarded as a mixture between FS and GW transmission and is thus an alternative to straight free-space

Table 6.2 Comparison Between Free-Space and Guided-Wave Interconnections
WG: Waveguide

Topic	Free-space	Guided-wave
2-D arrangement	Almost natural	Possible
Channel intersections	Yes, lossless	Yes, low loss achievable; no in case of fibers
Channel fan-out	Relatively easy	More complicated
Optical paths	Dynamic reconfiguration possible	Fixed
Maximum channel density	Decreasing with distance; interference through diffraction, misalignment	Independent of distance; usually free from crosstalk
Volume consumption	V_{FS}	$V_{GW} < V_{FS}$ possible [92]
Bandwidth limit	Practically none (carrier frequency)	Modal or chromatic dispersion
Propagation losses	None; only during beam manipulations	Negligible (silica fibers) up to distance-limiting
In-/out-coupling losses	None	Yes, to and from WG
Additional (lossy) optics required?	Yes	Not necessarily (butt coupling possible)
Cost of waveguiding medium	For free	Inexpensive (replicated channel WGs) up to rather costly (2-D arrangements)
Alignment and stability	Very critical; active corrections desired	More easy; passive

interconnections [93]. In [94] advances toward such a kind of backplane are reported. VCSEL and photodetector arrays are hybridly integrated with microlens arrays required for beam collimation and focusing. Holographic gratings serve for beam deflection and splitting, which thus provides signal broadcasting, however, the available power budget limits the allowable number of boards. Clearly a tradeoff between interconnect density and interconnect distance has been identified in accordance with the fifth topic in Table 6.2. Distances on the order of 10 cm already required about 1 mm device spacing. Data rates of up to 2 Gbit/s on the other hand were solely limited by the performance of the active optoelectronic components.

In [95], an equivalent planar optics [96, 97] wave propagation approach is employed to implement a neural-type data processing system, where optoelectronic chips are directly attached to the substrate, which thus resembles the functionality of a miniaturized optical backplane.

6.3.5. DIRECT INTERBOARD INTERCONNECTION

Direct free-space board-to-board data transfer, as in Fig. 6.10(e), altogether avoiding board edge connections, is rated a viable short-term commercial opportunity for optics within a high-performance computing system [98]. The advantage of such a scenario lies in the design freedom to place massively parallel I/O blocks next to transmitting and receiving integrated circuits (which might be found anywhere on the board) without the need to run numerous high-speed electrical signals over long distances to the board edge. VCSEL arrays as the key element will be able to satisfy the projected per-channel bandwidth in the several GHz range at least up to the year 2010. Because individual boards in mainstream computer systems have a positioning tolerance exceeding 1 mm, establishing and maintaining the alignment of the free-space optical links obviously imposes a problem with no known solution to date. The viability of active alignment [99, 99a] that might be achieved by, e.g., beam deflection through micro-electromechanical systems (MEMS) -type elements is at least a cost issue.

6.3.6. EXAMPLE BACKPLANE SYSTEM

A particularly practical approach to an optical backplane based on free-space coupling at the board-to-backplane interface is presented in Fig. 6.12 [85, 72].

The beam emitted from a standard 850-nm VCSEL-based transmitter module is collimated at the PCB edge, focused onto a metalized micromirror by a lens on the backplane, and deflected into a multimode polymer step-index waveguide for guided-wave propagation within the backplane.

6. Optical Backplanes, Board and Chip Interconnects   239

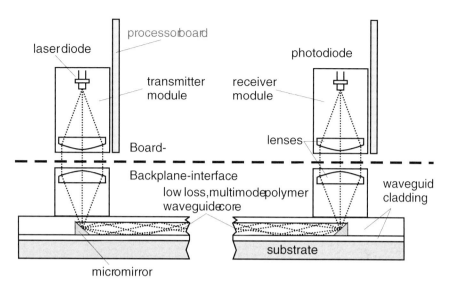

Fig. 6.12 DaimlerChrysler optical backplane concept relying on expanded beam free-space coupling to the processor boards and polymer channel waveguiding within the backplane [85].

The reverse process takes place at the output side, where the beam is focused onto a photodiode in a corresponding receiver module. The system is targeted especially for avionic applications so that immunity to electromagnetic interference is a major advantage of an optical solution. Other benefits of the system include:

- Elimination of optical connectors;
- Large lateral alignment tolerances of $+/-0.5$ mm for 1 dB excess loss; these are reached solely by aid of the remaining electrical pins further required for power and low-speed control lines;
- Low link losses and flexible signal routing.

The polymer waveguides depicted in Fig. 6.13 (top) have a cross-sectional area of about $250 * 200$ μm^2, a 0.35 numerical aperture, an estimated bit rate-distance product of 10 Gbit $*$ m/s (limited by intermodal rather than chromatic dispersion), and can be deposited on a wide range of reasonably smooth substrates such as aluminum, glass, PMMA, PCB materials, and even flexible foils. Attenuation coefficients as low as 0.03 dB/cm have been measured, allowing the integration of couplers and splitters, which is rather impractical with optical fibers. Figure 6.13 (center) shows an

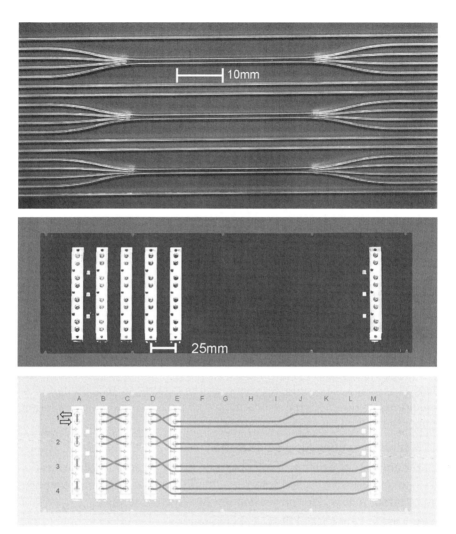

Fig. 6.13 Straight polymer waveguides and 4-to-1 combiners and corresponding splitters on a glass substrate (top) and example backplane fabricated by DaimlerChrysler with attached optical connectors (center) and various waveguide routings, as separately illustrated in the bottom part.

example backplane with the optical routing indicated in the bottom part, where incorporated waveguide crossings serve to realize ring networks with different numbers of nodes. A bidirectional backplane has been fabricated with a total loss of 4 dB including Fresnel losses, waveguide curvature, micromirror reflectivity, as well as a $90°$ crossing contributing 0.8 dB.

By deflection through 90° mirrors, integration of power splitters and a mode mixing region, a 4 ∗ 4 star network has also been implemented. Data transmission at 1 Gbit/s has been achieved over sufficiently long distances of 55 cm. System upgrade to 2.5 Gbit/s has been shown to be feasible using available VCSEL technology and 300-μm-diameter metal-semiconductor-metal (MSM) photodetectors [99b]. Environmental system tests including accelerated aging, vibration, and sensitivity to dust and condensation have all been successfully passed.

An obvious drawback of this rather matured system is the relatively low I/O density enforced by the several millimeter expanded beam diameter. This inhibits direct mapping of electrical lines into optical channels but requires electrical multiplexing and demultiplexing on the boards, which was judged acceptable for the present application, given the numerous remaining advantages.

6.4. Optical Board Interconnects

When discussing optical backplanes in the preceding section we touched on the subject of board interconnects. Even if a large-bandwidth optics-assisted data transfer along the backplane was guaranteed, the limitations of high-speed electrical signaling mentioned at the beginning of Sect. 6.3 would still apply on the individual printed circuit boards. Certainly, increasing the layer count of the PCBs is costly, an alternative differential signaling scheme will increase the number of IC pins, and moreover neither strategy is at all future-proof. The question thus is if photonics could also help to circumvent the intraboard communication bottleneck appearing at the horizon.

6.4.1. INTERCONNECT REQUIREMENTS AND CHOICE OF TECHNOLOGIES

Again there are several options available. Free-space transmission either parallel to the board surface or in a zig-zag manner using external optics overlaying the board would interfere with the space requirements of electronic devices or of the adjacent boards. In addition, distances of up to several tens of centimeters would make alignment extremely critical and accordingly strongly reduce the channel density. Any kind of waveguide put on top of the board would likewise impose undesired restrictions on the placement of other components. Waveguide integration into the PCB thus appears to be the method of choice, where channel waveguides have a clear edge over optical fibers owing to planar batch fabrication capability. Polymer waveguides are most widely investigated for this purpose. An approach to be accepted in the industrial environment should ideally be fully compatible with today's PCB manufacturing sequence. In particular the following requirements have to be met. The waveguide layer has to withstand high pressure (applied during lamination) at elevated temperature without deterioration, especially of the optical properties. High temperatures also occur during the soldering step. Lead-free soldering in an infrared lamp-heated oven will increase the temperature to about $200°$C for a period of up to 20 s, which accordingly determines the minimum glass transition temperature of the polymers. Preferably the optical layer should be placed in the center of the PCB to avoid undesired warp and at the same time reduce the temperature load. The optoelectronic modules have to be picked and placed onto the board by the same machines as the electronic devices, which requires some kind of passive self-alignment. The attenuation coefficient of the waveguides should not much exceed 0.1 dB/cm in order to keep the propagation losses within an acceptable 5 dB limit even for transmission distances of up to 50 cm. Because the polymers usually become less transparent at longer wavelengths, and red-emitting lasers suffer from inherent problems as mentioned in Sect. 6.2.3, the 850-nm wavelength region, already well established in datacom, is preferable. On the other hand, continuously decreasing operating voltages of integrated circuits make the design of laser drivers more challenging and thus speak in favor of devices operating in the 1.3 or 1.55 μm wavelength region.

6.4.2. APPROACHES TO INTRABOARD WAVEGUIDE INTERCONNECTS

For the interfacing of transmitter and receiver elements with the waveguides essentially three options are discussed, as illustrated in Fig. 6.14. In the first case (a) transceiver modules are placed on top of the board, where light from the VCSEL is guided down to the waveguide and back up to the photodetector at the receiving end. Deflections by $90°$ can be implemented either in the guiding pin or the waveguide itself. In the second option (b) laser and PD are displaced from the transceivers such that they butt-couple with the waveguide. The third case (c) envisions thin-film optoelectronic devices fully embedded into the optical layer, where electrical contacting is provided through the conventional multilayer structure.

6. Optical Backplanes, Board and Chip Interconnects 243

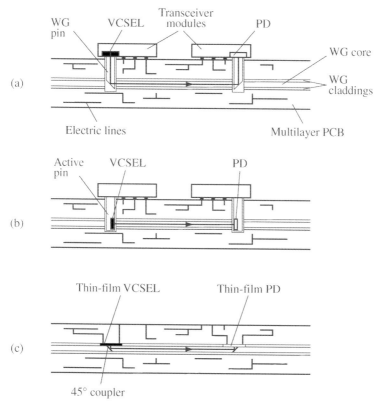

Fig. 6.14 Options for interfacing optoelectronic devices to the waveguide (WG) layer in a hybrid electrical/optical printed circuit board. Indirect coupling via waveguide pins (a), direct butt-coupling (b), and board-embedded one-dimensional VCSEL and photodetector (PD) arrays (c).

The three approaches in Fig. 6.14 are all investigated in ongoing projects [100, 101]. For option (a), waveguide structures are formed by hot-embossing of a polymer foil followed by core filling with an optical adhesive of higher refractive index. Sealing with an upper cladding layer and a common lamination step complete the PCB fabrication [102, 103]. Deflecting 45° mirrors are realized in the same replication step followed by selective metalization for highly reflecting surfaces. First optical data transmission experiments over prototype boards are reported in [104, 24]. A total of 20 trapezoidal shape waveguides of 125 μm average width are arranged on a 250-μm pitch. Minimum waveguide attenuation of about 0.5 dB/cm has already been reduced to about 0.1 dB/cm in subsequent board generations. Data rates up to 10 Gbit/s from a 930-nm wavelength VCSEL have been sent over some-cm long samples without internal beam deflection. Bit error rates of 10^{-11} at -9.5 dBm minimum received optical power have been measured by coupling the light into a 50-μm core diameter silica MMF pigtail of a 10 Gbit/s-compatible InGaAs pin-photodiode based receiver. The apparent size mismatch between waveguide core and pick-up fiber and associated modal noise effects increase the power penalty and lead to undesired channel-to-channel variations. These problems are already known from high-speed POF experiments discussed in Sect. 6.2.3 and could be alleviated through waveguide core size reductions, however, jeopardizing favorable alignment tolerances.

Figure 6.15 shows the advanced concept of data transmission over a waveguide-integrated printed circuit board as promoted by Infineon Technologies. Parallel optical link (PAROLI) transceivers are modified to

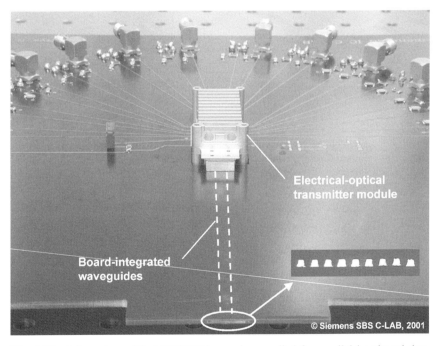

Fig. 6.15 Infineon's modified PAROLI transceiver applied for parallel intraboard data transmission over integrated polymer waveguides [104a].

6. Optical Backplanes, Board and Chip Interconnects

permit light output or input through the bottom instead of the front side of the module. A detailed description of the optical coupling concept relying on micromechanics-assisted passive alignment can be found in [104a]. Target bit rates of 2.5 Gbit/s should be achievable with waveguides of $70 * 70$ μm^2 cross-sectional area, whereas the photodetectors might demand smaller waveguides for data rates approaching 10 Gbit/s. In addition to intraboard connections, board-to-backplane and board-to-fiber concepts are also being explored. Apart from hardware realizations, much emphasis is put on the design process that has to lay the foundations for a seamless integration of optoelectronics into PCB manufacturing [100].

An approach toward option (b) in Fig. 6.14 is followed in [105] using a very similar fabrication sequence for the optical layer. Waveguides with $60 * 60$ μm^2 size cores and 0.3 numerical aperture embedded in a PCB with four electrical layers have been demonstrated as well as butt-coupling of a 4-channel, 250-μm pitch VCSEL array attached to a carrier board that is vertically inserted into the slotted hybrid PCB. Both separate transmitter and receiver boards with light out- or in-coupling over the board edge, respectively, are presented in [105a], where expected data rates of 1.25 Gbit/s per channel are limited by the available VCSEL and photodetector technology.

Fully embedding the optoelectronic devices into the PCB as in Fig. 6.14(c) is a particularly attractive vision. Ideally, the board performance is considerably improved by buried optical interconnects without any extra space consumption. In [101] essential components such as thin-film VCSEL and MSM photodetector arrays as well as 45° total internal reflection or tilted grating microcouplers are presented and in [106] first coupling results between waveguide and photodetector arrays are shown. Polymer waveguide formation in this case is accomplished by direct laser beam writing and the deflecting mirrors are fabricated in a reactive ion etching process. Possible concerns, e.g., regarding laser heat-sinking or the general compatibility to conventional PCB manufacturing, need to be addressed in future.

In all cases discussed before, first polymer waveguide implementations usually consist of straight parallel data lines. As the technology matures, more complicated waveguide configurations incorporating bends, crossings, splitters, etc., as well as multiple optical layers, might become feasible, where the design limits will be set by the available optical power budget.

Among the concrete applications for optical board-level signaling, optical synchronous clock distribution has attracted a lot of attention. Global synchrony as required in many architectures will be increasingly difficult to maintain electrically in future processing systems running at multi-GHz clock frequencies. Optics, on the other hand, might help reduce the clock skew as well as the associated dissipated power, which can amount to more than a quarter of a chip's total heat generation. The latter benefit is related to the property of a photon beam to provide large signal fanout independent of data rate and without parasitic reflections caused by impedance mismatch. In [107] laminated POF circuits are employed for clock delivery, taking advantage of sub-1-mm radius fiber bending capability. Although a 64-node network with favorable bandwidth and loss figures has been demonstrated, fiber insertion into holes in the PCB and their termination appears to be rather labor-intensive. A 1-to-48 clock signal distribution via an H-tree-shaped polymer waveguide circuit based on 1-to-2 splitters, 90° waveguide bends, and 45° total internal reflection waveguide mirrors is reported in [101]. An optical layer of this kind is aimed at replacing electrical circuit board layers, building up the conventional network from strongly bandwidth-limited, delay equalized transmission lines. Since a remaining electrical clock distribution is only needed over markedly reduced distances, considerable performance improvements of future supercomputer generations might be attainable. A more complete overview of optical clock distribution experiments can be obtained from refs. [63–83] in [108].

6.5. Optical Chip Interconnections

Contributing to the interconnection between separate electronic chips or even within a single chip is the greatest challenge optical datacom has set upon to date. Usually reference is made to the Semiconductor Industry Association's roadmaps [109], which predict a severe bottleneck for off-chip and global on-chip interconnections to be encountered within a period of roughly one decade. These insufficiencies especially lead to high power consumption and obstructed processor-memory interaction. Certainly numerous physical reasons exist to use photons instead of electrons for interconnection, which have been meticulously listed in [108, 109a] and on principle also apply at the interchip level. Regarding the rapid advances witnessed in the fields of optoelectronics and microoptics, many voices prophesy a bright future even for mm-distance optical datacom, generally accompanied by wishful thinking that no major or only well predictable progress ought to occur in the electronic domain over the extrapolated timeframe. In this section we will first look into the problems preventing the current use of optoelectronics on a silicon chip and then introduce some interconnect approaches chosen for custom-type processing systems.

6.5.1. ROADBLOCKS PREVENTING MAINSTREAM USE OF OCIs

This short section shall be devoted to the open question of when optical chip interconnects (OCIs), either inter- or intrachip, will be used in *mainstream* high-performance computer systems. The bold answer given in [98] simply says *most likely not before the year 2010 and perhaps never*. Because the arguments behind this statement are worthy of consideration for photonics-centered people, they shall be briefly summarized here:

- The SIA roadmap predicts continued performance scaling over the next decade so that chip designers will put every effort into achieving these goals;
- Even when excluding unforeseen inventions, conventional electronics technology still has a lot of remedies at hand: apart from interlevel materials with low dielectric constant (so-called low-K dielectrics), copper wiring, and additional metal layers (all of which are already pursued), in particular architectural changes reducing interconnect requirements, asynchronous designs mitigating the clock skew problem, and perhaps transistor-based repeaters integrated into the metalization layers might provide solutions;
- Prevailing conservative thinking (enforced by economic pressure) within the community in our case demands that photonics has to promise (or better, prove) extraordinary benefits before it can entice design people to trade valuable chip real estate for driver and receiver circuits that might bring along additional heat and noise generation;
- Substantial changes in the VLSI industry would be necessary to ensure optimized post-processing of dies for virtually seamless integration of optoelectronic devices, alignment and packaging of optics, and finally repeated testing;
- Optoelectronic components need to operate reliably from 70 up to 100°C ambient temperature in order not to unacceptably increase the systems' failure rates;

- The modification of computer-aided design tools to incorporate lasers and photodetectors would require dedicated long-term efforts that have not even started today;
- In mainstream systems optical interconnects will be accepted with only minor cost penalty.

According to the aforementioned boundary conditions, conventional processor design labs indeed appear as closed societies with "Photons Keep Out" signs attached to their entrance gates. With or without optics, bandwidth and latency are the two fields desiring the most progress. While in the former case, dense two-dimensional arrangements of optoelectronic devices hold quite some promise for huge amounts of interchip data throughput, possible reductions of signal latency are much less obvious when considering signal propagation through additional driver and receiver circuits. In fact, latency of an optical interconnect has been identified as a crucial factor determining the application area so that opportunities for OCIs are found in system architectures that are still uncommon as of today. More information on this subject has been collected, e.g., in [110].

6.5.2. DEMONSTRATOR REALIZATIONS EMPLOYING OCIs

In what follows, very few examples of system demonstrators aimed at exploiting the new degree of freedom gained from optical chip interconnects through vertical signal I/O via the third spatial dimension will be given. In order to subdivide the field we distinguish between free-space and guided-wave approaches.

6.5.2.1. Free-Space OCIs

In [110a] the importance of two-dimensional parallel optical interconnects is clearly emphasized. Various backplane- and board-type demonstrators have been built where a great deal of effort has been spent to relieve the alignment issue, which is most critical in free-space systems and necessitates detailed tolerance analyses. In particular, this is achieved by modularization of components requiring most accurate adjustments and in-situ measurements of alignment errors. Tradeoffs between macro- (one bulk lens for all channels), micro- (one lens per channel), and mini- (or clustered; small array of channels per lens) optical systems relaying between the optoelectronic device planes are also found in [110a] and [16].

6. Optical Backplanes, Board and Chip Interconnects

A macrooptical free-space system for global interconnection of a 4 ∗ 4 array of smart pixel arrays distributed over an area of about 7 ∗ 7 cm² is reported in [111]. The arrays each contain 64 VCSELs and photodetectors arranged in dense 4 ∗ 4 clusters and are placed with less than 10-μm lateral tolerance on a common multi-chip module. Mutual optical interconnection of the chips is achieved in a reflective configuration increasing the demonstrator volume to about one liter. The prototype from [111] in principle could be able to provide up to 128 Gbit/s interconnect bandwidth, where favorable scaling properties show potential for several Tbit/s throughput.

In [112] OCIs serve to take the first demonstrator steps toward a parallel processing system. Again, global connectivity is obtained through free-space optics and, as a special feature, the optical path is dynamically reconfigurable with spatial light modulators. In all previous cases proof-of-principles have been achieved and remarkable progress has been made in component integration. However, the optical systems still have a bulky appearance, demanding continued efforts in the fields of microoptics and optomechanics.

A radical proposal toward parallel computing is presented in [113] and explored in the U.S. 3D-OESP (3D OptoElectronic Stacked Processors) project [114, 114a]. In this case, three-dimensional stacks of processing layers are envisioned, intimately interconnected by intermediate optoelectronic layers. Although by this means, a high degree of compactness is achieved and alignment seems to be simplified, thermal management of such a system might become a rather serious issue.

A unique approach to short-distance interchip communication is described in [115]. According to the schematic in Fig. 6.16, the optical

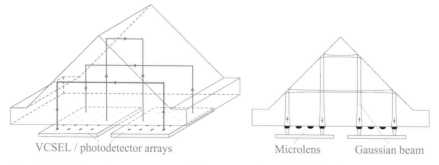

Fig. 6.16 Schematic arrangement of a free-space micooptical bridge for connecting two-dimensional laser/photodetector arrays located on the same electronic chip or on neighboring chips (left) and cross-sectional view illustrating diffraction-limited propagation and beam collimation and focusing with integrated microlenses (right).

pathway consists of a microoptical bridge integrating necessary functions of beam focusing and deflection. In the given demonstration, the optical pathway is assembled from two PMMA parts that are fabricated through a deep proton lithography prototyping technology. Application of injection molding for very low-cost replication has been shown to be feasible, permitting the use of alternative polymers with, e.g., higher glass transition temperatures. A wave propagation analysis relying on Gaussian beams yields a maximum transmission length of 5.2 mm (21 mm) with 100 μm (200 μm) diameter microlenses, corresponding to a peak channel density of 10^4/cm^2 ($2.5 * 10^3$/cm^2) for perfectly aligned systems. Clearly, the achievable channel data rate is determined by space and power consumption of the active components rather than by any inherent bandwidth limit (see Table 6.2). Owing to a 500-μm thickness restriction of the proton irradiation technique, only a $2 * (2 * 8)$ bridge with 250-μm channel pitch could be demonstrated. Despite its elegance, in some possible applications a drawback of this rather simple interconnect scheme might be experienced in the restrictions imposed by the regular placement of transmitter and receiver elements.

6.5.2.2. Two-Dimensional Waveguide Options for OCIs

Optical chip interconnection makes sense only if it can provide a large number of I/O channels distributed over the entire chip area. Consequently, one-dimensional waveguide configurations in the form of fiber ribbons as in Sect. 6.2.1, or for intraboard communication in Sect. 6.4.2, are inadequate. Because mechanical flexibility is usually desired, we concentrate on optical fiber-type waveguides, which essentially exist as ordered fiber bundles (OFBs) or fiber image guides (FIGs; also denoted imaging fiber bundles, IFBs), an example of both is shown in Fig. 6.17. Either type can be realized from silica glass or polymer optical fibers. On the left-hand side of Fig. 6.17, step-index POFs with 120- or 62-μm core diameter and 125-μm outer cladding diameter are arranged on a 250-μm pitch. Proper positioning is achieved through fiber insertion into a precision-drilled hole plate or stacking of individual U-grooved plates holding one fiber ribbon each [116]. As seen in the photograph, advantage is taken from the very high bending flexibility of POFs to fabricate a compact, small headroom 128-fiber connector based on 700-μm diameter MT guide pins for alignment. The experimentally determined bandwidth–distance product of a 120-μm diameter POF exceeds 2 GHz $*$ m. In [117] it was thus possible to transmit 10 Gbit/s data from a 980-nm VCSEL over a 17-cm-long interchip pathway incorporating an array of $2 * 8$ POFs.

6. Optical Backplanes, Board and Chip Interconnects

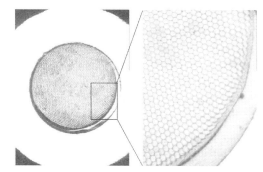

Fig. 6.17 Examples of two-dimensional optical pathways. Terminated ordered fiber bundle with 128 POFs arranged in four 4 ∗ 8 clusters with 250-μm pitch (left) [116, 128] and 1 mm diameter POF image guide (right) with detailed view of the hexagonal arrangement of about 15-μm diameter fibers [117] (© 2001 IEEE).

Lower attenuation than with PMMA material can be achieved with glass-type OFBs, as demonstrated in [118]. The fabrication procedure is able to yield rigid data conduits of various shapes [119] or flexible bundles where individual fibers are separated in a leaching process except for typically 1 to 3 mm-long rigid ends composed of fused all-glass material. Fiber image guides are manufactured similarly to OFBs, however, the stacking and drawing process is further repeated, which typically results in 2000 to 15000 per square millimeter fiber densities and variable core diameters in the order of 10 μm [120]. Traditionally FIGs have been used in flexible endoscopes and medical imaging systems. Loss and image resolution issues of FIGs for use in bit-parallel optical interconnects are studied in [121] in detail, where alternative fabrication methods are also outlined. In FIG-based links, a single data channel is supported by several waveguides. Because the fiber and source/detector array configurations generally have different pitches, at least 10 fibers per channel should be involved in order to avoid inhomogeneous power distributions resulting from a Moiré-type sampling effect. In [122] 6 ∗ 6 optical channels from a 125-μm spacing 980 nm VCSEL array have been transmitted over a 1-m-long FIG at 1 Gbit/s data rate, where the total insertion loss was as low as 1.1 dB and the optical cross-talk remained below −27 dB. In [123] a similar experiment is carried out at 250 Mbit/s with ten 840-nm VCSEL signals and the tradeoff between signal uniformity and channel density as functions of spot size is investigated. Other data are 1.93 mm diagonal length of the hexagonal-shape FIG, 1.35 m fiber length, 3 dB loss, and < −27 dB cross-talk. Further recent experiments relevant for chip-level interconnections carried out with FIGs are reported in [120].

Thin slices from a rigid FIG are known as fiber optic face plates [119]. Since with proper choice of the glass materials a wide range between 0.1 and up to (or even exceeding) unity for the fiber numerical aperture can be achieved [118], face plates can be employed, e.g., as artificial substrates essentially capturing and guiding the entire Lambertian-shaped light cone emitted from a two-dimensional LED array [124].

As seen in the example in Fig. 6.17, fiber image guides can also be composed of plastic optical fibers, where waveguide diameters of about 15 to 40 μm usually are somewhat larger compared to their silica-based counterparts. In [117], a 52-cm-long and 1-mm-diameter image-POF from Fig. 6.17 is used for single-channel data transmission with a 980-nm VCSEL. Unfortunately, the PMMA material shows an attenuation peak right at this wavelength (see Fig. 6.8) so that the 5 dB fiber loss restricted the maximum data rate to 8 Gbit/s. On the other hand, this value is also close to the 3-dB bandwidth limit of about 3 GHz $*$ m of the given step-index image fiber. POF image guides have been employed for various board-level interconnections in [125]. By embedding multiple FIG layers into a circuit board, where each bundle contains sharp 90° bends for in-board routing and data I/O via the board surface, the usual limitation of board-integrated waveguides to just one-dimensional arrays as outlined in Sect. 6.4 can be overcome. However, due to lack of compatibility with established PCB fabrication and a relatively large thickness of 8 mm for a realized three-layer circuit, placement on top of a processing board, similar to the schematic in Fig. 6.1, is deemed more practical for a real application. In addition to compact high-density point-to-point interconnections, promising approaches to incorporate branching nodes into FIG circuits are also reported in [125].

6.5.2.3. Guided-Wave System Demonstrator

Two-dimensional optical links are characterized by a high spatial regularity of data channels. It is intuitively clear that electronic processing systems potentially benefit more from optical chip interconnections the higher the similarity is between the architectures of optical link and electronic chip. In [95] it is emphasized that only appropriate, namely fine-grained and massively parallel, computing architectures rather that the traditional von Neumann implementation, can exploit the features of short-distance optical interconnections. In particular this characteristic is met by neural and reconfigurable computing structures. Field-programmable gate arrays (FPGAs) in which each unit cell is enhanced by an optical input and output node [126] have

6. Optical Backplanes, Board and Chip Interconnects

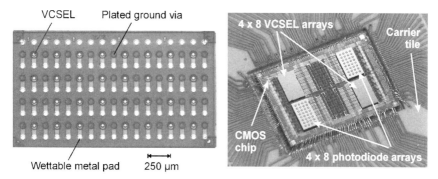

Fig. 6.18 Top-side view of a bottom-emitting 4 ∗ 8 elements VCSEL array [132] (left) and VCSEL and photodiode arrays flip-chip mounted onto an optoelectronic FPGA chip as part of an optical interconnect demonstrator within the European OIIC project [128] (right).

been chosen as a vehicle for an OCI system demonstrator in the European Commission-funded Optically Interconnected Integrated Circuits (OIIC) project [127–129]. The project is part of the EC's 1996 to 2000 Advanced Research Initiative in Microelectronics (MEL-ARI OPTO) cluster. A number of the achievements attained therein have been collected in [130]. The compilation of a technology roadmap dedicated to optoelectronic interconnects for integrated circuits has been an accompanying activity of the cluster [131]. The goal of the OIIC project is to demonstrate direct high-density optical interconnection between two-dimensional optoelectronic transmitter and receiver arrays hybridized onto silicon CMOS circuits. The left-hand part of Fig. 6.18 shows a VCSEL chip incorporating a total of 32 emitters in a rectangular 4 ∗ 8 grid [132]. The array is designed for bottom emission through the transparent GaAs substrate at wavelengths in the 980 nm regime and features coplanar, flip-chip bondable p- and n-contacts contained in a $250 * 250\ \mu m^2$ size unit cell. For 6 μm active diameter VCSELs, where the emission area is defined by selective oxidation of an AlAs layer, lasing occurs in several transverse modes. Typical operation characteristics include average threshold currents and voltages of 0.8 mA and 1.45 V, respectively, optical output powers of 1 mW at only 2 mA driving current, and peak power conversion efficiencies of 37% for about 3 mW output [24]. Devices of 3 μm active size have shown single-mode emission up to about 3 mW optical output power at 30 dB sidemode suppression ratio and have been employed for error-free 12.5 Gbit/s data transmission over silica multi- and single-mode fibers, respectively [132, 23]. Inherently the VCSEL arrays are thus able to deliver an interconnect bandwidth density of 20 Tbit/s/cm^2.

The right-hand side of Fig. 6.18 displays the photograph of a carrier tile holding an optoelectronic FPGA chip with interleaved analog driver and receiver circuitry, onto which two VCSEL and two InP based pin-photodetector arrays have been heterogeneously integrated. As of today, flip-chip bonding is the most realistic and advanced method for interfacing silicon VLSI with optoelectronics [133] and has thus been chosen in the given project. Of the VCSEL arrays, only the anti-reflection coated backside of the GaAs substrates is seen in Fig. 6.18, whereas the InP substrates of the photodiode arrays have been recessed in order to suppress optical cross-talk. The hardware in the right-hand side of Fig. 6.18 is the result of a major collaborative effort in which the CMOS IC was designed by IMEC-Ghent University and ETH Zurich, the optoelectronic arrays were fabricated by University of Ulm and ETH Zurich, the CMOS post-processing was done by IMEC, and the flip-chip mounting and packaging was accomplished by Marconi Caswell. Much effort has been devoted in the OIIC project to devise a manufacturable approach for system packaging. In fact it has been pointed out that the lack of proper interface technology of optoelectronic devices with electrical drivers and the optical system currently is the limiting factor rather than individual device performance [134]. Fig. 6.19 shows how the terminated ordered fiber bundle from Fig. 6.17 is employed for guided-wave 128-channel parallel interconnection between different chips in a rather compact way. Proper interfacing between optoelectronic chip and fiber connector is achieved with an all-passive alignment strategy [135] and light coupling is done without any additional optics. In the final system demonstrator on the right-hand side

Fig. 6.19 Detailed view of the OIIC system demonstrator's optical coupling unit with one right-angle 128-channel POF connector from Fig. 6.17 in place (left) and photograph of the entire demonstrator board, where optical connection is established between two of the three FPGA chips (right) [127a, 128].

of Fig. 6.19, three FPGA chips are mutually interconnected in order to implement a virtually three-dimensionally stacked processing architecture [136]. As a typical value for state-of-the-art commercial FPGAs, the system is targeted for 80 Mbit/s per channel data rate. Another testbed within the OIIC project pursues two-dimensional CMOS-integrated transceiver arrays for 1 Gbit/s channel data rate [137], where 2.5 Gbit/s operation is already entirely feasible with today's technology [138] and even 10 Gbit/s speeds are projected to impose no fundamental physical problem.

After briefly discussing some specific features of a current optical chip interconnect demonstrator, general comments on the operation wavelength finally seem to be appropriate. No standardization efforts are being made yet in the field of OCIs, so project teams in principle are free with their choice of wavelength. Owing to their high degree of maturity, advantages concerning the power budget, and ease of flip-chip integration, 980-nm devices very much lend themselves to incorporation into prototype systems. On the other hand, the 850-nm wavelength region has long been standardized for optical datacom over multimode fiber so that it appears highly desirable to merge these two worlds. Because the naturally used GaAs substrate is highly absorptive at 850 nm, the main challenge lies in the fabrication of bottom-emitting VCSEL arrays where, however, several feasible routes have already been devised. Among those are epitaxial growth on a transparent AlGaAs substrate [139], wafer bonding to a new host substrate [140, 141, 88] (the latter applied in [111]), or incorporation of light outcoupling holes [24]. Complete substrate removal after flip-chip bonding, leaving only a thin film of epitaxially grown material or even individual islands, has been demonstrated with LEDs [124], optical modulators [133] and VCSELs [142].

6.6. Conclusion

In this chapter we have dealt with ultra-short-reach optical interconnects ranging from the interframe down to the interchip level. Most remarkably, the vast majority of experimental demonstrations nowadays is based on vertical-cavity surface-emitting laser devices, which can thus be considered the key enabling technology. High-performance intersystem links almost routinely employ well-established graded-index silica multimode fibers and approach 10 Gbit/s per channel signaling rates. Continued development of high-bandwidth plastic optical fibers might bring about further cost savings in some application areas. Within a computer box, competition between traditional electrical and emerging optical interconnects is much more intense. Guided-wave approaches relying on polymer channel waveguides integrated into hybrid electrical/optical backplanes and printed circuit boards have already reached high levels of maturity. Free-space interconnects could offer some additional advantages but alignment and stability requirements will most probably be met in custom design and special purpose rather than in mainstream signal processing systems. Optical chip interconnects in a long-term view hold great promise for the realization of new computing architectures exploiting massively parallel three-dimensional off-chip data transport. Over a shorter timeframe, the focused efforts that are being made in the fields of heterogeneous integration and packaging will at least result in very compact transceiver modules for space-parallel two-dimensional data communication. In any case the prospects of cost-effective optical interconnection over increasingly shorter distances are bright and it will be exciting to witness the expected rapid progress in all aspects of this highly dynamic discipline.

ACKNOWLEDGMENTS

My sincere thanks are due to Casimer DeCusatis for giving me the opportunity to contribute to this handbook. I am grateful to Elmar Griese, Jon Hall, Max Kicherer, Roger King, Jörg-R. Kropp, Felix Mederer, Jörg Moisel, and Andreas Neyer, who have made available numerous valuable illustrations from their research work. Gerlinde Meixner, Hildegard Mack, and Christine Bunk were indispensable in saving precious time with the preparation of the manuscript. Finally I very much appreciate Karl Joachim Ebeling's continuous support and encouragement.

References

1. *SPIE Proceedings Conference Series,* "Optoelectronic Interconnects VIII," Proc. SPIE **4292** (2001); "VII" **3952A** (2000); "VI" **3632** (1999); "V" **3288** (1998); "IV" **3005** (1997); "III" **2400** (1995); "II" **2153** (1994); **1849** (1993).
2. *Conference Series*, "Optics in Computing," *Proc.* SPIE **4089** (June 2000) Quebec City, Canada; OSA Technical Digest Series (April 1999) Snowmass, Colo.; *Proc.* SPIE **3490** (June 1998) Brugge, Belgium; OSA Technical Digest Series (March 1997) Incline Village, Nev.; preceding conferences were titled "Optical Computing."
3. *Proc. IEEE* **88** (2000). Special issue on "Optical Interconnections for Digital Systems."
4. *IEEE J. Sel. Top. Quantum Electron.* 5: 2 (1999). Issue on "Smart Photonic Components, Interconnects, and Processing."
5. *IEEE J. Sel. Top. Quantum Electron.* 2: 1 (1996). Issue on "Smart Pixels."
6. *Appl. Opt.* 39: 5 (2000). Feature issue on "Optics in Computing."

7. *Appl. Opt.* 37: 2 (1998). Feature issue on "Massively Parallel Processing Using Optical Interconnects."
8. *Appl. Opt.* 35: 8 (1996). Feature issue on "Optical Computing."
9. Lee, S. H. Ed. (1997). "Selected papers on optical interconnects and packaging." *SPIE Milestone Series*, **MS 142**.
10. DeCusatis, C., E. Maass, D. P. Clement, and R. C. Lasky (Eds.). 1998. *Handbook of Fiber Optic Data Communication.* San Diego: Academic Press.
11. Tocci, C., and H. J. Caulfield (Eds.). 1994. *Optical Interconnections: Foundations and Applications.* Boston: Artech House.
12. Towe, E. (Ed.). (2000). "Heterogeneous integration." *SPIE Critical Reviews of Optical Science and Technology*, **CR76**.
13. Husain, A., and M. Fallahi (Eds.). (1998). "Heterogeneous integration: Systems on a chip." *SPIE Critical Reviews of Optical Engineering*, **CR70**.
14. Chen, R. T., and P. S. Guilfoyle (Eds.). 1996. "Optoelectronic interconnects and packaging." *SPIE Critical Reviews of Optical Science and Technology*, **CR62**.
15. Wilmsen, C., H. Temkin, and L. A. Coldren (Eds.). 1999. *Vertical-Cavity Surface-Emitting Lasers.* Cambridge, England: Cambridge University Press.
15a. Li, E. H., and K. Iga (Eds.). *Vertical-Cavity Surface-Emitting Laser Devices.* Heidelberg, Germany: Springer Verlag, in press.
15b. Cheng, J., and N. K. Dutta (Eds.). 2000. *Vertical-Cavity Surface-Emitting Lasers: Technology and Applications.* Amsterdam, The Netherlands: Gordon and Breach Publishing.
16. Tooley, F. A. P. (1996). "Challenges in optically interconnecting electronics." *IEEE J. Sel. Top. Quantum Electron.* 2: 3–13.
17. Tatum, J. A., A. Clark, J. K. Guenter, R. A. Hawthorne, and R. H. Johnson. 2000. Commercialization of Honeywell's VCSEL technology. In "Vertical-Cavity Surface-Emitting Lasers IV," (K. D. Choquette, and C. Lei, Eds.), *Proc.* SPIE **3946**: 144–151.
18. Crow, J. Sept. 1996. "Parallel fiber bus technology — A cost performance breakthrough." In *Proceeding 22nd Europ. Conf. on Opt. Commun., ECOC*, 2: 47–54. Oslo, Norway.
19. Drögemüller, K., D. Kuhl, J. Blank, M. Ehlert, T. Kraeker, J. Höhn, D. Klix, V. Plickert, L. Melchior, I. Schmale, P. Hildebrandt, M. Heinemann, F. P. Schiefelbein, L. Leininger, H.-D. Wolf, T. Wipiejewski, and A. Ebberg. May 2000. "Current progress of advanced high speed parallel optical links for computer clusters and switching systems." In *Proc. 50th Electron. Comp. & Technol. Conf., ECTC*, pp. 1227–1235. Las Vegas, Nev.
19a. Karstensen, H., L. Melchior, V. Plickert, K. Drˆgem ller, J. Blank, T. Wipiejewski, H.-D. Wolf, J. Wieland, G. Jeiter, R. Dal'Ara, and M. Blaser. May 1998. Parallel optical link (PAROLI) for multichannel gigabit rate

interconnections. In *Proc.* "48th Electron. Comp. & Technol. Conf., ECTC," pp. 747–754. Seattle, Wash.

20. Mueller, C., M. Donhove, S. Kilcoyne, T. Lowes, R. Martin, and C. Theorin. 2000. nLIGHTEN™ parallel optical modules. In "Vertical-Cavity Surface-Emitting Lasers IV," (Choquette, K. D., and C. Lei, Eds.), *Proc.* SPIE **3946**: 20–28.

21. Jönsson, J., M. Ghisoni, S. Hatzikonstantinidou, A. Kullander-Sjöberg, A. Risberg, R. Stevens, K. Streubel, J. Sveijer, and R. M. Von Würtemberg. 2000. Reliable vertical-cavity components for multimode data communication. In "Vertical-Cavity Surface-Emitting Lasers IV," (K. D. Choquette, and C. Lei, Eds.), *Proc.* SPIE **3946**: 144–151.

22. Satake, T., T. Arikawa, P. W. Blubaugh, C. Parsons, and T. K. Uchida. May 1994. "MT multifiber connectors and new applications." In *Proc. 44th Electron. Comp. & Technol. Conf., ECTC*, pp. 994–999. Washington, DC.

23. Wiedenmann, D., R. King, C. Jung, R. Jäger, R. Michalzik, P. Schnitzer, M. Kicherer, and K. J. Ebeling. 1999. "Design and analysis of single-mode oxidized VCSEL's for high-speed optical interconnects." *IEEE J. Sel. Top. Quantum Electron.* 5: 503–511.

24. Michalzik, R., K. J. Ebeling, M. Kicherer, F. Mederer, R. King, H. Unold, and R. Jäger. May 2001. "High-performance VCSELs for optical data links." *IEICE Trans. Electron.* **E84-C**.

24a. Greshishchev, Y. M., P. Schvan, J. L. Showell, M.-L. Xu, J. J. Ojha, and J. E. Rogers. Feb. 2000. "A fully integrated SiGe receiver IC for 10 Gb/s data rate." In *Proc. IEEE Int. Solid-State Circuits Conf., ISSCC*, paper MP 3.2. San Francisco, Calif.

25. See URL http://grouper.ieee.org/groups/802/3/ae/

26. Aronson, L. B., B. E. Lemoff, L. A. Buckman, and D. W. Dolfi. 1998. "Low-cost multimode WDM for local area networks up to 10 Gb/s." *IEEE Photon. Technol. Lett.* 10: 1489–1491.

27. Grann, E. B., K. Herrity, B. C. Peters, and B. Wiedemann. 2000. "Datacom applications for new VCSEL technologies." In *Vertical-Cavity Surface-Emitting Lasers IV*, (K. D. Choquette, and C. Lei, Eds.). *Proc.* SPIE **3946**: 165–169.

28. Lemoff, B. E., L. A. Buckman, A. J. Schmit, and D. W. Dolfi. May 2000. "A compact, low-cost WDM transceiver for the LAN." In *Proc. 50th Electron. Comp. & Technol. Conf., ECTC*, pp. 711–716. Las Vegas, Nev.

28a. Cunningham, D. G. Sept./Oct. 2001. "The status of the 10-Gigabit Ethernet standard." In *Proc. 27th Europ. Conf. on Opt. Commun., ECOC*, 3: 364–367. Amsterdam, The Netherlands.

29. Penty, R. V., M. Webster, A. B. Massara, and I. H. White. May 2000. "Physical layer strategies for 10 Gigabit Ethernet." In *Proc. 50th Electron. Comp. & Technol. Conf., ECTC*, pp. 487–490. Las Vegas, Nev.

30. Zhao, X., and F.-S. Choa. Sept. 2000. "10 Gb/s multimode fiber transmissions over any (loss-limited) distances using adaptive equalization techniques." In *Proc. 26th Europ. Conf. on Opt. Commun., ECOC*, 3: 57–58. Munich, Germany.
31. Seifert, R. 1998. *Gigabit Ethernet: Technology and Applications for High-Speed LANs.* Reading, Mass.: Addison Wesley.
32. Cunningham, D. G., and W. G. Lane. 1999. *Gigabit Ethernet Networking.* Indianapolis, Ind.: Macmillan Technical Publishing.
33. Raddatz, L., I. H. White, D. G. Cunningham, and M. C. Nowell. 1998. "An experimental and theoretical study of the offset launch technique for the enhancement of the bandwidth of multimode fiber links." *J. Lightwave Technol*. 16: 324–331.
34. George, J. E., and P. F. Kolesar. Dec. 1999. "10-Gigabit Ethernet development for LAN cabling system well underway." *Lightwave*: 87–89. The data have been compiled from http://grouper.ieee.org/groups/802/3/z/public/presentations/july1996/ANfibsur.txt
35. Welch, F. Nov. 2000. "Applications and opportunities for optoelectronics in high speed LANs." In *Proc. IEEE Lasers and Electro-Opt. Soc. Ann. Meet., LEOS*, 1: 290–291. Rio Grande, Puerto Rico.
36. Senior, J. M. 1985. *Optical Fiber Communications: Principles and Practice.* New York: Prentice Hall.
37. Michalzik, R., R. King, G. Giaretta, R. Jäger, and K. J. Ebeling. 2000. VCSEL arrays for 10 Gb/s multimode fiber optical interconnects. In "Optoelectronic Interconnects VII" (S. Tang, and J. P. Bristow, Eds.), *Proc.* SPIE **3952A**: 124–133.
38. Michalzik, R., G. Giaretta, A. J. Ritger, and Q. L. Williams. Nov. 1999. "10 Gb/s VCSEL based data transmission over 1.6 km of new generation 850 nm multimode fiber." *IEEE Lasers and Electro-Opt. Soc. Ann. Meet., LEOS*, postdeadline paper PD1.6. San Francisco, Calif.
39. Giaretta, G., R. Michalzik, and A. J. Ritger. May 2000. "Long distance (2.8 km), short wavelength (0.85 μm) data transmission at 10 Gb/sec over new generation high bandwidth multimode fiber." *Conf. on Lasers and Electro-Opt., CLEO*, postdeadline paper CPD13. San Francisco, Calif.
40. Peters, F. H., D. J. Welch, V. Jayaraman, M. H. MacDougal, J. D. Tagle, T. A. Goodwin, J. E. Schramm, T. D. Lowes, S. P. Kilcoyne, K. R. Nary, J. S. Bergey, and W. Carpenter. 2000. "10 Gb/s VCSEL-based data links." In *Vertical-Cavity Surface-Emitting Lasers IV*, (K. D. Choquette, and C. Lei, Eds.), *Proc.* SPIE **3946**: 152–164.
41. Michalzik, R., G. Giaretta, K. W. Goossen, J. A. Walker, and M. C. Nuss, Sept. 2000. "40 Gb/s coarse WDM data transmission with 825 nm wavelength VCSELs over 310 m of high-performance multimode fiber." In *Proc. 26th Europ. Conf. on Opt. Commun., ECOC*, **4**: 33–34. Munich, Germany.

42. Marcou, J. (Ed.). 1997. *Plastic Optical Fibres: Practical Applications.* Chichester, U.K.: J. Wiley & Sons.
43. *Proc. International Plastic Optical Fibres Conference*, Berlin, Germany (Oct. 1998).
44. *Proc. 8th International POF Conference*, Makuhari Messe, Chiba, Japan (July 1999).
45. *Proc. The International POF Technical Conference*, Cambridge, Mass. (Sept. 2000).
45a. *Proc. 10th International Plastic Optical Fibres Conference 2001*, Amsterdam, The Netherlands (Sept. 2001).
46. Koganezawa, K., and T. Onishi. Sept. 2000. "Progress in perfluorinated GI POF, Lucina^{TM}." In *Proc. The International POF Technical Conference*, pp. 19–21. Cambridge, Mass. See also Onishi, T., H., Murofushi, Y. Watanabe, Y. Takano, R. Yoshida, and M. Naritomi. Oct. 1998. "Recent progress of perfluorinated GI POF." In *Proc. 7th International Plastic Optical Fibres Conference '98*, pp. 39–42. Berlin, Germany.
47. Ishigure, T., and Y. Koike. Sept. 2000. "Potential bit rate of GI-POF link." In *Proc. The International POF Technical Conference*, pp. 14–18. Cambridge, Mass.
48. Yaseen, M., S. D. Walker, and R. J. S. Bates. Feb. 1993. "531-Mbit/s, 100-m all-plastic optical-fiber data link for customer premises network applications." In *Proc. Conf. on Optical Fiber Commun. / Int. Conf. on Integrated Optics and Optical Fiber Commun., OFC/IOOC*, pp. 171–172. San Jose, Calif.
49. Kicherer, M., F. Mederer, H. Unold, K. J. Ebeling, S. Lehmacher, E. Griese, and M. Naritomi. June 2000. "VCSEL based high-speed data transmission over polymer optical fibers and circuit board integrated waveguides." In *Proc. International Optoelectronics Symposium*, pp. 7–8. Kyoto, Japan.
50. Schnitzer, P., M. Grabherr, R. Jäger, F. Mederer, R. Michalzik, D. Wiedenmann, and K. J. Ebeling. 1999. "GaAs VCSEL's at $\lambda = 780$ and 835 nm for short-distance 2.5-Gb/s plastic optical fiber data links." *IEEE Photon. Technol. Lett.* 11: 767–769.
51. Li, W., G. Khoe, H. v.d. Boom, G. Yabre, H. de Waardt, Y. Koike, S. Yamazaki, K. Nakamura, and Y. Kawaharada. Dec. 1998. "2.5 Gbit/s transmission over 200 m PMMA graded index polymer optical fibre using a 645 nm narrow spectrum laser." In *Proc. IEEE Lasers and Electro-Opt. Soc. Ann. Meet., LEOS*, 2: 279–298. Orlando.
52. Khoe, G. D. July 1999. "Exploring the use of GIPOF systems in the 640 nm to 1300 nm wavelength area." In *Proc. 8th International POF Conference '99*, pp. 36–43. Makuhari Messe, Chiba, Japan.
53. Li, W., G. D. Khoe, H. P. A. v.d. Boom, G. Yabre, H. de Waardt, Y. Koike, M. Naritomi, and N. Yoshihara. July 1999. "Record 2.5 Gbit/s 550 m GI

POF transmission experiments at 840 and 1310 nm wavelength." In *Proc. 8th International POF Conference '99*, pp. 60–63. Makuhari Messe, Chiba, Japan.
54. Giaretta, G., F. Mederer, R. Michalzik, W. White, R. Jäger, G. Shevchuk, T. Onishi, M. Naritomi, R. Yoshida, P. Schnitzer, H. Unold, M. Kicherer, K. Al-Hemyari, J. A. Valdmanis, M. Nuss, X. Quan, and K. J. Ebeling. Sept. 1999. "Demonstration of 500 nm-wide transmission window at multi-Gb/s data rates in low-loss plastic optical fiber." In *Proc. 25th Europ. Conf. on Opt. Commun., ECOC*, 2: 240–241. Nice, France.
55. Mederer, F., R. Jäger, P. Schnitzer, H. Unold, M. Kicherer, K. J. Ebeling, M. Naritomi, and R. Yoshida. 2000. "Multi-Gb/s graded-index POF data link with butt-coupled single-mode InGaAs VCSEL." *IEEE Photon. Technol. Lett*. 12: 199–201.
56. Giaretta, G., W. White, M. Wegmuller, and T. Onishi. 2000. "High-speed (11 Gb/sec) data transmission using perfluorinated graded-index polymer optical fibers for short interconnects (<100 m)." *IEEE Photon. Technol. Lett*. 12: 347–349.
57. Grimes, G. J. Sept. 2000. "Switching applications of POF." In *Proc. The International POF Technical Conference*, pp. 35–37. Cambridge, Mass.
58. Lehman, J. A., R. A. Morgan, D. Carlson, M. Hagerott Crawford, and K. D. Choquette. 1997. "High-frequency modulation characteristics of red VCSELs." *Electron. Lett*. 33: 298–300.
59. Stevens, R., A. Risberg, R. Schatz, R. M. Von Würtemberg, B. Kronlund, M. Ghisoni, and K. Streubel. 2000. High-speed visible VCSEL for POF data links. In "Vertical-Cavity Surface-Emitting Lasers IV," (K. D. Choquette, and C. Lei, Eds.), Proc. SPIE **3946**: 88–94.
60. Lambkin, J. D., T. Calvert, B. Corbett, J. Woodhead, S. M. Pinches, A. Onischenko, T. E. Sale, J. Hosea, P. van Daele, K. Vandeputte, A. van Hove, A. Valster, J. G. McInerney, and P. A. Porta. 2000. Development of a red VCSEL-to-plastic fiber module for use in parallel optical data links. In "Vertical-Cavity Surface-Emitting Lasers IV," (K. D. Choquette, and C. Lei, Eds.), *Proc*. SPIE **3946**: 95–105.
61. Choquette, K. D., R. P. Schneider, M. Hagerott Crawford, K. M. Geib, and J. J. Figiel. 1995. "Continuous wave operation of 640-660 nm selectively oxidised AlGaInP vertical-cavity lasers." *Electron. Lett*. 31: 1145–1146.
62. White, W. R., M. Dueser, W. A. Reed, and T. Onishi. 1999. "Intermodal dispersion and mode coupling in perfluorinated graded-index plastic optical fiber." *IEEE Photon. Technol. Lett*. 11: 997–999.
63. Ishigure, T., M. Kano, and Y. Koike. Sept. 1999. "Which is serious factor to the bandwidth of GI POF: mode dependent attenuation or mode coupling." In *Proc. 25th Europ. Conf. on Opt. Commun., ECOC*, 1: 280–281. Nice, France.

64. Koike, Y. Sept. 2000. "Progress in GI-POF: status of high speed plastic optical fiber and its future prospect." In *Proc. The International POF Technical Conference*, pp. 1–5. Cambridge, Mass. See also Koike, Y. Oct. 1998. "POF—From the past to the future." In *Proc. 7th International Plastic Optical Fibres Conference '98*, pp. 1–8. Berlin, Germany.
65. Seidl, D., P. Merget, J. Schwarz, J. Schneider, R. Weniger, and E. Zeeb. Oct. 1998. "Application of POFs in data links of mobile systems." In *Proc. 7th International Plastic Optical Fibres Conference '98*, pp. 205–211. Berlin, Germany.
66. Ziemann, O., L. Giehmann, P. E. Zamzow, H. Steinberg, and D. Tu. Sept. 2000. "Potential of PMMA based SI-POF for Gbps transmission in automotive applications." In *Proc. The International POF Technical Conference*, pp. 44–48. Cambridge, Mass.
67. Khoe, G. D., W. Li, P. K. v. Bennekom, G. Yabre, H. P. A. v.d. Boom, H. de Waardt, A. H. E. Breuls, G. Kuyt, and P. J. T. Pleunis. July 1999. "Gigabit/s transmission via large core graded-index silica optical fibers in comparison with graded index polymer optical fibers." In *Proc. 8th International POF Conference '99*, pp. 192–195. Makuhari Messe, Chiba, Japan.
68. Bristow, J. 1996. Intra computer optical interconnects: progress and challenges. In "Optoelectronic Interconnects and Packaging," (R. T. Chen, and P. S. Guilfoyle, Eds.), SPIE Critical Reviews of Optical Science and Technology **CR62**: 318–326.
69. Zhou, G., Y. Zhang, and W. Liu. 2000. "Optical fiber interconnection for the scalable parallel computing system." *Proc. IEEE* 88: 849–855.
70. Melchior, L., and J. R. Kropp. May 1996. "A high density optical backplane connector." In *Proc. 46th Electron. Comp. & Technol. Conf., ECTC*, Orlando.
70a. Infineon Optical Backplane Connection/Module (OBC-M) with SMC Backplane Connector Plug (SMC BP) for PAROLI.
71. Yoshikawa, T., and H. Matsuoka. 2000. "Optical interconnections for parallel and distributed computing." *Proc. IEEE* 88: 849–855.
72. Rode, M., J. Moisel, O. Krumpholz, and O. Schickl. Sept. 1997. "Novel optical backplane board-to-board interconnection." In *Proc. 11th Int. Conf. on Integrated Optics and Opt. Fibre Commun. & 23rd Europ. Conf. on Opt. Commun., IOOC-ECOC*, 2: 228–231. Edinburgh, U.K.
73. Tan, Q., J. Vandewege, G. De Pestel, P. Vetter, and F. Migom. Sept. 1996. "2.5 Gb/s/mm optical fiber interconnections." In *Proc. 22nd Europ. Conf. on Opt. Commun., ECOC*, pp. 2.55–2.58. Oslo, Norway.
74. Bacher, W., P. Bley, and H. O. Moser. 1996. "Potential of LIGA technology for optoelectronic interconnects." In "Optoelectronic Interconnects and Packaging," (R. T. Chen, and P. S. Guilfoyle, Eds.), SPIE Critical Reviews of Optical Science and Technology **CR62**: 442–460.

75. Parker, J. W. 1991. "Optical interconnection for advanced processor systems: a review of the ESPRIT II OLIVES program." *J. Lightwave Technol.* 9: 1764–1773.
76. Cannell, G. J. Sept. 1996. "Standardized packaging and interconnection for inter- and intra-board optical communication (SPIBOC)." In *Proc. 22nd Europ. Conf. on Opt. Commun., ECOC*, pp. 3.249–3.252. Oslo, Norway.
77. Peall, R. G., H. F. Priddle, M. C. Geear, B. Shaw, A. Briggs, P. M. Harrison, A. Schmid, M. Bitter, J. Wieland, O. Zorba, and R. Harcourt. 1996. "12 × 2.5 Gbit/s receiver array module." *Electron. Lett.* 32: 682–683.
78. Hibbs-Brenner, M., and J. Lehman. Dec. 1998. "In search of the 1 meter cost competitive optical interconnect." In *Proc. IEEE Lasers and Electro-Opt. Soc. Ann. Meet., LEOS*, 2: 381–382. Orlando.
79. Kurokawa, T., N. Takato, and Y. Katayama. 1980. "Polymer optical circuits for multimode optical fiber systems." *Appl. Opt.* 19: 3124–3129.
80. Eldada, L., and L. Shacklette. 2000. "Advances in polymer integrated optics." *IEEE J. Sel. Top. Quantum Electron.* 6: 54–68.
81. Liu, Y. S. 1998. "Lighting the way in computer design." *IEEE Circuits & Devices.* 14: 23–31.
82. Liu, Y. S., R. J. Wojnarowski, W. A. Hennessy, J. P. Bristow, Y. Liu, A. Peczalski, J. Rowlette, A. Plotts, J. Stack, J. Yardley, L. Eldada, R. M. Osgood, R. Scarmozzino, S. H. Lee, and V. Uzguz. 1996. Polymer optical interconnect technology (POINT)—optoelectronic packaging and interconnects for board and backplane applications. In "Optoelectronic Interconnects and Packaging," (R. T. Chen, and P. S. Guilfoyle, Eds.), SPIE Critical Reviews of Optical Science and Technology **CR62**: 405–414.
83. Booth, B. L. 1989. "Low loss channel waveguides in polymers." *J. Lightwave Technol.* 7: 1445–1453.
84. Guttmann, J., H.-P. Huber, O. Krumpholz, J. Moisel, M. Rode, R. Bogenberger, and K.-P. Kuhn. 1999. "19″ polymer optical backplane." In *Proc. 25th Europ. Conf. on Opt. Commun., ECOC*, 1: 354–355. Nice, France.
85. Moisel, J., J. Guttmann, H.-P. Huber, O. Krumpholz, M. Rode, R. Bogenberger, and K.-P. Kuhn. 2000. "Optical backplanes with integrated polymer waveguides." *Opt. Eng.* 39: 673–679.
86. Sakamoto, T., H. Tsuda, M. Hikita, T. Kagawa, K. Tateno, and C. Amano. 2000. "Optical interconnection using VCSELs and polymeric waveguide circuits." *J. Lightwave Technol.* 18: 1487–1492.
87. Krauss, T. F., and R. M. De La Rue. 1999. "Photonic crystals in the optical regime—past, present and future." *Progress in Quantum Electron.* 23: 51–96.

88. Liu, Y., E. M. Strzelecka, J. Nohava, M. K. Hibbs-Brenner, and E. Towe. 2000. "Smart-pixel array technology for free-space optical interconnects." *Proc. IEEE* 88: 764–768.
89. Szymanski, T. H., and V. Tyan. 1999. "Error and flow control in terabit intelligent optical backplanes." *IEEE J. Sel. Top. Quantum Electron.* 5: 339–352.
90. Szymanski, T. H., and H. S. Hinton 1996. "Reconfigurable intelligent optical backplane for parallel computing and communications." *Appl. Opt.* 35: 1253–1268.
91. Iga, K., Y. Kokubun, and M. Oikawa. 1984. *Fundamentals of Microoptics: Distributed-Index, Microlens, and Stacked Planar Optics.* Tokyo, Japan: Academic Press.
92. Li, Y., and J. Popelek. 2000. "Volume consumption comparisons of free-space and guided-wave optical interconnections." *Appl. Opt.* 39: 1815–1825.
93. Chen, R. T., C. Zhao, and T.-H. Oh. 1996. Performance-optimized optical bidirectional backplane bus for multiprocessor systems. In "Optoelectronic Interconnects and Packaging," (R. T. Chen, and P. S. Guilfoyle, Eds.), SPIE Critical Reviews of Optical Science and Technology **CR62**: 299–317.
94. Kim, G., X. Han, and R. T. Chen. 2000. "An 8-Gb/s optical backplane bus based on microchannel interconnects: design, fabrication, and performance measurements." *J. Lightwave Technol.* 18: 1477–1486.
95. Fey, D., W. Erhard, M. Gruber, J. Jahns, H. Bartelt, G. Grimm, L. Hoppe, and S. Sinzinger. 2000. "Optical interconnects for neural and reconfigurable VLSI architectures." *Proc. IEEE* 88: 838–848.
96. Jahns, J. 1994. "Planar packaging of free-space optical interconnections." *Proc. IEEE* 82: 1623–1631.
97. Sinzinger, S., and J. Jahns. 1999. *Microoptics.* Wiley-VCH: Weinheim, Germany.
98. Lytel, R., H. L. Davidson, N. Nettleton, and T. Sze. 2000. "Optical interconnections within modern high-performance computing systems." *Proc. IEEE* 88: 758–763.
99. Hirabayashi, K., T. Yamamoto, and S. Hino. 1998. "Optical backplane with free-space optical interconnections using tunable beam deflectors and a mirror for bookshelf-assembled terabit per second class asynchronous transfer mode switch." *Opt. Eng.* 37: 1332–1342.
99a. Boisset, G. C., G. Robertson, and H. S. Hinton. 1995. "Design and construction of an active alignment demonstrator for a free-space optical interconnect." *IEEE Photon. Technol. Lett.* 7: 676–678.
99b. Moisel, J. Sept./Oct. 2001. Optical backplane. In *Proc. 27th Europ. Conf. on Opt. Commun., ECOC*, 3: 254–255. Amsterdam, The Netherlands.

100. Griese, E. 2000. Optical interconnections on printed circuit boards. In "Optics in Computing 2000," (R. A. Lessard, and T. V. Galstian, Eds.), *Proc.* SPIE **4089**: 60–71.
101. Chen, R. T., L. Lin, C. Choi, Y. J. Liu, B. Bihari, L. Wu, S. Tang, R. Wickmann, B. Picor, M. K. Hibbs-Brenner, J. Bristow, and Y. S. Liu. 2000. "Fully embedded board-level guided-wave optoelectronic interconnects." *Proc. IEEE* 88: 780–793.
102. Lehmacher, S., A. Neyer, and F. Mederer. Sept./Oct. 2001. "Polymer optical waveguides integrated in printed circuit boards." In *Proc. 27th Europ. Conf. on Opt. Commun., ECOC*, 3: 302–303. Amsterdam, The Netherlands.
103. Lehmacher, S., and A. Neyer. 2000. "Integration of polymer optical waveguides into printed circuit boards." *Electron. Lett.* 36: 1052–1053.
104. Kicherer, M., F. Mederer, R. Jäger, H. Unold, K. J. Ebeling, S. Lehmacher, A. Neyer, and E. Griese. Sept. 2000. "Data transmission at 3 Gbit/s over intraboard polymer waveguides with GaAs VCSELs." In *Proc. 26th Europ. Conf. on Opt. Commun., ECOC*, 3: 289–290. Munich, Germany.
104a. Griese, E., A. Himmler, K. Klimke, J.-R. Kropp, S. Lehmacher, A. Neyer, and W. Süllau. July/Aug. 2001. "Self-aligned coupling of optical transmitter and receiver modules to board-integrated optical multimode waveguides." In "Micro- and Nano-optics for Optical Interconnection and Information Processing," (M. R. Taghizadeh, H. Thienpont, and G. E. Jabbour, Eds.), *Proc.* SPIE **4455**, paper 32.
105. Krabe, D., F. Ebling, N. Arndt-Staufenbiel, G. Lang, and W. Scheel. May 2000. "New technology for electrical/optical systems on module and board level: the EOCB approach." In *Proc. 50th Electron. Comp. & Technol. Conf., ECTC*, pp. 970–974. Las Vegas, Nev.
105a. Bargiel, S., F. Ebling, H. Schröder, H. Franke, G. Spickermann, E. Griese, A. Himmler, C. Lehnberger, L. Oberender, G. Mrozynski, D. Steck, E. Strake, and W. Süllau. Apr. 2001. "Electrical-optical circuit boards with 4-channel transmitter and receiver modules." In *Proc. 6th Workshop Optics in Computing Technology*, ORT 2001, pp. 17–27, Paderborn, Germany.
106. Liu, Y., L. Lin, C. Choi, B. Bihari, and R. T. Chen. 2001. "Optoelectronic integration of polymer waveguide array and metal-semiconductor-metal photodetector through micromirror couplers." *IEEE Photon. Technol. Lett.* 13: 355–357.
107. Li, Y., J. Popelek, L. Wang, Y. Takiguchi, T. Wang, and K. Shum. 1999. "Clock delivery using laminated polymer fibre circuits." *J. Opt. A: Pure Appl. Opt.* 1: 239–243.
108. Miller, D. A. B. 2000. "Rationale and challenges for optical interconnects to electronic chips." *Proc. IEEE* 88: 728–749.
109. The Semiconductor Industry Association's (SIA's) *National Technology Roadmap for Semiconductors*, 1999 edition. See URL http://public.itrs.net/

109a. Miller, D. A. B. 1997. "Physical reasons for optical interconnection." *Int. J. Optoelectron.* 11: 155–168.
110. Collet, J. H., D. Litaize, J. Van Campenhout, C. Jesshope, M. Desmulliez, H. Thienpont, J. Goodman, and A. Louri. 2000. "Architectural approach to the role of optics in monoprocessor and multiprocessor machines." *Appl. Opt.*, 39: 671–682.
110a. Plant, D. V., and A. G. Kirk. 2000. "Optical interconnects at the chip and board level: challenges and solutions." *Proc. IEEE*, 88: 806–818.
111. Haney, M. W., M. P. Christensen, P. Milojkovic, G. J. Fokken, M. Vickberg, B. K. Gilbert, J. Rieve, J. Ekman, P. Chandramani, and F. Kiamilev. 2000. "Description and evaluation of the FAST-Net smart pixel-based optical interconnection prototype." *Proc. IEEE* 88: 819–828.
112. McArdle, N., M. Naruse, H. Toyoda, Y. Kobayashi, and M. Ishikawa. 2000. "Reconfigurable optical interconnections for parallel computing." *Proc. IEEE* 88: 829–837.
113. Ludwig, D., J. Carson, and L. Lome. 1996. Opto-electronic interconnects for 3 dimensional wafer stacks. In "Optoelectronic Interconnects and Packaging," (R. T. Chen, and P. S. Guilfoyle, Eds.), SPIE Critical Reviews of Optical Science and Technology, **CR62**: 366–389.
114. Esener, S., and P. Marchand. 1998. 3-D optoelectronic stacked processors: design and analysis. In "Optics in Computing, OC '98," (P. Chavel, D. A. B. Miller, and H. Thienpont, Eds.), Proc. SPIE **3490**: 541–545.
114a. Esener, S., P. Marchand, V. Ozguz, Y. Liu, D. Huang, and X. Zheng. Nov. 1999. "Packaging optoelectronic stacked processors and free-space optical interconnects." In *Proc. IEEE Lasers and Electro-Opt. Soc. Ann. Meet., LEOS*, 1: 94–95. San Francisco, Calif.
115. Thienpont, H., C. Debaes, V. Baukens, H. Ottevaere, P. Vynck, P. Tuteleers, G. Verschaffelt, B. Volckaerts, A. Hermane, and M. Hanney. 2000. "Plastic microoptical interconnection modules for parallel free-space inter- and intra-MCM data communication." *Proc. IEEE*, 88: 769–779.
116. Neyer, A., B. Wittmann, and M. Jöhnck. 1999. "Plastic-optical-fiber-based parallel optical interconnects." *IEEE J. Sel. Top. Quantum Electron*. 5: 193–200.
117. Mederer, F., R. Jäger, J. Joos, M. Kicherer, R. King, R. Michalzik, M. Riedl, H. Unold, K. J. Ebeling, S. Lehmacher, B. Wittmann, and A. Neyer. May 2001. "Improved VCSEL structures for 10 Gigabit-Ethernet and next generation optical-integrated PC-boards." In *Proc. 51st Electron. Comp. & Technol. Conf., ECTC*. Orlando.
118. Cryan, C. V. 1998. "Two-dimensional multimode fibre array for optical interconnects." *Electron. Lett*. 34: 586–587.
119. A product of Schott Communications Technologies, Inc., see URL http://www.opticalinterconnect.com/

120. Chiarulli, D. M., S. P. Levitan, P. Derr, R. Hofmann, B. Greiner, and M. Robinson. 2000. "Demonstration of a multichannel optical interconnection by use of imaging fiber bundles butt coupled to optoelectronic circuits." *Appl. Opt.* **39**: 698–703.
121. Li, Y., T. Wang, H. Kosaka, S. Kawai, and K. Kasahara. 1996. "Fiber-image-guide-based bit-parallel optical interconnects." *Appl. Opt.* **35**: 6920–6933.
122. Kosaka, H., M. Kajita, Y. Li, and Y. Sugimoto. 1997. "A two-dimensional optical parallel transmission using a vertical-cavity surface-emitting laser array module and an image fiber." *IEEE Photon. Technol. Lett.* **9**: 253–255.
123. Maj, T., A. G. Kirk, D. V. Plant, J. F. Ahadian, C. G. Fonstad, K. L. Lear, K. Tatah, M. S. Robinson, and J. A. Trezza. 2000. "Interconnection of a two-dimensional array of vertical-cavity surface-emitting lasers to a receiver array by means of a fiber image guide." *Appl. Opt.* **39**: 683–689.
124. Heremans, P., M. Kuijk, S. Peeters, R. Windisch, D. Filkins, and S. Borghs. Sept. 1999. "Thin-film 850 nm light emitters on fiber optic face plate." In *Proc. 25th Europ. Conf. on Opt. Commun., ECOC*, **2**: 304–305. Nice, France.
125. Li, Y., J. Ai, and J. Popelek. 2000. "Board-level 2-D data-capable optical interconnection circuits using polymer fiber-image guides." *Proc. IEEE* **88**: 794–805.
126. Van Campenhout, J., H. van Marck, J. Depreitere, and J. Dambre. 1999. "Optoelectronic FPGA's." *IEEE J. Sel. Top. Quantum Electron.* **5**: 306–315.
127. OIIC's full project name is *Generic Approach to Manufacturable Optoelectronic Interconnects for VLSI Circuits*. See URL http://www.intec.rug.ac.be/oiic/ Brunfaut, M., W. Meeus, J. Van Campenhout, R. Annen, P. Zenklusen, H. Melchior, R. Bockstaele, L. Vanwassenhove, J. Hall, B. Wittmann, A. Neyer, P. Heremans, J. Van Koetsem, R. King, H. Thienpont, and R. Baets. July/Aug. 2001. "Demonstrating optoelectronic interconnect in a FPGA based prototype system using flip chip mounted 2D arrays of optical components and 2D POF-ribbon arrays as optical pathways." In *Micro- and Nano-optics for Optical Interconnection and Information Processing* (M. R. Taghizadeh, H. Thienpont, and G. E. Jabbour, Eds.), *Proc. SPIE* **4455**: paper 23.
128. Vanwassenhove, L., R. Bockstaele, R. Baets, M. Brunfaut, W. Meeus, J. Van Campenhout, J. Hall, H. Melchior, A. Neyer, J. Van Koetsem, R. King, K. Ebeling, and P. Heremans. March 2001. "Demonstration of 2-D plastic optical fibre based optical interconnect between CMOS IC's." In *Proc. Optical Fiber Commun. Conf., OFC*, pp. WDD74-1–WDD74-3. Anaheim, CA.
129. Vanwassenhove, L., R. Baets, M. Brunfaut, J. van Campenhout, J. Hall, K. J. Ebeling, H. Melchior, A. Neyer, H. Thienpont, R.,Vounckx, J. van Koetsem, P. Heremans, F.-T. Lentes, and D. Litaize. May 2000.

"Two-dimensional optical interconnect between CMOS IC's." In *Proc. 50th Electron. Comp. & Technol. Conf., ECTC*, pp. 231–237. Las Vegas, Nev.

130. Neefs, H. (Ed.). 2000. "Optoelectronic interconnects for integrated circuits — Achievements 1996-2000." Luxembourg, Office for Official Publications of the European Communities, Belgium: See http://www.cordis.lu/esprit/src/melop-rm.htm

131. "Optoelectronic interconnects for integrated circuits." (Sept. 1999) MEL-ARI Technology Roadmap, 2nd Ed. Luxembourg, Office for Official Publications of the European Communities, Belgium: See http://www.cordis.lu/esprit/src/melop-rm.htm

132. King, R., R. Michalzik, D. Wiedenmann, R. Jäger, P. Schnitzer, T. Knödl, and K. J. Ebeling. 1999. 2D VCSEL arrays for chip-level optical interconnects. In "Optoelectronic Interconnects VI," (J. P. Bristow, and S. Tang, Eds.), Proc. SPIE **3632**: 363–372.

133. Krishnamoorthy, A. V., and K. W. Goossen. 1998. "Optoelectronic-VLSI: photonics integrated with VLSI circuits." *IEEE J. Sel. Top. Quantum Electron*. 4: 899–912.

134. Tooley, F. 1998. Optical interconnects do not require improved optoelectronic devices. In "Optics in Computing, OC '98," (P. Chavel, D. A. B. Miller, and H. Thienpont, Eds.), *Proc. SPIE* **3490**: 14–17.

135. Hall, J. P., M. J. Goodwin, J. J. Lewandowski, E. M. Thom, and D. J. Robbins. 1998. "Packaging of VCSEL, MC-LED and detector 2-D arrays." In *Proc. 48th Electron. Comp. & Technol. Conf., ECTC*, pp. 778–782. Seattle, Wash.

136. Brunfaut, M., J. Depreitere, W. Meeus, J. M. Van Campenhout, H. Melchior, R. Annen, P. Zenklusen, R. Bockstaele, L. Vanwassenhove, J. Hall, A. Neyer, B. Wittmann, P. Heremans, J. Van Koetsem, R. King, H. Thienpont, and R. Baets. Nov. 1999. "A multi-FPGA demonstrator with POF-based optical area interconnect." In *Proc. IEEE Lasers and Electro-Opt. Soc. Ann. Meet., LEOS*, 2: 625–626. San Francisco, Calif.

137. King, R., R. Michalzik, R. Jäger, K. J. Ebeling, R. Annen, and H. Melchior. 2001. 32-VCSEL channel CMOS-based transmitter module for Gb/s data rates. In "Vertical-Cavity Surface-Emitting Lasers V," (K. D. Choquette, and C. Lei, Eds.), *Proc.* SPIE **4286**: in press.

138. Madhavan, B., and A. F. J. Levi. 1998. "Low-power 2.5 Gbit/s VCSEL driver in 0.5 μm CMOS technology." *Electron. Lett*. 34: 178–179.

139. Ohiso, Y., K. Tateno, Y. Kohama, A. Wakatsuki, H. Tsunetsugu, and T. Kurokawa. 1996. "Flip-chip bonded 0.85-μm bottom-emitting vertical-cavity laser array on an AlGaAs substrate." *IEEE Photon. Technol. Lett*. 8: 1115–1117.

140. Choquette, K. D., K. M. Geib, B. Roberds, H. Q. Hou, R. D. Twesten, and B. E. Hammons. 1998. "Short wavelength bottom-emitting vertical cavity lasers fabricated using wafer bonding." *Electron. Lett.* 34: 1404–1405.
141. Lin, C.-K., and P. D. Dapkus. 2001. "Uniform wafer-bonded oxide-confined bottom-emitting 850-nm VCSEL arrays on sapphire substrates." *IEEE Photon. Technol. Lett.* 13: 263–265.
142. Pu, R., C. Duan, and C. W. Wilmsen. 1999. "Hybrid integration of VCSEL's to CMOS integrated circuits." *IEEE J. Sel. Top. Quantum Electron.* 5: 201–208.

Chapter 7 | Parallel Computer Architectures Using Fiber Optics

David B. Sher
Mathematics/Statistics/CMP Dept., Nassau Community College, Garden City, New York 11530

Casimer DeCusatis
IBM Corporation, Poughkeepsie, New York 12601

7.1. Introduction

In this chapter, we present an overview of parallel computer architectures and discuss the use of fiber optics for clustered or coupled processors in this context. Presently, a number of computer systems take advantage of commercially available fiber optic technology to interconnect multiple processors, thereby achieving improved performance or reliability for a lower cost than if a single large processor had been used. Optical fiber is also used to distribute the elements of a parallel architecture over large distances; these can range from tens of meters to alleviate packaging problems to tens of kilometers for disaster recovery applications. Not all of the parallel computer architectures discussed in this chapter use fiber optic connectivity, but they are presented as a context for those systems that are currently using optics. We will give a few specific examples of parallel computer systems that use optical fiber, in particular the Parallel Sysplex architecture from IBM. Other applications do not currently use optical fiber, but they are presented as candidates for optical interconnect in the near future, such as the Power-Parallel supercomputers which are part of the Advanced Strategic Computing Initiative (ASCI). Many of the current applications for fiber optics in this area use serial optical links to share data between processors, although this is by no means the only option. The use of parallel optical interconnects for intra-machine applications was discussed in Vol. 1, Chapter 10, and other schemes including plastic optics, optical backplanes, and free space optical interconnects are discussed in Vol. 2, Chapter 6. Toward the end of the chapter, we also provide some speculation concerning machines that have not yet been designed or built but which serve to illustrate potential future applications of optical interconnects. Because this is a rapidly developing area, we will frequently cite Internet references where the latest specifications and descriptions of various parallel computers may be found.

Computer engineering often presents developers with a choice between designing a computational device with a single powerful processor (with additional special-purpose coprocessors) or designing a parallel processor device with the computation split among multiple processors that may be cheaper and slower. There are several reasons why a designer may choose a parallel architecture over the simpler single processor design. Before each reason, and other categorizing methods in this chapter we will have a letter code, A, which we will use to categorize architectures we describe in other sections of the chapter.

A. **Speed** – There are engineering limits to how fast any single processor can compute using current technology. Parallel architectures can exceed these limits by splitting up the computation among multiple processors.
B. **Price** – It may be possible but prohibitively expensive to design or purchase a single processor machine to perform a task. Often a parallel processor can be constructed out of off-the-shelf components with sufficient capacity to perform a computing task.
C. **Reliability** – Multiple processors means that a failure of a processor does not prevent computation from continuing. The load from the failed processor can be redistributed among the remaining ones. If the processors are distributed among multiple sites, then even catastrophic failure at one site (due to natural or man-made disaster, for example) would not prevent computation from continuing.
D. **Bandwidth** – Multiple processors means that more bus bandwidth can be processed by having each processor simultaneously use parts of the bus bandwidth.
O. **Other** – Designers may have other reasons for adding parallel processing not covered above.

Current parallel processor designs were motivated by one or more of these needs. For example, the parallel sysplex family was motivated by reliability and speed, the Cray XMP was primarily motivated by speed, the BBN butterfly was designed with bandwidth considerations in mind, and the transputer family was motivated by price and speed.

After a designer has chosen to use multiple processors he must make several other choices:

- Which processors?
- How many processors?
- Which network topology?

The product of the speed of the processors and the number of processors is the maximal processing power of the machine (for the most part unachievable in real life). The effect of network topology is subtler.

7.1.1. NETWORK TOPOLOGY

Network topologies control communication between machines. While most multiprocessors are connected with ordinary copper-wired buses, we believe that fiber optics will be the bus technology of the future. Topology controls how many computers may be necessary to relay a message from one processor to another. A poor network topology can result in bottlenecks where all the computation is waiting for messages to pass through a few very important machines. Also, a bottleneck can result in unreliability with failure of one or few processors causing either failure or poor performance of the entire system.

Four kinds of topologies have been popular for multiprocessors. They are

- E. **Full connectivity** using a crossbar or bus. The historic C.mmp processor used a crossbar to connect the processors to memory (which allowed them to communicate). Computers with small numbers of processors (like a typical parallel sysplex system or tandem system) can use this topology but it becomes cumbersome with large (more than 16) processors because every processor must be able to simultaneously directly communicate with every other. This topology requires a fan in and fan out proportional to the number of processors, making large networks difficult.
- F. **Pipeline** where the processors are linked together in a line, and information primarily passes in one direction. The CMU Warp

processor was a pipelined multiprocessor and many of the first historical multiprocessors, the vector processors, were pipelined multiprocessors. The simplicity of the connections and the many numerical algorithms that are easily pipelined encourage people to design these multiprocessors. This topology requires a constant fan in and fan out, making it easy to lay out large numbers of processors and add new ones.

G. **Torus and Allied topologies** where an N processor machine requires $\sqrt{N}$ processors to relay messages. The Goodyear MPP machine was laid out as a torus. Such topologies are easy to layout on silicon so multiple processors can be placed on a single chip and many such chips can be easily placed on a board. Such technology may be particularly appropriate for computations that are spatially organized. This topology also has constant fan in and fan out. Adding new processors is not as easy as in pipelined processors but laying out this topology is relatively easy. Because of the ease of layout sometimes this layout is used on chips and then the chips are connected in a hypercube.

H. **Hypercube and Butterfly topologies** have several nice properties that have lead to their dominating large-scale multiprocessor designs. They are symmetric so no processor is required to relay more messages than any other is. Every message need only be relayed through $\log(N)$ processors in an N processor machine and messages have multiple alternate routes, increasing reliability under processor failure and improving message routing and throughput. Transputer systems and the BBN butterfly were some of the first multiprocessors that adapted this type of topology. This topology has a logarithmic fan out and that can complicate layout when the size of the processor may grow over time. There is an alternative topology called cube-connected cycles that has the same efficient message passing properties as the hypercube topology but constant fan out, easing layout considerably.

X. **Exotic** – There are a variety of less popular but still important topologies one can use on their network.

The more efficient and fast the bus technology is, the simpler the topology can be. A really fast bus can simply connect all the processors in a machine together by using time multiplexing giving $|N|$ slots for every possible connection between any two of the N processors.

7.1.2. COMPUTING TASKS

The primary computing task for the machine under consideration has a major effect on the network topology. Computing tasks fall into three general categories.

I. **Heavy computational tasks** – these tasks require much more computation than network communication. Some examples of this task are pattern recognition (SETI), code breaking, inverse problems, and complex simulations such as weather prediction and hydrodynamics.
J. **Heavy communication tasks** – these tasks involve relatively little computation and massive amounts of communication with other processors and with external devices. Message routing is the classic example of these tasks. Other such tasks are data base operations and search.
K. **Intermediate or mixed tasks** – these tasks lie between the above or are mixtures of the above. An example of an intermediate task is structured simulation problems, such as battlefield simulation. These simulations require both computation to represent the behavior and properties of the objects (like tanks) and communication to represent interaction between the objects. Some machines may be designed for a mixture of heavy computation and heavy communication tasks.

Historically, supercomputers focused on heavy computation tasks, particularly scientific programming, and mainframes focused on heavy communication tasks, particularly business and database applications.

7.2. Historical and Current Processors

This section uses the preceding categorization to illuminate the purposes and abilities of parallel processors that existed in the past and parallel processors that are in current use.

7.2.1. SIMD VS MIMD

When processors were simpler and memory expensive it made sense to have a single set of instructions stored in a global memory to control all the processors of a large parallel processor. For example, a parallel machine might have a 256 by 256 array of small processors connected in a torus topology and each processor would have one pixel of an image. All the processors would execute the same instructions supplied from an outside memory. For example, all the processors might sum their pixel values with their neighbors to the north, south, east, and west in turn. Then the processors would divide their result by 5. The machine would in these 6 instructions have blurred the image digitally. The MPP was designed to work in this manner. Such systems are called SIMD for Single Instruction (stream) Multiple Data (streams).

As memory became cheaper and more efficient and bandwidth more expensive, large parallel machines began to store instruction memory with the processors. Each processor in the machine executed its own instruction stream. This limited the synchronization required between the processors of the system. It also meant that the bandwidth of the system could be devoted entirely to data. Such systems are called MIMD for Multiple Instruction (streams) Multiple Data (streams). Such systems are more flexible than SIMD because the processors can engage in a separate part of an algorithm. Also such a machine can easily be partitioned, simulating several smaller machines. Every current parallel processing system is MIMD.

7.2.2. *HISTORICAL PARALLEL PROCESSORS*

Computer	Code	Description
IBM 2938 Array Processor	AFI	A programmable digital signal processor, it proves very popular in the petroleum industry
CDC~7600	AFI	Pipelined supercomputer
Pluribus	DJ	ARPAnet switch nodes
STARAN	AJ	a 4 × 256 1-bit PE array processor using associative addressing and a FLIP-network
IBM 3838 array processor	AFI	a general-purpose digital signal processor
C.mmp	E	crossbar connecting minicomputers to memories
ICL DAP		world's first commercial massively parallel computer

(*Continued*)

Computer	Code	Description
KL10 TOPS-10	E	up to three CPUs supported, but a customer built a five-CPU system
BBN Butterfly	DHJ	68000s connected through multistage network to disjoint memories, giving appearance of shared memory
Cray-X/MP	AEI	Up to four pipelined processors with shared memory
MPP	AGI	Large array of very small processors
Transputer	BXK	Extremely flexible small fast processors that can be connected in a wide variety of network topologies
NCube	AGI	NCube/10 hypercube using custom processors
Cray-2	AEI	four background processors, a single foreground processor, and a 4.1 nsec clock cycle
Connection Machine CM-1	OGK	up to 65536 single-bit processors connected in hypercube
Warp	AGI	programmable bit-slice VLIW systolic array
CM-2 Connection Machine	OGK	64k single-bit processors connected in hypercube, plus 2048 Weitek floating point units
iWarp	AGI	on-chip integrated communication, long-lived connections, systolic and low-latency communication, and 512~kbyte to 2~Mbyte per processor

7.2.3. CURRENT PARALLEL PROCESSORS

Computer	Code	Description
KLAT2, Kentucky Linux Athlon Testbed 2	BXI	64+2 700 MHz Athlon cluster using a variety of system hardware and software performance tricks, including a 264-NIC + 9 switch implementation of the new network topology
Atipa ATserver 6000	BEI	Fully connected high performance processors using fast optical ethernet, running LINUX
Berkeley Millennium	OXI	The UC Berkeley Millennium project aims to develop and deploy a hierarchical campus-wide "cluster of clusters" to support advanced applications in scientific computing, simulation, and modeling
Columbia (Riken/BNL) QCDSP	AGI	a series of computers ranging from 64-node, 3 Gflops machines to a 8,192-node, 400 Gflops machine and a 12,288-node, 600 Gflops machine designed for quantum field theory applications
Avalon	AEI	Avalon is a 140 processor Alpha Beowulf cluster constructed entirely from commodity personal computer technology. It uses fast ethernet for full connectivity.
Compaq AlphaServers	BDEJ	Large systems with multiple high-powered alpha computers connected through a crossbar switch largely devoted to network serving
Cray T3E	AGI	The CRAY T3ETM series efficiently scales performance and price/performance from tens to thousands of processors and up to 2.4 trillion calculations per second (teraflops). The Cray T3E-1200E distributed-memory parallel processing

(Continued)

Computer	Code	Description
		system follows the successful Cray T3ETM system with twice the performance and four times the memory.
Cray T932	AI	Small number of very fast processors and specialized connectivity hardware with unknown network topology. Uses some optical messaging between memory and CPUs.
CSPI MultiComputer 2741	AEI	The 2741 MultiComputer, a high-performance processor module, combines Myrinet gigabit-per-second network technology with the PowerPC architecture to deliver an open and expandable solution for your high-performance computing applications.
Fujitsu VPP700	AEI	Crossbar connected high-performance computing elements
Fujitsu VPP5000	AEI	Crossbar connected high-performance computing elements
Hitachi SR2201	AEI	The SR2201 uses Hitachi's new high-performance RISC chips, and has a high level of scalability. The SR2201 high-end model ranges from 32 up to a maximum of 2,048 processors, and compact model ranges from 8 up to a maximum of 64 processors that are able to meet a wide range of customer requirements. The SR2201 uses a crossbar switch network arrangement for high-speed message communication, and a pseudo vector processing function that provides a major boost in performance of large-scale scientific calculations.
Hitachi SR8000	AEI	Multidimensional crossbar connected high-performance computing

(*Continued*)

7. Parallel Computer Architectures Using Fiber Optics 279

Computer	Code	Description
HP 9000 Superdome	CEJ	The highly scalable SuperDome platform supports system configurations from 2 to 64 processors. SuperDome's cell-based hierarchical crossbar architecture can be configured as one large symmetric multiprocessor or as several independent partitions. SuperDome users can repartition the system according to business or application needs.
IBM RS/6000 S80	CDXJ	Up to 24-way symmetric multiprocessor with state-of-the-art RS64 III copper microprocessors. Since the memory is a resource accessed by all processors, it's not possible to share a single bus and get enough bandwidth to feed 24 RS64III microprocessors. Rather, groups of 6 microprocessors share a dedicated crossbar port to memory, running at 150 MHz, to avoid delays in memory access.
IBM RS/6000 SP	CAHK	This system is a distributed memory, multinode computer designed for speed, scalability, flexibility, manageability, and availability. Configurations range in size from 1 to 128 nodes.
IBM Parallel Sysplex	BCXK	One key customer business objective was to reduce the total cost of computing for S/390 systems. A key business objective was to address the increasing customer demands for improved application availability, not only in terms of failure recovery, but for the more important reduction of planned outage times.
NEC SX-5	AEI	High-performance shared memory parallel vector system — multinode systems scale to 5 Tflops

(Continued)

Computer	Code	Description
SGI 2000	CHJ	Very powerful processors with custom networking hardware designed in a modified hypercube to be servers.
SGI 3800	CEJ	With the industry's most advanced NUMA architecture from SGI, you can configure your SGI Origin 3800 system up to a single 512-processor shared memory system, or use partitioning to divide it into as many as 32 partitions and run them as a tightly coupled cluster.
SKY HPC-1	AXI	SKY has capitalized on its traditional expertise in embedded systems to build scalable high-performance computers (HPC) for applications where supercomputers once excelled. SKY's history of building high-performance, scalable, embedded computers designed for minimal space and power has now been applied to HPC programs, which require flexibility, ease of programming and speed.
Sun Ultra HPC 4500	CXJ	Highly available data center powerhouse for mission-critical applications. The system offers mainframe-class availability features, including full hardware redundancy, concurrent maintenance, online upgrades, and Hot CPU Upgrades. Additionally, this system offers mainframe-class resource management with fully fault isolated, Dynamic System Domains.
Sun Enterprise 10000	CXJ	The Sun Enterprise™ 10000 server has been designed specifically to meet the challenges of today's data centers — with Dynamic System Domains and an innovative, high-performance crossbar interconnect.

(*Continued*)

7. Parallel Computer Architectures Using Fiber Optics 281

Computer	Code	Description
Tsukuba CP-PACS/2048	AXI	Scientific computing machine designed with a large number of processing units connected in a cube with layered crossbars in all three dimensions
U-Tokyo GRAPE-6	AXI	The GRAPE-6 system consists of a host computer and 12 clusters. In each cluster 16 processor boards are connected to a host processor through two-level tree network. Each processor is designed specifically to solve gravity problems.

Computer	Web Site	References
KLAT2, Kentucky Linux Athlon Testbed 2	http://aggregate.org/KLAT2/	http://aggregate.org/KLAT2/InTheNews/
Atipa ATserver 6000	http://www.atipa.com/store/clustering/	
Berkeley Millennium	http://now.cs.berkeley.edu/Millennium	http://www.millennium.berkeley.edu/proj/Vineyard/pubs.html
Columbia (Riken/BNL) QCDSP	http://www.phys.columbia.edu/~cqft/qcdsp.htm	
Avalon	http://cnls.lanl.gov/avalon	http://loki-www.lanl.gov/papers/
Compaq AlphaServers	http://www5.compaq.com/alphaserver/technology/index.html	http://www5.compaq.com/alphaserver/whitepapers.html
Cray T3E	http://www.cray.com/products/systems/crayt3e/	http://www.cray.com/products/systems/crayt3e/papers.html
Cray T932	http://www.cray.com/products/systems/crayt90/932.html	

(*Continued*)

Computer	Web Site	References
CSPI MultiComputer 2741	http://www.cspi.com/multi-computer/datasheets/2741/description.htm	
Fujitsu VPP700	http://www.fujitsu.co.jp/hypertext/Products/Info_process/hpc/vpp-e/index.html	
Fujitsu VPP5000	http://www.fujitsu.co.jp/hypertext/Products/Info_process/hpc/vpp5000e/index.html	
Hitachi SR2201	http://www.hitachi.co.jp/Prod/comp/hpc/eng/sr1.html	
Hitachi SR8000	http://www.hitachi.co.jp/Prod/comp/hpc/eng/sr82e.html#TEC	
HP 9000 Superdome	http://www.unixservers.hp.com/solutions/alwayson/overview_features/index.html	http://www.unixservers.hp.com/highend/superdome/infolibrary/index.html
IBM RS/6000 S80	http://www.rs6000.ibm.com/hardware/enterprise/s80.html	http://www.rs6000.ibm.com/resource/technology/#64bit
IBM RS/6000 SP	http://www.rs6000.ibm.com/hardware/largescale/SP/index.html	http://www.rs6000.ibm.com/resource/technology/index.html#sp
IBM Parallel Sysplex	http://www.research.ibm.com/journal/sj/362/nick.html	http://www.ibm.com/search?lv=c&o=10&i=0&t=0&n=100&v=10&lang=en&cc=us&q=sysplex&realm=ibm
NEC SX-5	http://www.hstc.necsyl.com/index.html	
SGI 2000	http://www.sgi.com/origin/2000/	http://www.sgi.com/origin/2000/datasheet.html

(*Continued*)

7. Parallel Computer Architectures Using Fiber Optics

Computer	Web Site	References
SGI 3800	http://www.sgi.com/origin/3000/3800.html	http://www.sgi.com/origin/3000/datasheet.html
SKY HPC-1	http://www.skycomputers.com/hardware/HPC1.html	http://www.skycomputers.com/technical/technical.html
Sun Ultra HPC 4500	http://www.sun.com/servers/midrange/e4500	http://www.sun.com/servers/wp.html
Sun Enterprise 10000	http://www.sun.com/servers/highend/10000/	http://www.sun.com/servers/wp.html
Tsukuba CP-PACS/2048	http://www.rccp.tsukuba.ac.jp/cppacs/architecture-e.html	http://www.rccp.tsukuba.ac.jp/cppacs/architecture-e.html
U-Tokyo GRAPE-6	http://grape.astron.s.u-tokyo.ac.jp/~makino/papers/gbp2000-full/node3.html#SECTION00030000000000000000	http://grape.astron.s.u-tokyo.ac.jp/grape/computer.htm

7.3. Detailed Architecture Descriptions

In this section we present descriptions of four of the architectures mentioned in the previous tables. Two of these machines are chosen primarily for their historical interest — the CMU Warp and the transputer — and three of these machines represent current architectures of great moment — Sky HPC-1, Tsukuba CP-PACS/2048, and Parallel Sysplex.

7.3.1. CMU WARP (LATER INTEL IWARP)

The Warp processor was designed in a project at Carnegie Mellon University (CMU). It was designed for computationally intensive parallel processing on large data sets as is common in signal and image processing. The name of the processor, Warp, refers to an image processing operation.

The processors were designed to operate in a pipeline, each processor receiving data from the previous processor and sending processed data on to the next processor in the system. This kind of processing avoids many synchronization and communication problems that occur in more complex network topologies.

Such processing sometimes has problems with latency (the time necessary to fill or exhaust the pipeline) but in the data-intensive processing that the Warp was designed for latency issues are small relative to the size of the data set and the required computation per data point.

Each Warp processor had a state-of-the-art floating-point coprocessor (pipelined also) and fast integer arithmetic. Each processor was controlled by a VLIW (very long instruction word) microprocessor that allowed maximal internal parallelism because the powerful components of this processor could be addressed and controlled separately.

This machine is an early example of a powerful MIMD processor. However, the topology of this machine and the processing expected of it are such that optical components and in particular optical buses are not necessary. It was designed to avoid the issues optical components are designed to address. A bibliography of papers related to the Warp processor is available at http://www.cs.cmu.edu/afs/cs/project/iwarp/archive/WWW-pages/iwarp-papers.html.

7.3.2. TRANSPUTER

The transputer was one of the more important parallel architectures and many such machines were sold to academic institutions and research groups. This computer actually had a language, Occam, designed for programming it. The transputer was a processor designed to easily connect with other transputers through four bidirectional I/O ports. These ports were serial, making the work of interconnection easy and allowing flexibility in network architectures. This allowed the design of a wide variety of network topologies using transputers. Each transputer was a state-of-the-art RISC (reduced instruction set computation) processor. While the transputer was not as computationally powerful as a typical Warp processor, it was much more flexible in its I/O capabilities.

The main idea of the transputer was to make it easy to write parallel software for it. Thus a program could be written that runs on a single processor that simulates a network of transputers. This program could be debugged on the processor and then run on the network. Thus programs generally did not have to worry about the exact size or shape of the multiprocessor. See [9] for a general description of the transputer and some interesting ideas for programming it.

Once again the transputer was designed to deal flexibly with the problems of interprocessor communication in a multiprocessor. A high-speed optical bus system can reduce these problems considerably or even render them moot through time multiplexing. Essentially, a sufficiently high bandwidth system can allow all the multiprocessors to communicate with any other multiprocessor in the system and share system resources at their maximum capacity.

7.3.3. SKY HPC-1

Sky's HPC-1 architecture is an example of a NUMA (Non-Uniform Memory Access) architecture. This is one of the most popular architectures around because of the simplification of programming tasks. Basically, every processor has access to all the memory of the machine. The processors and their memory are connected together using a high-speed network. Most of the systems previously described, including the HPC-1, use crossbar switches to maximize the network bandwidth.

The HPC-1 network architecture uses packet-based communications to optimize memory operations. In a packet-based architecture, such as SKYchannel, both the routing information and the data payload travel together in one transmission or packet. Examples of other packet-based communications include the fixed-size packets (or cells) of ATM and the variable length packets of Sun's XDBus.

Many of the SKYchannel advantages are partly linked to the use of high-speed FIFOs (packet queues).

- Latency issues with FIFOs were solved by providing a mechanism for early cut-through. Some systems use a technique called "worm-hole routing," but other solutions are available.
- Split transactions have been added to reduce the contention at the destination node. The actual connectivity of a HPC-1 is a hierarchical crossbar system in which there is a master crossbar-connected subsystem whose processors lie on the subsystem crossbar. This connectivity suffices for the machine sizes SKY provides.

All of the technological innovations of the SKY architecture can be applied with great effect to a system designed around a high-speed optical bus. Such a system would still be packet-based and still use the FIFOs but would use the bus instead of crossbars to route the packets between processors.

7.3.4. TSUKUBA CP-PACS/2048

This machine is an interesting example of modern thought in parallel machine design. It was designed by collaboration between physicists and computer scientists as a high-speed machine devoted to difficult problems in physics. The details of the machine design are particularly available because the machine was designed for academic rather than commercial purposes.

This machine is designed as a cube of processors, each processor participating in three crossbar switches. Thus the processor with coordinates 5,4,3 can instantly communicate with any processor that shares any of the coordinates; for example it is connected to **5**,7,1 and to 6,**4**,9 and to 7,2,**3**.

Each processor is an extremely high-speed numerical computation device that at one time would have been called a supercomputer itself. The Tsukuba CP-PACS/2048 has 2048 of these processors tightly coupled together using the network described previously.

Data transfer on the network is made through Remote DMA (Remote Direct Memory Access), in which processors exchange data directly between their respective user memories with a minimum of intervention from the operating system. This leads to a significant reduction in the startup latency, and a high throughput.

A well-balanced performance of CPU, network, and I/O devices supports the high capability of CP-PACS for massively parallel processing (614Gflops).

We believe that in the future high speed optical buses can replace the crossbar switches of this architecture, making it more flexible, and easier to maintain.

7.3.5. PARALLEL SYSPLEX AND GDPS

High-end computer systems running over metropolitan area networks (MANs) are proving to be a near-term application for multi-terabit communication networks. Large computer systems require dedicated storage area networks (SANs) to interconnect with various types of direct attach storage devices (DASD), including magnetic disk and tape, optical storage devices, printers, and other equipment. This has led to the emergence of client–server-based networks employing either circuit or packet switching, and the development of network-centric computing models. In this approach, a high bandwidth, open protocol network is the most critical resource in a computer system, surpassing even the processor speed in its importance to overall performance. The recent trend toward clustered, parallel computer architectures to enhance performance has also driven the requirement for high bandwidth fiber optic coupling links between computers. For example, large water-cooled mainframe computers using bipolar silicon processors are being replaced by smaller, air-cooled servers using complementary metal oxide semiconductor (CMOS) processors. These new processors can far surpass the performance of older systems because of their ability to couple together many central processing units in parallel. One widely adopted architecture for clustered mainframe computing is known as a Geographically Dispersed Parallel Sysplex (GDPS). In this section, we will describe the basic features of a GDPS and show how this architecture is helping to drive the need for high-bandwidth dense-wavelength division multiplexing (DWDM) networks.

In 1994, IBM announced the Parallel Sysplex architecture for the System/390 mainframe computer platform (note that the S/390 has recently been rebranded as the IBM eServer z series). This architecture uses high-speed fiber optic data links to couple processors together in parallel [1–4], thereby increasing capacity and scalability. Processors are interconnected via a coupling facility, which provides data caching, locking, and queuing services; it may be implemented as a logical partition rather than a separate physical device. The gigabit links, known as InterSystem Channel (ISC), HiPerLinks, or Coupling Links, use long-wavelength (1300-nm) lasers and single-mode fiber to operate at distances up to 10 km with a 7 dB link budget (HiPerLinks were originally announced with a maximum distance of 3 km, which was increased to 10 km in May 1998). If good quality fiber is used, the link budget of these channels allows the maximum distance to be increased to 20 km. When HiPerLinks were originally announced, an optional interface at 531 Mbit/s was offered using short-wavelength lasers on MM fiber. The 531 Mbit/s HiPerLinks were discontinued in May 1998 for the G5 server and its follow-ons. A feature is available to accommodate operation of 1 Gbit/s HiPerLinks adapters on multimode fiber, using a mode conditioning jumper cable at restricted distances (550 meters maximum).

The physical layer design is similar to the ANSI Fibre Channel Standard, operating at a data rate of 1.0625 Gbit/s, except for the use of open fiber control (OFC) laser safety on long-wavelength (1300 nm) laser links (higher order protocols for ISC links are currently IBM proprietary). Open fiber control is a safety interlock implemented in the transceiver hardware; a pair of transceivers connected by a point-to-point link must perform a handshake sequence in order to initialize the link before data transmission occurs. Only after this handshake is complete will the lasers turn on at full optical power. If the link is opened for any reason (such as a broken fiber or unplugged connector) the link detects this and automatically deactivates the lasers on both ends to prevent exposure to hazardous optical power levels. When the link is closed again, the hardware automatically detects this condition and reestablishes the link. The HiPerLinks use OFC timing corresponding to a 266 Mbit/s link in the ANSI standard, which allows for longer distances at the higher data rate. Propagating OFC signals over DWDM or optical repeaters is a formidable technical problem, which has limited the availability of optical repeaters for HiPerLinks. OFC was initially used as a laser eye safety feature; subsequent changes to the international laser safety standards have made this unnecessary, and it has been discontinued on the most recent version of z series servers. The 1.06 Gbit/s HiPerLinks will continue to support OFC in order to interoperate with installed equipment; this is called "compatibility mode." There is also a 2.1 Gbit/s HiPerLink channel, also known as ISC-3, which does not use OFC; this is called "peer mode."

Because all the processors in a GDPS must operate synchronously with each other, they all require a multimode fiber link to a common reference clock known as a Sysplex Timer (IBM model 9037). The sysplex timer provides a time of day clock signal to all processors in a sysplex; this is called an External Timing Reference (ETR). The ETR uses the same physical layer as an ESCON link, except that the data rate is 8 Mbit/s. The higher level ETR protocol is currently proprietary to IBM. The timer is a critical component of the Parallel Sysplex; the sysplex will continue to run with degraded performance if a processor fails, but failure of the ETR will disable the entire sysplex. For this reason, it is highly recommended that two redundant timers be used, so that if one fails the other can continue uninterrupted operation of the sysplex. For this to occur, the two timers must also be synchronized with each other; this is accomplished by connecting them with two separate, redundant fiber links called the Control Link Oscillator (CLO). Physically, the CLO link is the same as an ETR link except that it carries timing information to keep the pair of timers synchronized. Note that because the two sysplex timers are synchronized with each other, it is possible that some processors in a sysplex can run from one ETR while others run from the second ETR. In other words, the two timers may both be in use simultaneously running different processors in the sysplex, rather than one timer sitting idle as a backup in case the first timer fails.

There are three possible configurations for a Parallel Sysplex. First, the entire sysplex may reside in a single physical location, within one data center. Second, the sysplex can be extended over multiple locations with remote fiber optic data links. Finally, a multi-site sysplex in which all data is remote, copied from one location to another, is known as a Geographically Dispersed Parallel Sysplex, or GDPS. The GDPS also provides the ability to manage remote copy configurations, automates both planned and unplanned system reconfigurations, and provides rapid failure recovery from a single point of control. There are different configuration options for a GDPS. The single site workload configuration is intended for those enterprises that have production workload in one location (site A) and discretionary workload (system test platforms, application development, etc.) in another location (site B). In the event of a system failure, unplanned site failure, or planned workload shift, the discretionary workload in site B will be terminated to provide processing resources for the production work from site A (the resources are acquired from site B to prepare this environment, and the critical workload is restarted). The multiple site workload configuration is intended for those enterprises that have production and discretionary workload in both site A and site B. In this case, discretionary workload from either site may be terminated to provide processing resources for the production workload from the other site in the event of a planned or unplanned system disruption or site failure.

Multi-site Parallel Sysplex or GDPS configurations may require many links (ESCON, HiPerLinks, and Sysplex Timer) at extended distances; an efficient way to realize this is the use of wavelength division multiplexing technology. Multiplexing wavelengths is a way to take advantage of the high bandwidth of fiber optic cables without requiring extremely high modulation rates at the transceiver. This type of product is a cost effective way to utilize leased fiber optic lines, which are not readily available everywhere and may be very high cost (typically the cost of leased fiber (sometimes known as dark fiber) where available is $300/mile/month). Traditionally, optical wavelength division multiplexing (WDM) has been widely used in telecom applications, but has found limited usage in datacom applications. This is changing, and a number of companies are now offering multiplexing alternatives to datacom networks that need to make more efficient use of their existing bandwidth. This technology may even be the first step toward development of all-optical networks. For Parallel Sysplex applications, the only currently available WDM channel extender that supports GDPS (Sysplex Timer and HiPerLinks) in addition to ESCON channels is the IBM 2029 Fiber Saver [5–8] as described in Chapter 5 (note that the 9729 Optical Wavelength Division Multiplexer also supported GDPS but has been discontinued; other DWDM products are expected to support GDPS in the future, including offerings from Nortel and Cisco). The 2029 allows up to 32 independent wavelengths (channels) to be combined over one pair of optical fibers, and extends the link distance up to 50 km point-to-point or 35 km in ring topologies. Longer distances may be achievable from the DWDM using cascaded networks or optical amplifiers, but currently a GDPS is limited to a maximum distance of 40 km by timing considerations on the ETR and CLO links (the sysplex timer documents support for distances up to only 26 km, the extension to 40 km requires a special request from IBM via RPQ 8P1955). These timing requirements also make it impractical to use TDM or digital wrappers in combination with DWDM to run ETR and CLO links at extended distances; this implies that at least 4 dedicated wavelengths must be allocated for the sysplex timer functions. Also note that since the sysplex timer assumes that the latency of the transmit and receive sides of a duplex ETR and CLO link are approximately equal, the length of these link segments should be within 50 m of each other. For this reason, unidirectional $1 + 1$ protection switching is not supported for DWDM systems using the 2029; only bidirectional protection switching will work properly. Even so, most protection schemes cannot switch fast enough to avoid interrupting the sysplex timer and HiPerLinks operation.

HiPerLinks in compatibility mode will be interrupted by their open fiber control, which then takes up to 10 seconds to reestablish the links. Timer channels will also experience loss of light disruptions, as will ESCON and other types of links. Even when all the links reestablish, the application will have been interrupted or disabled and any jobs that had been running on the sysplex will have to be restarted or reinitiated, either manually or by the host's automatic recovery mechanisms depending on the state of the job when the links were broken. For this reason, it is recommended that continuous availability of the applications cannot be ensured without using dual redundant ETR, CLO, and HiPerLinks. Protection switching merely restores the fiber capacity more quickly, it does not ensure continuous operation of the sysplex in the event of a fiber break.

To illustrate the use of DWDM in this environment, consider the construction of a GDPS between two remote locations for disaster recovery,

IBM Parallel Sysplex Architecture

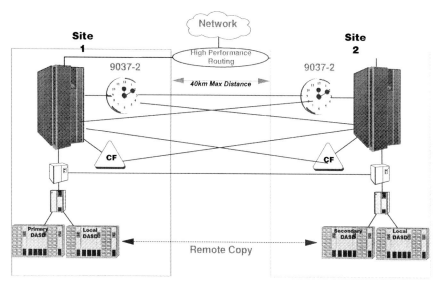

Fig. 7.1

as shown in Fig. 7.1. There are four building blocks for a Parallel Sysplex; the host processor (or Parallel Enterprise Server), the coupling facility, the ETR (Sysplex Timer), and disk storage. Many different processors may be interconnected through the coupling facility, which allows them to communicate with each other and with data stored locally. The coupling facility provides data caching, locking, and queuing (message passing) services. By adding more processors to the configuration, the overall processing power of the sysplex (measured in millions of instructions per second or MIPS) will increase. It is also possible to upgrade to more powerful processors by simply connecting them into the sysplex via the coupling facility. Special software allows the sysplex to break down large database applications into smaller ones, which can then be processed separately; the results are combined to arrive at the final query response. The coupling facility may either be implemented as a separate piece of hardware, or as a logical partition of a larger system. The HiPerLinks are used to connect a processor with a coupling facility. Because the operation of a Parallel Sysplex depends on these links, it is highly recommended that redundant links and coupling facilities be used for continuous availability.

Thus, in order to build a GDPS, we require at least one processor, coupling facility, ETR, and disk storage at both the primary and secondary locations, shown in Fig. 7.1 as site A and site B. Recall that one processor may be logically partitioned into many different sysplex system images; the number of system images determines the required number of HiPerLinks. The sysplex system images at site A must have HiPerLinks to the coupling facilities at both site A and B. Similarly, the sysplex system images at site B must have HiPerLinks to the coupling facilities at both site A and B. In this way, failure of one coupling facility or one system image allows the rest of the sysplex to continue uninterrupted operation. A minimum of two links is recommended between each system image and coupling facility. Assuming there are S sysplex system images running on P processors and C coupling facilities in the GDPS, spread equally between site A and site B, the total number of HiPerLinks required is given by

$$\# \text{ HiPerLinks} = S * C * 2 \tag{7.1}$$

In a GDPS, the total number of inter-site HiPerLinks is given by

$$\text{inter-site } \# \text{ HiPerLinks} = S * C \tag{7.2}$$

The Sysplex Timer (9037) at site A must have links to the processors at both site A and B. Similarly, the 9037 at site B must have links to the processors at both site A and B. There must also be two CLO links between the timers at sites A and B. This makes a minimum of four duplex inter-site links, or eight optical fibers without multiplexing. For practical purposes, there should never be a single point of failure in the sysplex implementation; if all the fibers are routed through the same physical path, there is a possibility that a disaster on this path would disrupt operations. For this reason, it is highly recommended that dual physical paths be used for all local and inter-site fiber optic links, including HiPerLinks, ESCON, ETR, and CLO links. If there are P processors spread evenly between site A and site B, then the minimum number of ETR links required is given by

$$\# \text{ ETR links} = (P * 2) + 2 \text{ CLO links} \tag{7.3}$$

In a GDPS, the number of inter-site ETR links is given by

$$\text{inter-site } \# \text{ ETR links} = P + 2 \text{ CLO links} \tag{7.4}$$

These formulas are valid for CMOS-based hosts only; note that the number of ETR links doubles for ES/9000 Multiprocessor models due to differences in the server architecture.

7. Parallel Computer Architectures Using Fiber Optics

In addition, there are other types of inter-site links such as ESCON channels to allow data access at both locations. In a GDPS with a total of N storage subsystems (also known as Direct Access Storage Devices or DASD), it is recommended that there be at least four or more paths from each processor to each storage control unit (based on the use of ESCON Directors at each site); thus, the number of inter-site links is given by

$$\text{inter-site \# storage (ESCON) links} = N * 4 \tag{7.5}$$

In addition, the sysplex requires direct connections between systems for cross-system coupling facility (XCF) communication. These connections may be provided by either ESCON channel-to-channel links or HiPerLinks. If coupling links are used for XCF signaling, then no additional HiPerLinks are required beyond those given by equations (7.1) and (7.2). If ESCON links are used for XCF signaling, at least two inbound and two outbound links between each system are required, in addition to the ESCON links for data storage discussed previously. The minimum number of channel-to-channel (CTC) ESCON links is given by

$$\text{\# CTC links} = S * (S - 1) * 2 \tag{7.6}$$

For a GDPS with S_A sysplex systems at site A and S_B sysplex systems at site B, the minimum number of inter-site channel-to-channel links is given by

$$\text{inter-site \# CTC links} = S_A * S_B * 4 \tag{7.7}$$

Because some processors also have direct local area network (LAN) connectivity via FDDI or ATM/SONET links, it may be desirable to run some additional inter-site links for remote LAN operation as well.

As an example of applying these equations, consider a GDPS consisting of two system images executing on the same processor and a coupling facility at site A, and the same configuration at site B. Each site also contains one primary and one secondary DASD subsystem. Sysplex connectivity for XCF signaling is provided by ESCON CTC links, and all GDPS recommendations for dual redundancy and continuous availability in the event of a single failure have been implemented. From eq. (7.1–7.7), the total number of inter-site links required is given by

# of inter-site links:

$$\begin{aligned}
\text{\# CTC links} &= S_A * S_B * 4 = 2 * 2 * 4 = 16 \\
\text{\# timer links} &= P + 2 = 2 + 2 = 4 \\
\text{\# HiPerLinks} &= S * C = 4 * 2 = 8 \\
\text{\# storage (DASD) links} &= N * 4 = 8 * 4 = 32
\end{aligned} \quad (7.8)$$

or a total of 60 inter-site links. Note that currently, only ESCON links may be used for the direct connection between local and remote DASD via the Peer-to-Peer Remote Copy (PPRC) protocols. Other types of storage protocols such as Fibre Channel or FICON may be used for the DASD connections. Note that any synchronous remote copy technology will increase the I/O response time, because it will take longer to complete a writing operation with synchronous remote copy than without it (this effect can be offset to some degree by using other approaches, such as parallel access to storage volumes). The tradeoff for longer response times is that no data will be lost or corrupted if there is a single point of failure in the optical network. PPRC makes it possible to maintain synchronous copies of data at distances up to 103 km; however, these distances can only be reached using either DWDM with optical amplifiers or by using some other form of channel extender technology. The performance and response time of PPRC links depends on many factors, including the number of volumes of storage being accessed, the number of logical subsystems across which the data is spread, the speed of the processors in the storage control units and processors, and the intensity of the concurrent application workload. In general, the performance of DASD and processors has increased significantly over the past decade, to the point where storage control units and processors developed within the past two years have their response time limited mainly by the distance and the available bandwidth. Many typical workloads perform several read operations for each write operation; in this case the effect of PPRC on response time is not expected to be significant at common access densities.

Similar considerations will apply to any distributed synchronous architecture such as Parallel Sysplex. In some cases, such as disaster recovery applications, where large amounts of data must be remotely backed up to a redundant storage facility, an asynchronous approach is practical. This eliminates the need for sysplex timers, and trades off continuous real-time data backup for intermittent backup; if the backup interval is sufficiently small, then the impact can be minimized. One example of this approach is the eXtended Remote Copy (XRC) protocols supported by FICON channels on a z series server. This approach interconnects servers and DASD between a primary and a backup location, and periodically initiates a remote copy of data from the primary to the secondary DASD. This approach requires fewer fiber optic links, and because it does not use a sysplex timer the distances can be extended to 100 km or more. The tradeoff with data integrity must be assessed on a case-by-case basis; some users prefer to implement XRC as a first step toward a complete GDPS solution.

The use of a parallel computing architecture over extended distances is a particularly good match with fiber optic technology. Channel extension is well known in other computer applications, such as storage area networks; today, mainframes are commonly connected to remote storage devices housed tens of kilometers away. This approach, first adopted in the early 1990s, fundamentally changed the way in which most people planned their computer centers, and the amount of data they could safely process; it also led many industry pundits to declare "the death of distance." Of course, unlike relatively low bandwidth telephone signals, performance of many data communication protocols begins to suffer with increasing latency (the time delay incurred to complete transfer of data from storage to the processor). While it is easy to place a long-distance phone call from New York to San Francisco (about 42 milliseconds round trip latency in a straight line, longer for a more realistic route), it is impossible to run a synchronous computer architecture over such distances. Further compounding the problem, many data communication protocols were never designed to work efficiently over long distances. They required the computer to send overhead messages to perform functions such as initializing the communication path, verifying it was secure, and confirming error-free transmission for every byte of data. This meant that perhaps a half dozen control messages had to pass back and forth between the computer and storage unit for every block of data, while the computer processor sat idle. The performance of any duplex data link begins to fall off when the time required for the optical signal to make one round trip equals the time required to transmit all the data in the transceiver memory buffer. Beyond this point, the attached processors and storage need to wait for the arrival of data in transit on the link, and this latency reduces the overall system performance and the effective data rate. As an example, consider a typical fiber optic link with a latency of about 10 microseconds per kilometer round trip. A mainframe available in 1995 capable of executing 500 million instructions per second (MIPS) needs to wait not only for the data to arrive, but also for 6 or more handshakes of the overhead protocols to make the round trip from the computer to the storage devices. The computer could be wasting 100 MIPS of work, or

20% of its maximum capacity, while it waits for data to be retrieved from a remote location 20 kilometers away. Although there are other contributing factors, such as the software applications and workload, this problem generally becomes worse as computers get faster, because more and more processor cycles are wasted waiting for the data. As this became a serious problem, various efforts were made to design lower latency communication links. For example, new protocols were introduced that required fewer handshakes to transfer data efficiently, and the raw bandwidth of the fiber optic links was increased from ESCON rates (about 17 Mbytes/s) to nearly 100 Mbyte/s for FICON links. But for very large distributed applications, the latency of signals in the optical fiber remains a fundamental limitation; DASD read and write times, which are significantly longer, will also show a more pronounced effect at extended distances.

While the performance of any large-scale computer system is highly application dependent, we can infer some of the effects caused by extended distances. For the case of I/O requests to DASD on an ESCON link, assume that at the primary site a typical storage read or write operation takes 3 ms. The latency of an inter-site fiber optic link is about 10 microseconds/km round trip; this must be multiplied by the inter-site distance and the number of acknowledgments required by the data link protocol to determine the impact of inter-site distance on performance. If we assume a conservative datacom protocol (such as ESCON) that requires 6 acknowledgments per operation, then at a distance of 40 km the additional delay is (10 microseconds/km/round trip) (40 km) (6 round trips) = 2.4 ms. The time required for a DASD read operation from site B to DASD in site A is then $3 + 2.4 = 5.4$ ms. Similarly, a data mirroring application might require a write operation to the DASD in site A that would then be remote copied to DASD in site B. This operation would take 3 ms for the local write, 2.4 ms latency, and 3 ms for the remote write, or 8.4 ms total. If the data must first be requested from site B before this operation can begin, this adds another 2.4 ms for a total of 10.8 ms. In a similar fashion, performance of all ESCON and HiPerLinks will degrade with distance; there is no general formula to predict this impact, it must be evaluated for each software application and datacom protocol individually.

7.4. Optically Interconnected Parallel Supercomputers

Latency is not only a problem for processor-to-storage interconnections, but also a fundamental limit in the internal design of very large computer systems. Today, many supercomputers are being designed to solve so-called

7. Parallel Computer Architectures Using Fiber Optics

"Grand Challenge" problems, such as advanced genetics research, modeling global weather patterns or financial portfolio risks, studying astronomical data such as models of the Big Bang and black holes, design of aircraft and spacecraft, or controlling air traffic on a global scale. This class of high risk/high reward problems is also known as "Deep Computing." A common approach to building very powerful processors is to take a large number of smaller processors and interconnect them in parallel. In some cases, a computational problem can be subdivided into many smaller parts, which are then distributed to the individual processors; the results are then recombined as they are completed to form the final answer. This is one form of asynchronous processing, and there are many problems that fall into this classification; one of the best examples is SETI@home, free software which can be downloaded over the Internet to any home personal computer. Part of the former NASA program, SETI (Search for Extra-Terrestrial Intelligence) uses spare processing cycles when a computer is idle to analyze extraterrestrial signals from the Arecibo Radio Telescope, searching for signs of intelligent life. There are currently over 1.6 million SETI@home subscribers in 224 countries, averaging 10 teraflops (10 trillion floating point operations performed per second) and having contributed the equivalent of over 165,000 years of computer time to the project. Taken together, this is arguably the world's largest distributed supercomputer, mostly interconnected with optical fiber via the Internet backbone.

More conventional approaches rely on large numbers of processors interconnected within a single package. In this case, optical interconnects offer bandwidth and scalability advantages, as well as immunity from electromagnetic noise, which can be a problem on high-speed copper interconnects. For these reasons, fiber optic links or ribbons are being considered as a next-generation interconnect technology for many parallel computer architectures, such as the PowerParallel and NUMA-Q designs. The use of optical backplanes and related technologies is also being studied for other aspects of computer design (see Chapter 6). To minimize latency, it is desirable to locate processors as close together as possible, but this is sometimes not possible due to other considerations, such as the physical size of the packages needed for power and cooling. Reliability of individual computer components is also a factor in how large we can scale parallel processor architectures. As an example, consider the first electronic calculator built at the University of Pennsylvania in 1946, ENIAC (Electronic Numeric Integrator and Computer), which was limited by the reliability of its 18,000 vacuum tubes; the machine couldn't scale beyond filling a room about 10 by 13 meters, because tubes would blow out faster than people could run from one end of the machine to the other replacing them. Although the reliability of individual components has improved considerably, modern-day supercomputers still require some level of modularity, which comes with an associated size and cost penalty.

A well-known example of Deep Computing is the famous chess computer, Deep Blue, that defeated grand master Gary Kasparov in May of 1997. As a more practical example, the world's largest supercomputer is currently owned and operated by the U.S. Department of Energy, to simulate the effects of nuclear explosions (such testing having been banned by international treaty). This problem requires a parallel computer about fifty times faster than Deep Blue (although it uses basically the same internal architecture). To accomplish this requires a machine capable of 12 teraflops, a level computer scientists once thought impossible to reach. Computers with this level of performance have been developed gradually over the years, as part of the Accelerated Strategic Computing Initiative (ASCI) roadmap; but the current generation, called ASCI White, has more than tripled the previous world record for computing power. This single supercomputer consists of hundreds of equipment cabinets housing a total of 8,192 processors, interconnected with a mix of copper and optical fiber cables through two layers of switching fabric. Because the cabinets can't be pressed flat against each other, the total footprint of this machine covers 922 square meters, the equivalent of 2 basketball courts. This single computer weighs 106 tons (as much as 17 full-size elephants) and had to be shipped to Lawrence Livermore National Labs in California on 28 tractor trailers. It's not feasible today to put the two farthest cabinets closer together than about 43 meters, and this latency limits the performance of the parallel computer system. Furthermore, ASCI White requires over 75 terabytes of storage (enough to hold all 17 million books in the Library of Congress), which may also need to be backed up remotely for disaster recovery; so, the effects of latency on the processor-to-storage connections are also critically important. Future ASCI programs call for building a 100 Teraflop machine by 2004.

7.5. Parallel Futures

Current Parallel Sysplex systems have been benchmarked at over 2.5 billion instructions per second, and are likely to continue to significantly increase in performance each year. The ASCI program has also set aggressive goals for future optically interconnected supercomputers. However, even these are not the most ambitious parallel computers being designed for future applications.

There is a project currently under way, led by IBM fellow Monty Denneau, to construct a mammoth computer nicknamed "Blue Gene" which will be dedicated to unlocking the secrets of protein folding. Without going into the details of this biotechnology problem, we note that it could lead to innumerable benefits, including a range of designer drugs, whole new branches of pharmacology, and gene therapy treatments that could revolutionize health care, not to mention lending fundamental insights into how the human body works. This is a massive computational problem, and Blue Gene is being designed for the task; when completed, it will be 500 times more powerful that ASCI White, a 12.3 petaflop machine — well over a quadrillion (10^{15}) operations per second, forty times faster than today's top 40 supercomputers combined. The design point proposes 32 microprocessors on a chip, 64 chips on a circuit board, 8 boards in a 6-foot-high tower, and 64 interconnected towers for a total of over 1 million processors. Because of improvements in packaging technology, Blue Gene will occupy somewhat less space than required by simply extrapolating the size of its predecessors; about 11×24 meters (about the size of a tennis court), with a worst-case diagonal distance of about 26 meters. However, the fast processors proposed for this design can magnify the effect of even this much latency to the point where Blue Gene will be wasting about 1.6 billion operations in the time required for a diagonal interconnect using conventional optical fiber. Further, a machine of this scale is expected to have around 10 terabytes of storage requirements, easily enough to fill another tennis court, and give a processor-to-storage latency double that of the processor-to-processor latency. Because of the highly complex nature of the protein folding problem, a typical simulation on Blue Gene could take years to complete, and even then may yield just one piece of the answer to a complex protein-folding problem.

While designs such as this have yet to be realized, they illustrate the increasing interest in parallel computer architectures as an economical means to achieving higher performance. Both serial and parallel optical links are expected to play an increasing role in this area, serving as both processor-to-processor and processor-to-storage interconnects.

ACKNOWLEDGMENTS

The terms System/390, S/390, OS/390, ES/9000, MVS, G5, Parallel Enterprise Server, 9037 Sysplex Timer, Enterprise Systems Connection, ESCON, Fibre Connection, FICON, Parallel Sysplex, HiPerLinks, Fiber Transport Services, FTS, Fiber quick connect, 9729 Optical Wavelength Division Multiplexer, Geographically Dipersed Parallel Sysplex, and GDPS are trademarks of IBM Corporation.

References

1. DeCusatis, C., D. Stigliani, W. Mostowy, M. Lewis, D. Petersen, and N. Dhondy. Sept./Nov. 1999. "Fiber optic interconnects for the IBM S/390 parallel enterprise server." *IBM Journal of Research and Development.* **43**(5/6):807–828; also see *IBM Journal of Research & Development*, special issue on "IBM System/390: Architecture and Design," 36:4 (1992).
2. DeCusatis, C., Jan./Feb. 2001. "Dense wavelength division multiplexing for parallel sysplex and metropolitan/storage area networks." *Optical Networks*, pp. 69–80.
3. DeCusatis, C., and P. Das, Jan. 31-Feb. 1, 2000. "Subrate multiplexing using time and code division multiple access in dense wavelength division multiplexing networks." *Proc. SPIE Workshop on Optical Networks*, Dallas, Tex., pp. 3–11.
4. "Coupling Facility Channel I/O Interface Physical Layer" (IBM document number SA23-0395), IBM Corporation, Mechanicsburg, Pa. (1994).
5. Bona, G.L., et al. December 1998. "Wavelength division multiplexed add/drop ring technology in corporate backbone networks." Optical Engineering, special issue on Optical Data Communication.
6. "9729 Operators Manual" (IBM document number GA27-4172), IBM Corporation, Mechanicsburg, Pa. (1996).
7. "IBM Corporation 9729 optical wavelength division multiplexer." *Photonics Spectra* (Special Issue: The 1996 Photonics Circle of Excellence Awards), vol. 30 (June 1996).
8. DeCusatis, C., D. Petersen, E. Hall, and F. Janniello. December 1998. "Geographically distributed parallel sysplex architecture using optical wavelength division multiplexing." *Optical Engineering*, special issue on Optical Data Communication.
9. Thiébaut, D., Parallel Programming in C for the Transputer, 1995, http://www.cs.smith.edu/~thiebaut/transputer/descript.html?clkd=iwm.

Relevant Web Sites:

http://www.npac.syr.edu/copywrite/pcw/node1.html is a parallel computing textbook.

http://www.gapcon.com/info.html is a list of the top 500 super computers (all of the current computers referenced were taken from this list).

http://compilers.iecc.com/comparch/article/93-07-068 is a timeline for the history of parallel computing.

Part 2 | The Future

Chapter 8 | Packaging Assembly Techniques

Ronald C. Lasky
Consultant, Medway, Massachusetts 02053

Adam Singer
Prashant Chouta
Cookson Performance Solutions, Foxborough, Massachusetts 02053

8.1. Packaging Assembly — Overview

The field of electronics packaging is a multidisciplinary field that requires sound knowledge of a wide array of subjects. A typical electronic package contains millions of transistors assembled in chips, resistors, diodes, capacitors, and other components. To perform the different logical functions these components are connected to each other to form functional entities. To function, these circuits are supplied with electric energy, some of which is converted into thermal energy. An electronic object will function correctly if the package has good electrical, mechanical, thermal, metallurgical, etc., properties. In summary, the package has to perform the following four main functions:

- Sound signal distribution and maintenance of signal integrity with minimum cross-talk
- Mechanical support to the entire assembly
- Provision of a path of least resistance to the flow of heat, enhancing heat dissipation properties
- Protection from the environment

A proper selection of materials, processes, assembly techniques, board design, component layout, operating procedures etc., will ensure a reliable product that is also easy to manufacture but also is reliable. It is also desirable that the packaging occupies a minimum of space and weight.

8.1.1. SURFACE-MOUNT TECHNOLOGY (SMT)

The electronics industry is continuously seeking faster and smaller products. This is a constant challenge from a design and assembly perspective. The need for products with higher functionality and reduced real estate space has resulted in the evolution of Surface-Mount Components. A typical Surface-Mount Technology process involves mounting electronic packages on the surfaces of the PCBs while conventional Through-Hole Technology entails the attachment of components through insertion into the through-holes on the board. SMT offers several inherent advantages: The components are small and can be mounted on either side of the board, they offer reduction in weight and size, more interconnections per unit area, better electrical properties, quicker signal propagation, and reduction in cost due to real estate savings. Through-hole components (THCs) are still often preferred for the robustness of their solder joint and for their ease of assembly due to generous pitches. However, even with the advantages offered by Surface-Mount Technology it is very unlikely that SMT will completely displace Through-Hole technology, therefore, in the present-day scenario, mixed technology, with surface-mount and through-hole parts on the same board, has gained popularity.

8.1.2. SMT VS THT

Changing the assembly method of components from THT to SMT entails a number of distinct modifications:

- The packaging design must change to accompany new land pattern configurations and new component styles having different lead types and shapes.
- The assembly manufacturing processes also change to accompany the new component styles.
  - Surface-mount components (SMCs) are more amenable to placement by automated equipment, thus resulting in faster throughput. SMCs are also normally soldered using a reflow soldering process versus THC wave soldering process.
  - Inspection processes are different because of SM solder joints.

- Testing changes are needed to accommodate the above differences.

8.1.3. PACKAGING HIERARCHY

The main levels of an electronic assembly are integrated circuits or chips, chip carrier, packages and printed circuit (PC) boards. The smallest functional unit of the assembly is the silicon chip, which provides the logic functions of the circuit. The chip is attached to a chip carrier by means of wires or solder; the resulting joints are also referred to as the first-level of interconnection. A plastic body is often molded around the chip. The resulting component is often called a first-level package or a component package — the most common examples of these packages are Ball Grid Arrays, Quad Flat Packs, and small outline integrated circuits (SOICs) shown in Fig. 8.1. The attachment of the chip carrier to the Printed Circuit Board (PCB) is usually done by means of solder. This is also known as the

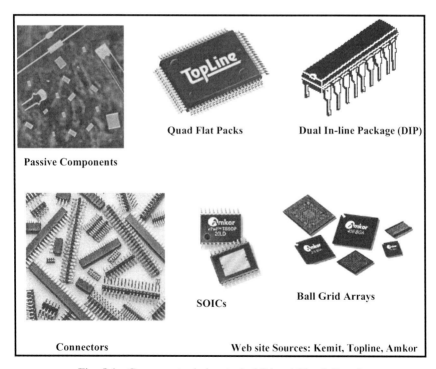

Fig. 8.1 Component mix in a typical Printed Circuit Board.

second-level interconnection. The eutectic solder composition of tin and lead is the most common material used for the second-level interconnection. The assembled PCB card may then be attached to a motherboard by means of connectors. Typically a PC card or board are referred to as second level packages.

8.1.4. PCB OR SECOND-LEVEL ASSEMBLY TECHNIQUES

The assembly process of a Printed Ciruit Board is dictated by the component mix, the board design, the technology available, and the assembly capability of the manufacturing site. Typically, it consists of five main steps as follows:

- Applying solder paste on to the conductive pads on the board, also known as stencil printing
- Placing components onto the board with the component leads on the solder paste
- Reflow/Wave Soldering
- Cleaning (Depends on type of solder paste), and
- Testing and Inspection

However, the actual process isn't as trivial as it sounds. To ensure a quality manufacturing process many different steps, such as adhesive dispensing, post printing paste inspection, and techniques like design of experiments and statistical process control may be required.

The surface mount components, when attached to the substrate, form three major types of SMT assembly — commonly referred to as Type 1, Type 2, and Type 3 as shown in Fig. 8.2. The figure also shows the different processing steps for each of the assembly methods.

> The Type 1 assembly contains only surface-mount components. This assembly can either be single-sided or double-sided.
>
> The Type 3 assembly contains passive surface-mount components like resistors and capacitors glued to the bottom and Through-hole components on the top side.
>
> The Type 2 assembly is a combination of Type 1 and Type 3 assembly. It generally does not contain any active surface-mount devices on the bottom.

8. Packaging Assembly Techniques   307

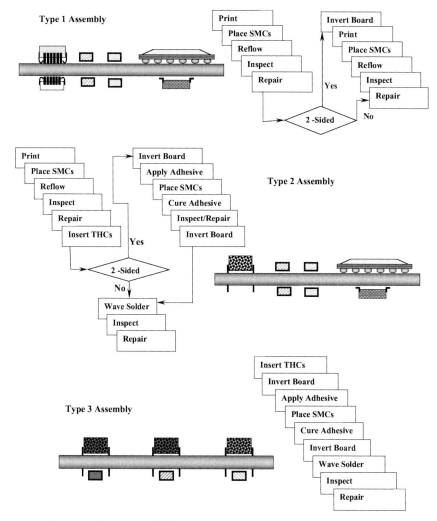

Fig. 8.2   Schematic of different types of assembly with their process flows.

Furthermore, in order to accommodate the complexities of the surface-mount components, additional reference designators like Type 1C, Type 2C, and Type 3C have been developed.

The following sections will describe the different types of first interconnects and package construction of a few common components.

8.1.4.1. Interconnection Methodologies for First-Level Packages

Connections between the chip and its package substrate are commonly performed by one of three technologies: wirebond, solder, or controlled collapse chip connection (C4), also called "flip chip (FC)," or "solder bump," and tape automated bonding (TAB), depending on the number and spacing of I/O connections on the chip and substrate as well as permissible cost.

8.1.4.2. Wirebond

Wirebonding is the most common chip-bonding technology, spanning the needs from consumer electronics to mainframes. The widespread use of wirebonding is based on its low cost per I/O and compliant wires resulting in a reliable joint. Moreover, the compliant wires also provide an excellent path for heat dissipation and provide strain relief to allow for the difference in thermal expansion between the chip and the module material. Finished wirebond connections are shown in Fig. 8.7. However, with the changing face of today's technological needs, this technology has been continuously pushed to the limits. There is a physical limit on the maximum number of wirebonds on the face of a die, as the interconnections are formed on the periphery of the die. This situation results in an increased size of the chip by a factor of three as compared to area array solder flip chip connections. As a result, for die with a high number of interconnections, area array technologies are preferred.

8.1.4.3. Thermocompression Wirebonding

Thermocompression bonding consists of heating and applying thermal and mechanical pressure to two joining bodies. This combined energy facilitates diffusion of the metals, thus resulting in a metallurgical bond between the two surfaces. The ductility of the wire also plays an important role in the formation of the bond. Thermocompression bonding of gold to aluminum metallization is one of the most extensively studied interdiffusion mechanisms due to its wide proliferation.

The chip terminal to wire interconnection is made by ball bonding, while interconnection of the wire to chip carrier circuitry is made by wedge bonding. This process is shown in Fig. 8.3.

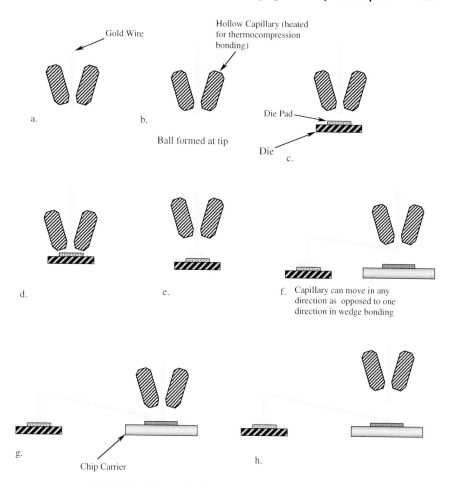

Fig. 8.3 Schematic of different stages in ball bonding.

8.1.4.4. Tape-Automated Bonding (TAB)

TAB is a concept for an interconnection in which the metallurgical bond is similar to a wirebond. TAB is a composite of printed circuit wires held in an area configuration by a thin organic substrate (usually polyimide). The flexible leads or the wires, typically copper, are made by photoprinting an evaporated or a plated layer of metal on a polyimide carrier. The joining of the entire set of leads is usually done by thermocompression. This approach allows a high lead count that cannot be achieved in either Dual Inline Package (DIP) or Quad Flat Packages (QFPs). Despite these advantages, TAB is dif-

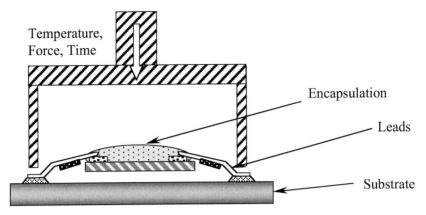

Fig. 8.4 TAB Outer Lead Bonding.

ficult to assemble to circuit boards. This difficulty has hurt its acceptance.

8.1.4.5. Flip Chip Technology

The need for achieving higher I/O per unit area as well as making the first level attachment truly SMT compatible has pushed the demand for area array interconnection at the first level. Flip chip technology uses the entire bottom surface area of the die for the purpose of interconnection. This area technology has several inherent advantages, as follows:

- More I/O per unit area
- SMT compatible
- Quicker signal propagation due to short transmission length
- Less signal distortion
- Economical bumping process, and
- SMT compatible assembly process.

The flip chip assembly process depends on the type of solder used for the bumping process.

For flip chips with high lead solder bumps, eutectic tin-lead solder paste is first printed onto the substrate pads, then the chips are placed and reflowed. For chips with eutectic solder bumps, there is no need to print additional solder paste. In this case the entire volume of solder is provided by the solder bump. Figure 8.5 shows a cross-section of the flip chip with eutectic solder bumps. The aluminum pad is usually covered with a thin

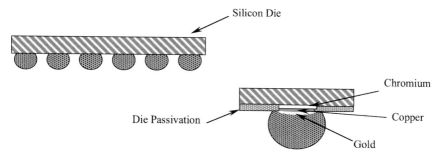

Fig. 8.5 Cross-section of a flip chip solder bump.

layer of nickel, which acts as a diffusion barrier, and then by a flash of gold. The solder bumps are usually formed by a metal, evaporation technique. Solder spheres are also widely used to bump the wafers, and recently stencil printing to "bump" has been demonstrated.

To accommodate the thermal mismatch between the die and the chip carrier, underfill is dispensed between the die and the chip carrier to even out the stresses created due to the different coefficients of thermal expansion. Flip chip assembly to the second-level package is shown in Fig. 8.6.

8.1.5. SURFACE MOUNT FIRST-LEVEL PACKAGES

Through-hole packages are gradually being replaced by surface-mount component packages, which offer inherent advantages like fine pitch, area array, and finally ease of manufacturing assembly. The significant driving force for the use of SMT is the reduced package size and cost, and improved board utilization. Hence, for these reasons, SMT has mostly replaced through-hole technology for PCB assembly. The advance of SMT can be attributed to better space utilization on the PCB board, thus leading to denser and smaller packages. The surface-mount packages have come a long way from their initial use of the package periphery for second-level interconnections to using the entire area underneath the package in area array packaging. The continuous push to pack more electronic functionality has driven the limits of leaded peripheral packages like QFPs. Eventually there will be a limit beyond which it would be physically impossible to manufacture and assemble such packages. This has given rise to area array packages, which offer more I/O per unit area. These packages also have greater spacing between interconnections thus enabling easier assembly.

Area array technology utilizes the whole area underneath the package

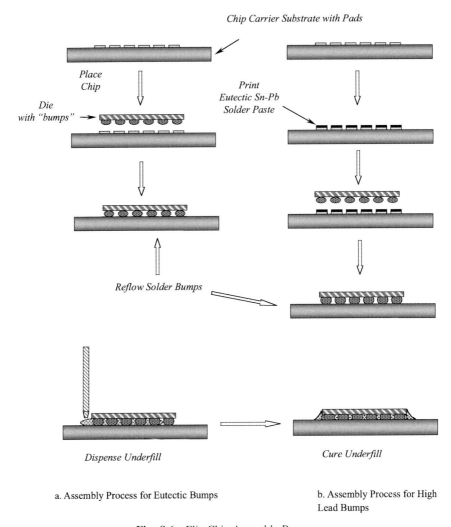

a. Assembly Process for Eutectic Bumps

b. Assembly Process for High Lead Bumps

Fig. 8.6 Flip Chip Assembly Process.

for interconnections, thus resulting in higher I/O density. The advantages of this technology are:

- Higher density of I/O interconnections
- Less real estate consumption
- Better electrical performance and reduced inductance
- High-speed, low-noise signal transmission

- Better thermal performance from the area array
- More component placement tolerance due to higher interconnection spacings
- The leads compliance are more robust
- Standardization

However, the drawbacks of this rapidly emerging technology are as follows:

- Increased difficulty in inspection of solder interconnections
- Possible difficulties in rework/repair
- Increase in board thickness and difficulty in routing to be able to connect to the denser interconnections

The most common and widely used type of area array packages is the Ball Grid Array (BGA) package; the following lines will describe the construction of a few of those packages.

8.1.5.1. Ball Grid Array Packages

BGA packages are rapidly gaining acceptance in the electronics industry as a low-cost, higher yielding alternative to fine pitch leaded packages. Conventional BGA technology emphasizes soldering directly between the chip carrier package and the interconnect substrate. It is also referred to as face-bonding or controlled collapse soldering. BGA packages truly reflect the advantages and disadvantages of area array interconnection mentioned previously.

BGAs are classified in a number of different ways. Based on the array of solder joints, BGAs can be classified as peripheral array, staggered array, depopulated array, or a full area array. However, the most convenient way of classifying BGA packages is based on the type of material of the chip-carrier substrate. Accordingly, their classification is briefly described next. Figure 8.7 illustrates the different described BGA packages.

8.1.5.2. Plastic Ball Grid Array (PBGA) Packages

These types of BGA packages have a glass-reinforced bismaleimide-triazine (BT) epoxy (i.e., plastic) chip carrier, hence the name PBGA. The die is mounted to the laminate chip carrier. The first-level interconnection within the package is predominantly thermosonic gold wire bonding. From there, copper traces are routed to an array of metal pads on the bottom of the

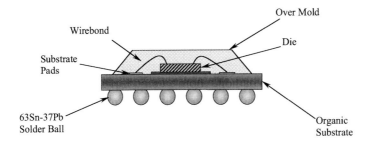

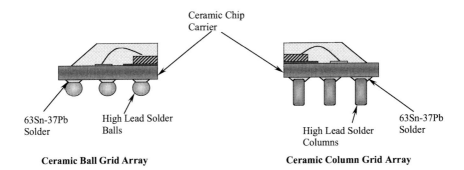

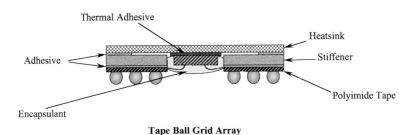

Fig. 8.7 Schematic of different BGA Packages.

printed circuit board. Vias in the form of plated through-holes run from the top surface to the bottom. A "glob-top" encapsulation or overmold is then performed to completely cover the chip, wires, and the substrate bond pads. Typically, eutectic tin-lead solder or eutectic tin-lead solder with 2%-silver is used for second-level interconnection.

8.1.5.3. Ceramic Ball Grid Array (CBGA) Packages

The chip-carrier substrate in this case is a ceramic instead of a BT resin epoxy-glass laminate. CBGAs give better hermetic sealing to the package as compared to PBGAs. The construction is similar to that of a PBGA, except that in many cases high-lead solder balls are used at the second-level interconnection.

The Ceramic Column Grid Array (CCGA) package consists of high-lead ceramic columns instead of the high-lead solder balls in the CBGA package. The increased standoff between the chip carrier and the substrate accounts for better thermal fatigue performance.

8.1.5.4. Tape Ball Grid Array (TBGA) Packages

The TBGA is characterized by a flexible polyimide tape, flip-chip stiffener, and high-temperature solder balls. The tape is a 0.002-inch thick polyimide layer with copper metallization on each side. The close proximity of signal traces to the ground plane provides a very low signal noise. Copper stiffener plates are bonded to the tape by an adhesive in order to give mechanical rigidity to the package.

8.1.5.5. Metal Ball Grid Array (MBGA) Packages

This package is featured with thin-film circuits deposited on the anodized aluminum substrate. The substrate, which is in direct contact with the die and circuitry, acts as a heat sink as well as the wiring substrate. The aluminum substrate is an effective shield against electromagnetic interferences.

8.2. Optoelectronic Packaging Overview

Optoelectronic packaging is an outgrowth of electronic packaging: Where possible, similar techniques and materials are employed, such as through-hole assembly and eutectic tin/lead solder joints for many components. However, the differences in the packaging needs of optoelectronics are great.

- Optical alignment requires greater precision than any electronics assembly (can be as precise as $+/-0.5$ micron);
- Hand soldering is used in most cases, since parts are fragile and many optical materials cannot be exposed to wave soldering or solder reflow ovens;

Table 8.1 **Three Different Levels of Optoelectronic Packaging**

	Component	*Module*	*System*
Materials	Unique hi-T solder	Mix of custom & standard solders	Standard solders
Chips	Bare die	Bare & packaged	Packaged
Precision	Highest (0.5–1 μm)	High (1–2 μm)	Low (PTH)
Automation	Little to none	Little to none	All automated soon
Testing	Simple	Simple to complex	Complex

- Bundles of fiber up to 10m long pose challenges in how assemblies can be handled;
- Generally, due to the young age of this industry as well as the proprietary nature of optoelectronics assembly to date, many needed standards have not yet been set.

Optoelectronic packaging can be grouped into three categories, each with different material and assembly requirements, as shown in Table 8.1.

Two other forms of optoelectronic or optical packaging exist that are not addressed in this chapter: at a level smaller than considered here (on-chip waveguides), and at a level larger than considered here (from one optical system to another via splicing fiber). On-chip issues are wafer fabrication issues, not necessarily packaging issues, and are beyond the scope of this chapter. System-to-system interconnect, while arguably packaging, consists largely of fiber splicing, which is well understood today.

The remainder of this chapter explains these three levels of packaging in more detail.

8.3. Component Level Optoelectronic Packaging

The assembly of optoelectronic components is similar to IC assembly:

- Parts being assembled are GaAs- or silicon-based nanostructures originally fabricated on a wafer
- Parts can be assembled by expensive, automated machines of great precision and handling delicacy in cleanrooms
- Parts can be assembled by inexpensive labor, most likely in Asia, with moderate precision and handling delicacy

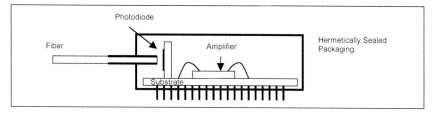

Fig. 8.8 Typical Optoelectronic Component (Receiver Shown Here).

8.3.1. ASSEMBLY ISSUES

At the component level, the highest precision in any form of manual assembly is required — ± 0.5 μm up to 1 μm where fiber alignment is concerned. Connections to the substrate can be wirebonds or flip chip connections, where 5 to 10 micron precision is required. Parts are soldered or brazed in place sequentially, by hand, so that they can be placed precisely. Care must be taken when placing a new part so that parts assembled previously (in the same housing) are not disturbed.

8.3.2. MATERIALS ISSUES

Solders and brazing materials tend to include nickel and gold, nickel for a mechanically strong connection (to prevent optical misalignment due to vibration or other induced stresses) and gold for a highly conductive connection (for the high frequencies — at or above 2.4 GHz — of optical communications). Optical connection materials (epoxies) need to be non-birefringent (not adversely affecting the optical signal in any way, such as polarization-dependent losses or wavefront distortion). Most material solutions cannot be subjected to stress, or else strain-induced birefringence is likely. Consequently, assembly procedures must take into consideration the stresses induced in optical materials.

Substrate materials tend to be ceramic, which is both capable of high-frequency operation and compatible with hermetic packaging. The housing is metal, usually Kovar.

8.4. Module Level Optoelectronic Packaging

Module assembly can combine both packaged components and unpackaged devices, and thus include the challenges of both component and system assembly. Modules usually have multiple fiber I/Os and many devices, usually 50–100 including passives.

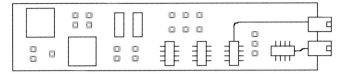

Fig. 8.9 Typical Optoelectronic Module.

8.4.1. ASSEMBLY ISSUES

In addition to the issues faced in component assembly, there are more devices, active and passive, that require a wide range of assembly techniques. For a module including high I/O devices, these devices might be placed and reflowed first, then the sensitive optoelectronic parts placed and hand soldered. Additionally, fiber cleaning and splicing may be needed. After assembly, testing of the module can be complex and expensive, using both optical and electronic testers to find defects.

8.4.2. MATERIALS ISSUES

In addition to the materials used in component assembly, modules require more traditional electronic materials such as solders. Proprietary fiber cleaning solutions fiber splicing are also needed. The substrate can be ceramic or organic, depending on the module's frequency (copper-clad organics reach limits at about 5 GHz) and hermeticity (organic substrates tend to fail hermeticity requirements) needs.

8.5. System Level Optoelectronic Packaging

System level packaging involves the assembly of optoelectronic modules and components onto a system board or backplane. It uses traditional electronic board assembly techniques with a few twists — mainly for fiber handling and for accommodating the materials within the optoelectronic modules and components.

8.5.1. ASSEMBLY ISSUES

Because optoelectronic components have high sensitivity to temperature and mechanical stress, the board must be reflowed before the optoelectronics can be manually placed. Then, special fiber handling techniques must be employed to prevent bending below a specified radius of curvature, which

8. Packaging Assembly Techniques

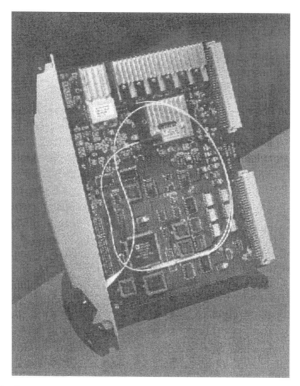

Fig. 8.10 Typical Optoelectronic System Board (Courtesy Prismark, used with permission).

is not yet standardized. Lastly, fiber splicing is likely, requiring proprietary fiber cleaning to minimize the signal loss due to the splice.

Assembly equipment companies will be introducing automated equipment for assembling optoelectronics at this level in the summer of 2001. Such equipment would combine a through-hole placement machine, modified for fiber handling, with robotic spot soldering.

After assembly, the testing systems are expensive ($1M+) and complex, requiring a room full of instrumentation and optoelectronics more advanced than the product being tested.

8.5.2. MATERIALS ISSUES

Low-temperature solders must be employed for the through-hole optoelectronic components and modules, to minimize the likelihood of reflowing solder connections within the module or component, and to prevent the materials within the optoelectronics from being subjected to temperature-induced stresses.

References

1. Seraphim, D. P., R. Lasky, and C. Y. Li. 1989. *Priniciples of Electronic Packaging*. New York: McGraw-Hill.
2. Tummala, R. R. and E. J. Rymaszewski. 1997. *Microelectronics Packaging Handbook*, 2nd ed. Boston: International Thomson Publishing.
3. Prasad, R. 1997. *Surface Mount Technology: Principles and Practice*, 2nd ed. Dordrecht, Netherlands: Kluwer Academic Publishers Group.
4. Harper, C. A. 1997. *Electronic Packaging & Interconnection Handbook*, 2nd ed. New York: McGraw-Hill.
5. Marcoux, P. P. 1992. *Fine Pitch Surface Mount Technology: Quality, Design, and Manufacturing Techniques*. Van Nostrand Reinhold.

Chapter 9 | InfiniBand—The Interconnect from Backplane to Fiber

Ali Ghiasi

Broadcom Corporation, Cupertino, California 95014

This chapter describes InfiniBand™ (IB) optical link interface. InfiniBand group was formed to facilitate development of a uniform interconnect from backplane to fiber. Prior to the development of IB nearly every computer manufacture developed a proprietary interconnect. Two earlier standardization efforts in this area were SCI (Scalable Coherent Interface) [1] and HiPPi6400 [2]; each attempted to develop an open high-performance interconnect. IB takes a further step in defining the high-performance System Area Interconnect, by eliminating the bottleneck inside the box. IB links scale from chip, backplane, IO Card, to fiber.

9.1. Introduction

Improvement in VLSI CMOS has enabled fabrication of more complex and faster processors, so that the I/O has now become the primary bottleneck [3]. CMOS devices operating at speeds greater than 10 Gb/s have now been demonstrated [4]. Existing wide electrical buses, sometimes 64–256 bit wide operating at a typical speed of 100 MHz, are becoming insufficient and impractical for the I/O need. A further complication with wide slow buses is the associated package size and cost. With increasing number of I/O additional routing channels are required to route the signals, which increases PCB stack-up layers and the total system cost. It was becoming impractical to increase bus width, and the natural solution was to

Table 9.1 **Performance of InfiniBand Link**

Link parameters		Raw bandwidth	
Link width	Signaling rate	Unidirectional	Bi-directional
1	2.5 Gb/s	250 MByte/s	500 MByte/s
4	2.5 Gb/s	1 GByte/s	2 GByte/s
12	2.5 Gb/s	3 GByte/s	6 GByte/s

increase the speed with broad availability of CMOS ASIC I/O operating at 2.5 Gb/s.

Early on, InfiniBand group studied two possible signaling schemes: Source Synchronous and Serial Link. Source Synchronous interfaces have been implemented in SCI with 0.5 GByte throughput [1] and HiPPi6400 at 1 Gbyte throughput [2]. To support backplane and long fiber applications one has to implement complex de-skew sequence and training similar to HiPPi6400. The alternative, based on serial link technology and the 8B/10B widely in use in Fiber Channel [5] and Gigabit Ethernet [6], provided robust symmetrical signaling, without the need for complex analog de-skew. To meet the High Performance Computing (HPC) requirements a single serial data stream near term from an ASIC couldn't meet the data throughput. A link layer was developed so the serial link could be scaled from 1, to 4, and to 12 lanes wide with each physical lane operating at 2.5 Gb/s. The 4 lanes implementation has been adapted in IEEE 802.3ae [7] as the basis for XAUI interface and similarly by Fiber Channel 10GFC [8].

Infiniband Link provides an interoperable interface with a raw bandwidth of 250 MBytes/s, 1 GByte/s, or 3 Gbyte/s as shown in Table 9.1. Figure 9.1 shows an example of a 4X-SX optical transceiver.

Through link negotiation a low-cost 1X wide 250 MByte/s link natively interfaces to a high throughput 12X wide 3 GByte/s link. In the next 2 to 3 years, the IB signaling rate is expected to increase to 5 Gb/s, while supporting a large application base protecting the investment via the interoperability.

9.2. Infiniband Link Layer

Infiniband link layer provides the connection between two InfiniBand protocol aware ports. Each physical link may have 1, 4, or 12 physical lanes. A physical link may be copper cable, copper backplane, or fiber optics.

9. InfiniBand—The Interconnect from Backplane to Fiber

Fig. 9.1 4X-SX Optical Transceiver (Courtesy of Alvesta Inc.).

Prior to any data transmission, links are trained, width negotiated, and deskewed.

InfiniBand allows connection of two protocol-aware devices with different widths. During link negotiation the common denominator is established between two protocol-aware devices, possibly a single lane wide. In the unforeseen circumstance of a lane failing, the link will renegotiate to a new lower common denominator[1] as long as lane 0 functions. For the case of two 12-lane-wide ports, a failure on lane 4 or above results in renegotiation to a 4-lane-wide link.

9.2.1. INFINIBAND PACKET FORMAT

A protocol-aware device sequentially byte stripes the data over 1, 4, or 12 lanes, with start of data packet always on lane 0. Each port operates from a clock with ± 100 PPM allowed transmission of 5 Kbyte of uninterruptible data. IB packets, including header, may be as long as 4608 bytes long, but the actual data portion is only 4096 bytes (4Kbyte) long.

InfiniBand link uses 8B/10B control characters for packet management [9]. Several key symbols are as listed in Table 9.2.

---

[1] When a link is established, only the discrete common denominators 1, 4, and 12 are supported.

Table 9.2 **Link Control Symbol**

COM	K28.5, Comma, PLL alignment symbol
SDP	K27.7, Start of Data Packet
SLP	K28.2, Start of Link Packet Delimiter
EGP	K29.7, END of Good Packet Delimiter
PAD	K23.7, Packet Padding Symbol
SKP	K28.0, Skip Symbol
EBP	K30.7, End of Bad Packet
Idle	Pseudo-Random Fill Pattern

The idle pattern is transmitted on all lanes as the fill pattern to reduce EMI. The pattern is a pseudo-random data generated by an 11th order LFSR $X^{11} + X^9 + 1$.

Typical packet format for a 4X link is shown in Fig. 9.2. All data transmission starts by start of data packets "SDP" and ends with "EGP" or "EBP". The Comma are transmitted for PLL byte alignment. A protocol-aware transmitter sends at least 3 rows of "SKP" to allow clock elasticity adjustment by a maximum of two retimes in the path. At the input of the protocol-aware device a transmission may have at least 1 row or at most 5 rows of SKP.

InfiniBand link layer stripes the data into a SerDes with 1, 4, or 12 channels for delivery over the fiber or copper media. Figure 9.3 shows the data stripping operation for a 4X link.

9.2.2. IB ELECTRICAL INTERFACE

IB electrical interconnect provides typical transmission over 20″ of FR4 Printed Circuit Board (PCB) trace including two connectors or over 17 m of Twin-ax copper cable. The signaling is based on 100 ohms differential impedance with common mode of 0.75 ± 0.25 V. The driver and receiver provide differential source and load terminations of 100 Ω. The minimum differential peak-to-peak amplitude is 1000 mV, receiver sensitivity is 175 mV, which provides 15 dB of dynamic range. To allow transmission over 20″ of PCB traces the pre-emphasis driver is specified [10].

IB electrical specification was designed to be system friendly by relaxing jitter specification to allow ASIC cell implementation. To allow ASIC

9. InfiniBand—The Interconnect from Backplane to Fiber

Lane	0	1	2	3
Time	SDP			
				EGP
	COM	COM	COM	COM
	SKP	SKP	SKP	SKP
	SKP	SKP	SKP	SKP
	SKP	SKP	SKP	SKP
	SLP			
				EGP
	SDP			
				EGP
	Idle	Idle	Idle	Idle
	Data	Data	Data	Data

Fig. 9.2 Shows structure of data format as it gets striped across 4 lanes.

implementation, IB and XAUI total transmitter jitters is specified at 0.35 (UI) instead of the typical 0.25 (UI). An InfiniBand Retimer or Repeater, sometimes referred to "IB Signal Conditioner" [9], often is required prior to direct connection of an IB electrical signal to an optical transceiver to meet optical jitter specification, due to the relaxed electrical jitter parameters.

A signal conditioner may be implemented with the use of CDR retiming or with additional FIFO to allow regeneration from a new reference. The Signal Conditioner may be placed on the board or integrated with the optoelectronic components into a solderable or pluggable assembly for

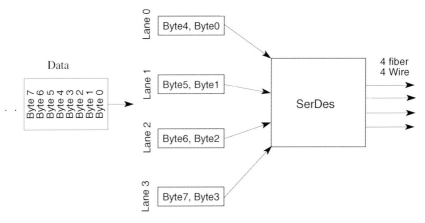

Fig. 9.3 Data Format Striped Across 4x Link.

attachment to an IB board. For complete IB electrical specification please see "IB High Speed Electrical" [10].

9.2.3. FIBER PLANT TECHNOLOGY SOLUTIONS

IB provides fiber plant options for 1X, 4X, and 12X wide links supporting broad application base from medium to high performance as listed in Table 9.3. Detailed specifications for three options are provided in section 1.5.

9.3. Optical Signal and Jitter Methodology

This section defines the characteristics of IB compliant optical signals and the measurement methodology. IB Optical transceiver is composed of an optical transmitter, an optical receiver, and an IB compliant Retimer. The purpose of the IB Retimer is to reduce relaxed IB electrical jitter levels to allow transmission over fiber optic links. To allow scalable architecture, IB electrical signal is designed to drive through 20″ of FR4 PCB trace with 2 connectors, but at the edge a Retimer must regenerate the signal prior to transmission over fiber.

A typical IB compliant link is shown in Fig. 9.4. IB electrical signal are regenerated prior to transmission into the optical transmitter at TP1. TP2 is the optical transmitter output at the optical receptacle. TP3 is IB compliant point after designated maximum fiber transmission length as specified in Table 9.3. TP4 is the output of the optical receiver prior to regeneration and

Table 9.3 **Fiber Optic Plant Options**

Link	Very short reach (VSR)	Longer reach
1X Wide		
Designation	IB-1X-SX	IB-1X-LX
Wavelength	850nm	1300nm
Connector	dual-LC[1]	dual-LC[1]
Worst-case operating range	2m-250m using 50/125μm 500MHz.km fiber	2m-10km with single mode fiber
	2m-125m using 62.5/125μm 200MHz.km fiber	
4X Wide		
Designation	IB-4X-SX	See Note 2
Wavelength	850nm	
Connector	single MPO	
Worst-case operating range	2m-125m using 50/125μm 500MHz.km fiber	
	2m-75m using 62.5/125μm 200MHz.km fiber	
12X Wide		
Designation	IB-12X-SX	
Wavelength	850nm	See Note 2
Connector	dual MPO	
Worst-case operating range	2m-125m using 50/125μm 500MHz.km fiber	
	2m-75m using 62.5/125μm 200MHz.km fiber	

[1] LC is the registered trademark of Lucent Technology.
[2] 4X wide and 12X wide LX may be specified in future versions of the specification.

transmission as an IB complaint electrical signal levels. It is foreseeable in a self-contained application with tighter electrical specification to interface directly to the optical transmitter and receiver without the use of Retimer. IB optical links are defined such that each lane has a BER $\leq 10^{-12}$ over lifetime, temperature, and operating range when driven through cable plant as specified.

The Gigabit Ethernet optical link model developed by IEEE 802.3z w estimates the link performance [6]. However, transmitter power and extinction ratio were replaced by Optical Modulation Amplitudes (OMA) [5] in the link model.

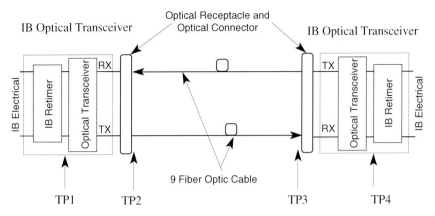

Fig. 9.4 IB Optical Link and the compliance point.

9.3.1. OPTICAL SIGNAL POLARITY AND QUIESCENT CONDITION

An electrical logic Zero level at the input of the optical transmitter generates a low level of optical power on the fiber and similarly a logic One generates a high level of optical power. If an electrical input to an Optical Transmitter is quiescent, then the optical power of that lane is not modulated. IB allows during quiescent period to leave transmitter DC optical power at average optical power.

9.3.2. OPTICAL TRANSMITTER MASK COMPLIANCE

The optical transmitter pulse shape characteristics are specified in the form of a compliance mask on the eye diagram of Fig. 9.5. The eye mask amplitude is normalized such that an amplitude of 0.0 represents logic ZERO and an amplitude of 1.0 represents logic ONE. The eye mask normalized time scale is in Unit Interval (UI), where 1UI is 400 ps at 2.5 Gb/s. An IB compliant transmitter must meet the eye mask as defined by the reference O/E converter.

This transmitter compliance mask is used to verify the overall response of the optical transmitter for rise time, fall time, pulse overshoot, pulse undershoot, and ringing. Compliance with the optical mask is a good indicator that deterministic effects are within generally acceptable limits, but it does not guarantee compliance with IB jitter specifications.

For uniform eye mask measurements, the optical transmitter signal is measured using an O/E converter with an equivalent fourth-order

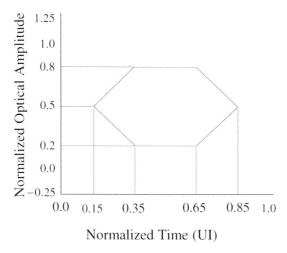

Fig. 9.5 Optical Transmitter Compliance Mask.

Bessel-Thompson response given by:

$$Hp = \frac{105}{105 + 105y + 45y^2 + 10y^3 + 4^4}$$

where

$$y = 2.114p, \qquad p = \frac{J\omega}{\omega_r}, \qquad \omega_r = 2\pi f_r, \qquad f_r = 1.875 MHz$$

The O/E converter filter response is based on ITU-T G.957 definition, which provides a physical implementation. The specified O/E converter is only intended to provide uniform measurement and does not represent the noise response of an IB Optical Receiver.

The reference O/E converter has a 3 dB frequency response of 1.88 GHz with equivalent response of the fourth order Bessel-Thompson. The reference O/E converter response must meet the parameter specified in Table 9.4, with tolerance not exceeding the specifications in Table 9.5.

9.3.3. RISE/FALL TIMES MEASUREMENT

Optical rise and fall times are specified based on unfiltered O/E converter waveforms. Some lasers have ringing or overshoot, which can reduce the accuracy of 20%–80% rise and fall time measurements. Therefore, a fourth-order Bessel-Thompson filter defined in the previous section is a convenient

Table 9.4 Equivalent Response of Reference O/E Converter

f/f_0	f/fr	Attenuation (dB)	Distortion (UI)
0.15	0.2	0.1	0
0.3	0.4	0.4	0
0.45	0.6	1.0	0
0.6	0.8	1.9	0.002
0.75	1.0	3.0	0.008
0.9	1.2	4.5	0.025
1.0	1.33	5.7	0.044
1.05	1.4	6.4	0.055
1.2	1.6	8.5	0.10
1.35	1.8	10.9	0.14
1.5	2.0	13.4	0.19
2.0	2.67	21.5	0.30

Table 9.5 Attenuation Tolerance of Reference O/E Converter

Reference frequency f/fr	Attenuation tolerance $\Delta a (dB)$
0.1 - 1.00	-0.0 to +0.5
1.00 - 2.00	+0.5 to +3.0

Note. Intermediate values of Δa **shall** be linearly interpolated on a logarithmic frequency scale.

filter for measurement of the rise and fall time. Because the limited response of the 4th-order Bessel-Thompson filter will adversely impact the measured rise and fall times, the following equation should be used to correct for the filter response:

$$T_{rise/fall} = \sqrt{(T_{rise/fall}\ measured)^2 - (T_{rise/fall}\ filter)^2}$$

For the purpose of rise and fall time measurement, 3 dB bandwidth of the fourth order Bessel-Thomson filter may be different from the reference O/E converter with 1.875 GHz bandwidth defined for eye mask compliance.

9. InfiniBand—The Interconnect from Backplane to Fiber

IB compliance definition for optical rise time and fall time is RMS of rise and fall times. The IEEE 802.3z Gigabit Ethernet optical link model analysis uses the larger of rise time and fall time, which provides an overly pessimistic analysis [11]. Therefore, IB optical specifications are defined using $T_{r/f}(RMS)$, which is the RMS mean of rise time and fall time as defined below:

$$T_{rise/fall}(RMS) = \sqrt{\frac{T_{rise}^2 + T_{fall}^2}{2}}$$

9.3.4. OPTICAL MODULATION AMPLITUDE

Optical Modulation Amplitude (OMA) is defined as the absolute difference between the optical power of a logic ONE level and the optical power of a logic ZERO level. *OMA* is related to Extinction Ratio (*ER*) measured in (dB) and Average Optical Power (P_{ave} measured in dBm) by the equation:

$$OMA = 2 \times 10^{P_{ave}/10} x \left[\frac{1 - 10^{-Er/10}}{1 + 10^{-Er/10}} \right]$$

OMA specification generally improves the optical transmitter yield as it allows a higher power laser with lower Extinction Ratio to pass.

9.3.5. OPTICAL JITTER SPECIFICATION

The IB jitter specification is based on the same methodology as the Fibre Channel [12] and the IEEE 802.3z Gigabit Ethernet standards [6]. Figure 9.4 shows jitter compliance test points TP1, TP2, TP3 and TP4 for an IB-1X optical link. Similar compliance points exist for the 4X and 12X link.

TP1 is the intermediate electrical signal at input of the optical transmitter with more strength than IB electrical specifications. TP2 is an IB compliant point located at the optical receptacle. Any measurement of optical signal at TP2 is recommended to be done with a 2 meter fiber jumper wrapped 10 turns around a 25.4 mm diameter mandrel to reduce high order mode. TP3 is located at the output of the optical fiber over the maximum specified length and adjacent to the Optical Receiver. TP4 is the output of the optical receiver electrical signal. TP1 and TP4 are not IB compliance point and physically may not be accessible.

Typical jitter values at TP1 and TP4 are listed in Table 9.6 for reference. The total jitter of an optical component is measured at BER of 10^{-12}

Table 9.6 Maximum Jitter of an Optical Link

InfiniBand link	Compliance point	DJ (UI)	DJ (ps)	TJ (UI)	TJ (ps)
1X-SX, 1X-LX	TP1	0.10	40	0.25	100
	TP2	0.23	92	0.46	184
	TP3	0.32	128	0.54	212
	TP4	0.40	160	0.70	280
4X-SX, 12X-SX	TP1	0.10	40	0.25	100
	TP2	0.25	100	0.48	192
	TP3	0.30	120	0.53	212
	TP4	0.40	160	0.70	280

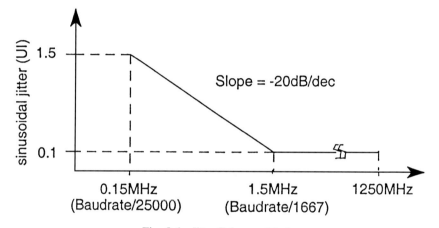

Fig. 9.6 Jitter Tolerance Mask.

with K28.5+, K28.5- test pattern. An IB optical port must provide BER of $\leq 10^{-12}$ under worst case data patterns and operating conditions. Additional test patterns with detail methodologies are defined in FC-MJS [12].

The total jitter specified at TP4 in Table 9.6 does not include any sinusoidal jitter (SJ) component. For all link types, the TP4 must tolerate total jitter of 0.75 UI^2 with addition of 0.10 UI of sinusoidal jitter (SJ) over a swept frequency from 1.5 MHz to 1250 MHz.

A noticeable difference between 1X link and 4X /12X links is the more relaxed jitter parameters specified for 4X and 12X links. More relaxed

---

[2] With additional 0.1 (UI) of SJ the DJ must be reduced by 0.05 (UI) to meet the TJ of 0.75 at TP4.

jitter specification increases manufacturing yield of array transmitters and receivers, but also reduces the length of 4X and 12X links by half. Due to practical implementation of pulling ribbon fiber limiting the length to 125 m for 4X and 12X link was not seen as a penalty.

9.3.6. JITTER METHODOLOGY

IB jitter methodology is based on Fibre Channel MJS [12]. MJS breaks down the jitter components into random jitter (RJ) and deterministic jitter (DJ). Random components of successive physical components are added in quadrature, where the DJs are added linearly. Deterministic jitter is composed of Duty Cycle Distortion (DCD), Sinusoidal Jitter (SJ), and Inter-symbol Interference (ISI). Total jitter is defined as

$$TJ = DJ + RJ = DJ + 14\sigma \quad \text{for BER of } 10^{-12}. \quad (9.1)$$

The optical jitter compliance points are TP2 and TP3, but values for TP1 and TP4 are provided for reference for the purpose of construction of SerDes, laser transmitter, and optical receiver. To determine the amount of deterministic jitter and random jitter that an Optical Transmitter under test adds from TP1 to TP2, the following analysis applies:

$$DJ\ (Transmitter) = DJ_2 - DJ_1$$
$$RJ(Transmitter) = \sqrt{(TJ_2 - DJ_2)^2 - (TJ_1 - DJ_1)^2}$$

where

$$DJ1 = DJ \text{ at TP1}$$
$$DJ2 = DJ \text{ at TP2}$$
$$TJ1 = TJ \text{ at TP1}$$
$$TJ2 = TJ \text{ at TP2}$$

A similar analysis can be made to TP3 and TP4.

9.3.7. BIT-TO-BIT SKEW

IB link provides up to 24 ns of skew at the receive input of the protocol-aware device between any two receive lanes. To allow interoperable as well as pluggable interfaces, each component has a generous skew allocation. Skew is defined as maximum differential bit-to-bit skew between any two

Table 9.7 **Maximum Optical Bit-to-Bit Skew Values**

Skew parameter	Maximum value	Note
Optical Cable Assembly	3.0ns	1
Transmitter	500ps	2
Receiver	500ps	3

[1] An optical cable assembly includes optical cable, optical connectors, and adaptors.
[2] Between any two physical lanes within a transmitter.
[3] Between any two physical lanes within a receiver.

lanes. All IB optical ports must limit maximum skew to values defined in Table 9.7.

9.4. Optical Specifications

This section defines detail specifications for 1X (1 Lane), 4X (4 Lanes), and 12X (12 Lanes) wide fiber optics transceivers. InfiniBand optical transceivers may be permanently attached on the circuit board or may be fabricated as part of the pluggables through the chassis.

9.4.1. 1X FIBER OPTIC SPECIFICATIONS

IB 1X optical link carries a duplex data as shown in Fig. 9.7, the signals are generated using one laser and one photodetector. Two classes of 1X links are defined, a 1X-SX based on 850 nm VCSEL with multimode fiber for short reach and a 1X-LX based on 1310 nm FP laser with single-mode fiber for operation up to 10 Km. IB optical parameters compromised the maximum distance achievable in favor of lower cost and flexibility.

Two fiber types are specified for multimode 1X-SX:

  i) 1X-SX/50 link: 500MHz.km $50 \mu m/125 \mu m$ Multimode fiber Range Up to 250 m
  ii) 1X-SX/62 link: 200MHz.km $62.5 \mu m/125 \mu m$ Multimode fiber Range Up to 125 m

The intermediate reach link, 1X-LX, operates in the 1300-nm wavelength region using single-mode (SM) fiber. The optical parameters are optimized

9. InfiniBand—The Interconnect from Backplane to Fiber

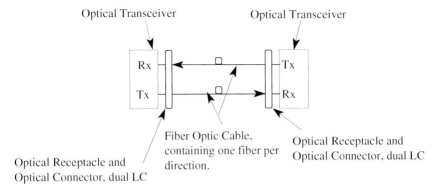

Fig. 9.7 1X-SX/LX Fiber Optic Link Configuration.

to allow a typical 1X-LX Optical Transceiver to operate with uncooled Fabry-Perot (FP) laser, Distributed Feedback (DFB) laser, or emerging 1310 nm VCSEL[13].

9.4.1.1. 1X-SX Optical Parameters

Optical parameters for 1X-SX link are listed in Table 9.8 [14]. The baud rate including 8B/10B coding is 2500 Mb/s. Optical passive loss accounts for fiber attenuation, connector, and splice loss. Total system penalty [11] with the addition of passive loss results in Link Power budget. Fiber plant and connectors specifications are described in section 9.5.

IB transmitter and receiver are specified with more relaxed jitter specification and shorter reach, but longer distance is possible with transmitter or receiver performing better than worst case. When IB optical transmitters are squelched for input electrical levels below V_{RSD} (Signal Threshold), as defined by IB electrical [10], then the output optical signal shall be less than 2 mW.

An Optical Receiver for a 1X-SX link must meet parameters specified at TP3 in Table 9.8. Conformance testing for a stressed receiver at TP3 is based on FC-PI methods of Annex A [5]. If the average received optical power on any Lane is less than -30 dBm then the corresponding high-speed electrical received output is squelched to less than 85 mV [10]. The electrical outputs of every lane with an optical power greater than -20 dBm is not squelched to allow testing for sensitivity.

Table 9.8 **Link Parameters for 1X-SX**

Link specifications	Unit	IB-1X-SX/50	IB-1X-SX/62.5	Notes
Nominal Signaling Rate	MBaud	2500	2500	
Rate tolerance	ppm	±100	±100	
Optical Passive Loss (Max)	dB	2.44	2	
Optical System Penalty (Max)	dB	3.56	4	
Total link power budget (Max)	dB	6	6	
Operating Distance (Max)	m	250	125	
Fiber Mode-field (Core) Diameter	μm	50	62.5	
Transmitter Specifications-TP2				
Type		Laser	Laser	
Center Wavelength (Max)	nm	860	860	
Center Wavelength (Min)	nm	830	830	
Average Launch Power (Max)	dBm	−4	−4	1
Optical Modulation Amplitude (Min)	mW	0.196	0.196	2
Rise/Fall Time RMS 20%–80% (Max)	ps	150	150	3
RIN12 (OMA) (Max)	dB/Hz	−117	−117	4
Receiver Specification-TP3				
Average Received Power (Max)	dBm	−1.5	−1.5	
Optical Modulation Amplitude (Min)	mW	0.05	0.05	2
Return Loss of the Receiver (Min)	dB	12	12	
Stressed Receiver Sensitivity OMA (Min)	mW	0.102	0.102	2,5

(*Continued*)

Table 9.8 continued

Link specifications	Unit	IB-1X-SX/50	IB-1X-SX/62.5	Notes
Stressed Receiver ISI Penalty (Max)	dB	2	2	5
Stressed Receiver DCD component of DJ at TP2 (Max)	ps	40	40	
Receiver Electrical 3 dB upper cutoff Frequency (Max)	GHz	2.8	2.8	6
Receiver Electrical 10 dB upper cutoff Frequency (Max)	GHz	6	6	6

[1] Must also comply to lesser of Class 1 laser safety limit CDRH or EN60825.
[2] Optical Modulation Amplitude are Peak-Peak.
[3] Rise and fall time are unfiltered response calculated as RMS of rise and fall time.
[4] Annex A4 [5]
[5] Annex A6 [5]
[6] Annex A7 [5]

9.4.1.2. 1X-LX Optical Parameters

The 1X-LX Optical link specification for transmitter, link, and the receiver is given below [14]. The link budget parameters for 1X-LX single-mode fiber optic links operating at 2.5Gbaud/s including 8B/10B are listed in Table 9.9. Fiber plant and connectors specifications are described in section 9.5.

IB optical transmitters are squelched for input electrical levels below V_{RSD} (Signal Threshold) as defined in by IB electrical [10], then the output optical modulation amplitude should not exceed 2.0 μW.

If the average received optical power on any Lane is less than -30 dBm then the corresponding high-speed electrical signalling output shall be squelched to less than V_{RSD} 85 mV (Signal Threshold) [10]. The electrical outputs of every lane with an optical power greater than -20dBm shall not be squelched to allow testing for receiver sensitivity.

To allow flexible laser transmitter design the RMS spectral width of 1X-LX was defined as a function of laser center wavelength. A 1X-lLX compliant transmitter RMS spectral width must lie below the Fig. 9.8 curve, calculated using the worst-case fiber at a given center wavelength for all operating conditions. For operating wavelengths less than 1312 nm fiber zero dispersion was assumed to be 1324 nm, and for wavelengths greater

Table 9.9 **Link Parameters for 1X-LX**

Link specifications	Unit	IB-1X-LX	Notes
Nominal Signaling Rate	MBaud	2500	
Rate tolerance	ppm	±100	
Optical Passive Loss (Max)	dB	6.64	
Optical System Penalty (Max)	dB	2.36	
Total link power budget (Max)	dB	9	
Operating Distance (Max)	m	10,000	
Fiber Mode-field (Core) Diameter	μm	9	
Transmitter Specifications-TP2			
Type		Laser	
Center Wavelength (Max)	nm	1270	
Center Wavelength (Min)	nm	1360	
Average Launch Power (Max)	dBm	−3	1
Optical Modulation Amplitude (Min)	mW	0.186	2,3
Rise/Fall Time RMS 20%–80% (Max)	ps	150	4
RIN12 (OMA) (Max)	dB/Hz	−120	5
Receiver Specification-TP3			
Average Received Power (Max)	dBm	−1.5	
Optical Modulation Amplitude (Min)	mW	0.0234	2
Return Loss of the Receiver (Min)	dB	20	
Stressed Receiver Sensitivity OMA (Min)	mW	0.0365	2
Stressed Receiver ISI Penalty (Max)	dB	0.58	6
Stressed Receiver DCD component of DJ at TP2 (Max)	ps	40	
Receiver Electrical 3 dB upper cutoff Frequency (Max)	GHz	2.8	7
Receiver Electrical 10 dB upper cutoff Frequency (Max)	GHz	6	7

[1] Must also comply to lesser of Class 1 laser safety limit CDRH or EN60825.
[2] Optical Modulation Amplitude are Peak-Peak.
[3] Figure 8 shows trade off between OMA and spectra width.
[4] Rise and fall time are unfiltered response calculated as RMS of rise and fall times.
[5] Annex A4 [5]
[6] Annex A6 [5]
[7] Annex A7 [5]

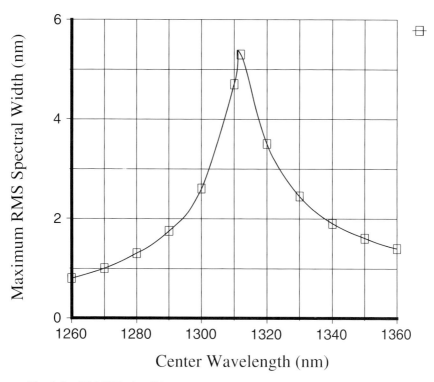

Fig. 9.8 1X-LX Trade-off between RMS spectral width with center wavelength.

than 1312 the zero-dispersion was assumed to be 1300 nm to allow for worst case link analysis.

9.4.2. 4X SYSTEM OVERVIEW

A 4X-SX optical link carries duplex data composed of 4 transmit and 4 receive signals as shown in Fig. 9.9. Connector is based on a single 12 position MPO[3], the first 4 positions carry the transmit signal, followed by 4 unused positions, and the last 4 positions are for the receive signal. To simplify manufacturing and commonality between 4X and 12X link, ribbon fiber with 12 fiber may be used but the middle 4 fibers do not carry IB signals. To allow interchangeability between 4X and 12X ribbon cable, the connector keys must face up on both sides of the cable.

[3] In the United States MPO is known as MTP, which is a registered trademark of USCONEC Inc.

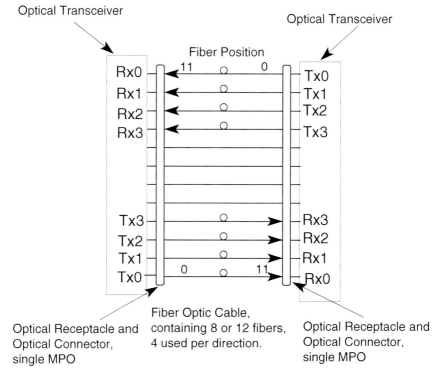

Fig. 9.9 4X-SX Fiber Optic Link Configuration.

The optical parameters and link distance were only specified up to 125m with 50/125 μm fiber and 75m with 62/125 μm fiber, selected based on the application need, cost, and ease of development, instead of maximum link distance possible. Two fiber types are specified for 4X-SX:

i) 4X-SX 50/ 500MHz.km 50μm/ 125μm MM fiber range up to 125 m
ii) 4X-SX 62/ 200MHz.km 62.5μm/ 125μm MM fiber range up to 75 m.

9.4.2.1. 4X-SX Optical Parameter

Optical parameters for 4X-SX link are listed in Table 9.10 [14]. The data rate including 8B/10B coding is 2.5 Gb/s. Fiber plant and connector specifications are described in section 9.5.

IB transmitter and receiver are specified with more relaxed jitter specification and shorter reach, but longer distance is possible with transmitter

9. InfiniBand—The Interconnect from Backplane to Fiber

Table 9.10 **Link Parameters for 4X-SX**

Link specifications	Unit	IB-4X-SX/50	IB-4X-SX/62.5	Notes
Nominal Signaling Rate	MBaud	2500	2500	
Rate tolerance	ppm	±100	±100	
Optical Passive Loss (Max)	dB	1.9	1.8	
Optical System Penalty (Max)	dB	2.9	3	
Total link power budget (Max)	dB	4.8	4.8	
Operating Distance (Max)	m	125	75	
Fiber Mode-field (Core) Diameter	μm	50	62.5	
Transmitter Specifications-TP2				
Type		Laser	Laser	
Center Wavelength (Max)	nm	860	860	
Center Wavelength (Min)	nm	830	830	
Average Launch Power (Max)	dBm	−4	−4	1
Optical Modulation Amplitude (Min)	mW	0.196	0.196	2
Rise/Fall Time RMS 20%–80% (Max)	ps	150	150	3
RIN12 (OMA) (Max)	dB/Hz	−117	−117	4
Receiver Specification-TP3				
Average Received Power (Max)	dBm	−1.5	−1.5	
Optical Modulation Amplitude (Min)	mW	0.05	0.05	2
Return Loss of the Receiver (Min)	dB	12	12	
Stressed Receiver Sensitivity OMA (Min)	mW	0.085	0.085	2,5
Stressed Receiver ISI Penalty (Max)	dB	0.9	0.9	5
Stressed Receiver DCD component of DJ at TP2 (Max)	ps	60	60	
Receiver Electrical 3 dB upper cutoff Frequency (Max)	GHz	2.8	2.8	6
Receiver Electrical 10 dB upper cutoff Frequency (Max)	GHz	6	6	6

[1] Must also comply to lesser of Class 1 laser safety limit CDRH or EN60825.
[2] Optical Modulation Amplitude are Peak-Peak.
[3] Rise and fall time are unfiltered response calculated as RMS of rise and fall time.
[4] Annex A4 [5]
[5] Annex A6 [5]
[6] Annex A7 [5]

or receiver performing better than worst case. Optical transmitter specifications for the 4X-SX link specified at TP2 are listed in Table 9.5. Optical modulation amplitude must not exceed 5.0 μW if any high-speed electrical input signals are less than the V_{RSD} (85 mV, Signal Threshold) [10].

Optical receiver parameters for the 4X-SX link specified at TP3 are listed in Table 9.5 [16]. If the average received optical power on any Lane is less than −26 dBm then the corresponding high-speed electrical signalling output is squelched to less than V_{RSD} (85 mV, Signal Threshold) [10]. The electrical outputs of every lane with an optical power greater than −18 dBm is not squelched to allow testing for sensitivity limits.

9.4.3. 12X-SX SYSTEM OVERVIEW

A 12X-SX optical link carries duplex 12X data composed of 12 transmit and 12 receive signals as shown in Fig. 9.10. Connector is based on single 12 position dual MPO[4]. The first connector carries 12 transmit signals and the second connector carries 12 receive signals. To allow commonality between 4X and 12X link the connector keys at both ends of a fiber optic cable face up.

The optical parameters and link distance were only specified up to 125m with 50/125 μm fiber and 75m with 62/125 μm fiber, range selected based on the application need, cost, and ease of development, instead of maximum link distance possible. Two fiber types are specified for 12X-SX:

  i) 4X-SX 50/ 500MHz.km 50μm/ 125μm MM fiber range of 2–125 m
  ii) 4X-SX 62/ 200MHz.km 62.5μm/ 125μm MM fiber range of 2–75 m.

9.4.3.1. 12X-SX Optical Parameters

Optical parameters for 12X-SX links are listed in Table 9.11 [14]. The data rate on each fiber, including 8B/10B coding, is 2.5 Gb/s. Fiber plant and connector specifications are described in section 9.5.

Optical transmitter specifications for the 12X-SX link parameters specified at TP2 are listed in Table 9.11. If any high-speed electrical input signals are less than the V_{RSD} (85 mV, Signal Threshold) [10] then the optical modulation amplitude must not exceed 5.0 μW.

---

[4] In the United States MPO is known as MTP, which is a registered trademark of USCONEC Inc.

9. InfiniBand—The Interconnect from Backplane to Fiber

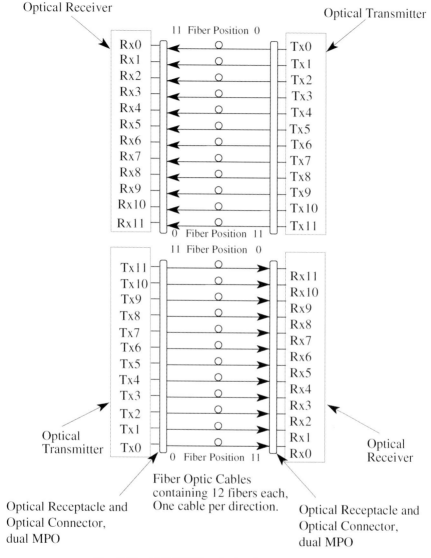

Fig. 9.10 12X-SX Fiber Optic Link Configuration.

Optical receiver parameters for the 12X-SX link specified at TP3 are listed in Table 9.11. If the average received optical power on any lane drops below −26 dBm then the corresponding high-speed electrical output is squelched to less than V_{RSD} (85 mV, Signal Threshold) [10]. The optical

Table 9.11 **Link Parameters for 12X-SX**

Link specifications	Unit	IB-4X-SX/50	IB-4X-SX/62.5	Notes
Nominal Signaling Rate	MBaud	2500	2500	
Rate tolerance	ppm	±100	±100	
Optical Passive Loss (Max)	dB	1.9	1.8	
Optical System Penalty (Max)	dB	2.9	3	
Total link power budget (Max)	dB	4.8	4.8	
Operating Distance (Max)	m	125	75	
Fiber Mode-field (Core) Diameter	μm	50	62.5	
Transmitter Specifications-TP2				
Type		Laser	Laser	
Center Wavelength (Max)	nm	860	860	
Center Wavelength (Min)	nm	830	830	
Average Launch Power (Max)	dBm	−4	−4	1
Optical Modulation Amplitude (Min)	mW	0.196	0.196	2
Rise/Fall Time RMS 20%–80% (Max)	ps	150	150	3
RIN12 (OMA) (Max)	dB/Hz	−117	−117	4
Receiver Specification-TP3				
Average Received Power (Max)	dBm	−1.5	−1.5	
Optical Modulation Amplitude (Min)	mW	0.05	0.05	2
Return Loss of the Receiver (Min)	dB	12	12	
Stressed Receiver Sensitivity OMA (Min)	mW	0.085	0.085	2,5
Stressed Receiver ISI Penalty (Max)	dB	0.9	0.9	5
Stressed Receiver DCD component of DJ at TP2 (Max)	ps	60	60	
Receiver Electrical 3 dB upper cutoff Frequency (Max)	GHz	2.8	2.8	6
Receiver Electrical 10 dB upper cutoff Frequency (Max)	GHz	6	6	6

[1] Must also comply to lesser of Class 1 laser safety limit CDRH or EN60825.
[2] Optical Modulation Amplitude are Peak-Peak.
[3] Rise and fall time are unfiltered response calculated as RMS of rise and fall time.
[4] Annex A4 [5]
[5] Annex A6 [5]
[6] Annex A7 [5]

9. InfiniBand—The Interconnect from Backplane to Fiber

receiver electrical output, if any lane has an optical power greater than −18 dBm, should not be squelched, to allow sensitivity testing.

9.5. Optical Receptacle and Connector

This section defines the optical receptacle and connector for the 1X link based on LC (LC® Lucent Technologies) and for the 4X/12X link based on MPO (also known as MTP in the U.S. (MTP® USCONEC)).

9.5.1. 1X CONNECTOR-LC

InfiniBand 1X port is defined based on Duplex LC optical connector conforming to ANSI/TIA/EIA 604-10 (FOCIS 10), Fiber Optic Connector Intermateability Standard, Type "LC". In addition it will comply with Fibre Channel Physical Interface (FC-PI), Revision 11.0. Figure 9.11 shows a duplex LC connector.

InfiniBand defines orientation for the 1X-SX and 1X-LX fiber optic transceivers to follow the convention of Fig. 9.12. When looking into the

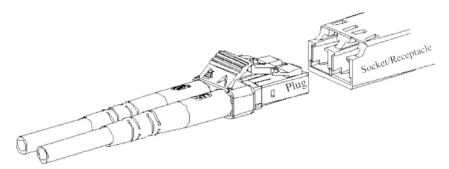

Fig. 9.11 Duplex LC Plug and Socket/Receptacle (Courtesy of Lucent Technology).

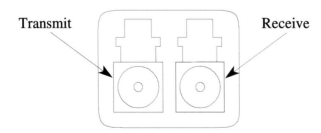

Fig. 9.12 1X-SX and 1X-LX Optical Receptacle Orientation Looking Into the Optical Port.

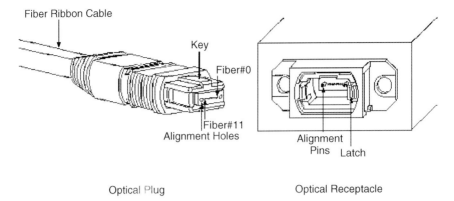

Fig. 9.13 MTP Optical Plug and Receptacle (Courtesy of USCONEC).

face plate, assuming keys are up, the transmitter is on the left and the receiver is on the right.

To allow ease of distiction between single-mode and multimode, beige color designates multimode and blue designates single-mode.

9.5.2. 4X-SX CONNECTOR-SINGLE MPO

A single 12 position MPO connector provides bi-directional full duplex optical connection with single mating action. The 4X-SX Optical Connector on each end of the Fiber Optic Cable consists of a female MPO plug conforming to IEC 1754-7-4 "Push/Pull MPO Female Plug Connector Interface". The female MPO connector is similar to male MPO, except it has no alignment pins. Figure 9.13 shows outline drawing of a MTP (In the U.S. MPO connector is known as MTP® USCONEC) connector plug and receptacle with push-pull coupling mechanism. Each MPO connector hold a rectangular 6.4 mm by 2.5 mm MT ferrule. The MT ferrule in a male MPO connector holds two precision alignment pins with 0.7 mm diameter.

The 4X-SX Optical Transceiver interface is a male MPO receptacle. The male MPO receptacle having two alignment pins complies to IEC 1754-7-5 and to IEC 1754-7-3; in addition it conforms to Push/Pull MPO Adapter Interface standard.

4X-SX Optical Transceivers follow the Transmit/Receive convention showed in Fig. 9.14. When looking into the optical receptacle with key

9. InfiniBand—The Interconnect from Backplane to Fiber 347

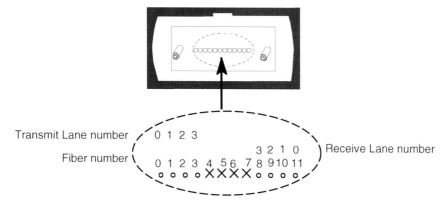

Fig. 9.14 Optical Receptacle Orientation Looking Into the Transceiver.

up, fibers are numbered from left to right as 0 through 11. The first 4 fiber positions from the left (fibers 0, 1, 2, 3) are the transmit optical lanes and the last 4 fibers from the left (fibers 8, 9, 10, 11) are the receive lanes. Fibers 0, 1, 2, 3 carry transmit Lanes 0, 1, 2, 3 signals respectively. Fibers 8, 9, 10, 11 carry receive Lanes 3, 2, 1, 0 signals respectively. The middle 4 fibers (fiber 4, 5, 6, 7) may be physically present but do not carry IB signals.

9.5.3. 12X-SX CONNECTOR - DUAL MPO

The 12X-SX link is defined based on dual MPO creating a full duplex bi-directional optical connection. The 12X-SX link optical connectors on each end of the cable consist of a two female MPO plug, conforming to IEC 1754-7-4 "Push/Pull Type MPO Female Plug Connector Interface." Each MPO connector plug holds a rectangular 12 position MT ferrule.

The 12X-SX optical port holds two male MPO receptacles, each with two fixed alignment pins conforming to IEC 1754-7-5 and IEC 1754-7-3 "Push/Pull MPO Adapter Interface" as shown in Fig. 9.15. The center-center spacing of the two MPO connectors is 20.0 mm ±0.5 mm to allow development of the duplex plug.

12X-SX optical transceivers follow the Transmit/Receive convention shown in Fig. 9.15. When looking into the optical receptacle with keys up, left receptacle carries the transmit signals and right receptacle carries the receive signals. The transmit receptacle fibers 0, 1, 2, 3, 4, 5, 6, 7, 8, 9, 10, 11 carry transmit Lanes 0, 1, 2, 3, 4, 5, 6, 7, 8, 9, 10, 11 respectively.

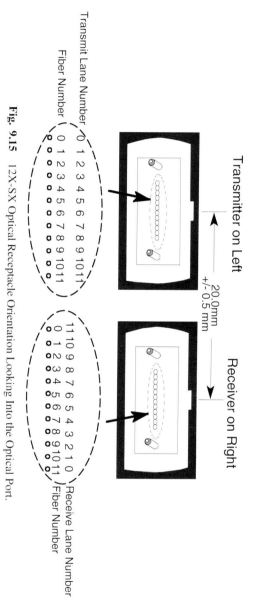

Fig. 9.15 12X-SX Optical Receptacle Orientation Looking Into the Optical Port.

The receive receptacle fibers 0, 1, 2, 3, 4, 5, 6, 7, 8, 9, 10, 11 carry receive Lanes 11, 10, 9, 8, 7, 6, 5, 4, 3, 2, 1, 0 respectively.

9.6. Fiber Optic Cable Plant Specifications

This section defines optical fiber cables specifications supporting 1X-SX, 1X-LX, 4X-SX, and 12X-SX varients.

9.6.1. OPTICAL FIBER SPECIFICATION

IB optical cables comply with the fiber cable specifications of Table 9.12 for the respective variant. Single Mode Fiber (SMF) conforms to TIA/EIA-492CAAA-98 "Dispersion-Unshifted Single-Mode Optical Fibers". Multimode Fiber (MMF) 50/125 μm conforms to TIA/EIA-492AAAB-98 "Detail Specification for 50-μm Core Diameter/125-μm Cladding Diameter Class Ia Graded-Index Multimode Optical Fibers" or IEC 60793-2 Type A1a. Multimode Fiber (MMF) 62.5/125 μm conforms to TIA/EIA-492AAAA-A-97 "Detail Specification for 62.5-μm Core Diameter/125-μm Cladding Diameter Class Ia Graded-Index Multimode Optical Fibers" or IEC 60793-2 Type A1b.

The MMF modal bandwidth specified in Table 9.12 is with overfilled launch condition measured in accordance with TIA2.2.1 method TIA/EIA/455-204-FOTP204. IB optical system penalties are calculated based on overfilled launch condition, which are gives lower fiber bandwidth. The specification under reasonable conditions is conservative; in practice better link performance is expected.

Optical passive loss for the cable plant must be verified by the methods of OFSTP-14A. Optical passive loss of a fiber optic cable is the sum of attenuation losses due to the fiber cable, connectors, and splices.

9.6.2. FIBER OPTIC CONNECTORS AND SPLICES

A fiber optic link may have one or more connectors and splices, provided the total passive loss conforms to the loss budget specified for each varient type. The loss of the fiber plant is verified by the methods of OFSTP-14A. The Optical Passive Loss of a Fiber Optic Cable is the sum of attenuation losses due to the fiber, connectors, and splices. A total loss budget of 1.5dB is assigned to optical connectors and splices.

Table 9.12 Optical Fiber Specifications

Parameters/Fiber core supported variant	SMF (9 μm) 1X-LX	MMF (50/125 μm) 1X-SX,4X-SX,12X-SX	MMF (62.5/125 μm) 1X-SX,4X-SX,12X-SX	Units
Conformance	TIA/EIA-492CAAA	TIA/EIA-492AAAB	TIA/EIA-492AAAA	N/A
Nominal Fiber Specification Wavelength	1310	850	850	nm
Fiber Cable Attenuation (Max)	0.5	3.5	3.5	dB/km
Conformance	TIA/EIA-492CAAA	TIA/EIA-492AAAB	TIA/EIA-492AAAA	N/A
Nominal Fiber Specification Wavelength	1310	850	850	nm
Modal Bandwidth with Overfilled Launch (Min)	not applicable	500	200	MHz·km
Zero Dispersion Wavelength λ_0	$1300 \leq \lambda_0 l \leq 1320$	$1295 \leq \lambda_0 \leq 1300$	$1320 \leq \lambda_0 l \leq 1365$	nm
Zero Dispersion Slope S0 (Max)	0.093	0.11 for $1300 \leq \lambda_0 l \leq 1320$ and $0.001(\lambda_0 - 1190)$ for $1295 \leq \lambda_0 \leq 1300$	0.11 for $1320 \leq \lambda_0 l \leq 1348$ and $0.001(1458 - \lambda_0)$ for $1348 \leq \lambda_0 \leq 1365$	ps/nm²·km

Fiber optic connectors and splices for all multimode variants (1X-SX, 4X-SX, 12X-SX) must meet a return loss of 20dB minimum as measured by the methods of FOTP-107 or equivalent. The single-mode 1X-LX variant must meet a return loss of 26dB minimum.

References

1. Ibel, M. Aug. 1997. "High-performance cluster computing using SCI." Hot Interconnect V.
2. HiPPi6400, Working Draft, NCITS T11.1, Project 1249-D, Rev 2.2.
3. Charlesworth, A. Aug. 1997. "Gigaplane-XB: Extending the ultra enterprise family." Hot Interconnect.
4. Momtaz, A. "Fully integrated SONET OC48 transceiver in standard CMOS." ISSC 2001, MP 5.2, pp. 76–77. In addition, private discussions.
5. Fibre Channel Physical Interface (FC-PI), NCITS Project 1235-D, Rev 11, Jan. 2001.
6. IEEE 802.3 Standard, "Carrier sense multiple access with collision detection (CSMA/CD) access method and physical layer specifications." 802.3z Gigabit Ethernet Section, 2000 Edition (ISO/IEC 8802-3:2000) IEEE Standard for Information Technology.
7. IEEE Draft 802.3ae/D2.3, "Media Acess Control (MAC) parameters, physical layer, and management parameters for 10 Gb/s operation." IEEE Standard for Information Technology, March 2001.
8. Fibre Channel 10 Gigabit (10GFC), NCITS Project 1413-D, Rev 1.1, May 2001.
9. "Link/Phy Interface." October 2000. Chapter 5 in *InfiniBand Architecture Specification*, Volume 2, Release 1.0.
10. "High Speed Electrical." October 2000. Chapter 6 in *InfiniBand Architecture Specification*, Volume 2, Release 1.0.
11. Cunnigham, D. June 1999. "Gigabit ethernet networking." Pearson Higher Education.
12. Fibre Channel Methodologies for Jitter Specification (MJS), NCITS Project 1230, Rev 10, June 1999.
13. Whitaker, T. July 2000. "Long Wavelength VCSELs Move Closer to Reality." *Compound Semiconductor* **6** (5): 65–67.
14. "Fiber Attachment." October 2000. Chapter 8 in *InfiniBand Architecture Specification*, Volume 2, Release 1.0.

Chapter 10 | New Devices for Optoelectronics: Smart Pixels

Barry L. Shoop
Andre H. Sayles
Photonics Research Center and Department of Electrical Engineering and Computer Science, United States Military Academy, West Point, New York 10996

Daniel M. Litynski
College of Engineering and Applied Sciences, Western Michigan University, Kalamazoo, Michigan 49008

Since the first edition of this chapter, the field of smart pixel technology has continued to make significant progress. This progress includes improvements to individual device performance, improvements to circuit performance, extensions to larger arrays, as well as applications to new areas. Over this period, a focus on vertical-cavity surface-emitting laser (VCSEL) technology has produced significant improvements in both individual device performance and array uniformity and performance. Furthermore, during this period, others outside of the smart pixel community have recognized the importance and potential of integrating electronic circuits with optoelectronic devices for a variety of application areas. The smart pixel community has also worked to expand to other non-traditional smart pixel applications including smart imaging systems, and applications based on digital signal processing. As a result of these two mutually supporting trends, the fields of Smart Photonics, Optoelectronic-VLSI, and Electronic-Enhanced Optics all have emerged, with support for progress in smart pixel technology.

The fundamental concepts of smart pixel technology presented in the first edition of this chapter remain valid and therefore, this revision will primarily reflect the current state-of-the-art in devices, applications, and demonstrations.

10.1. Historical Perspective

Optics has long held the promise of high-speed, high-throughput, parallel information processing and distribution. The focus of its early applications was on analog signal processing techniques such as the optical Fourier transform, matrix-matrix and matrix-vector processors, and correlators. During this period, optics was used almost exclusively for front-end, pre-processing of wide-bandwidth, high-speed analog signals, which were subsequently processed using digital electronic techniques. Although work continues on special-purpose analog applications, much of the recent optics work has focused on semiconductor devices for logic, switching, and interconnection applications.

Prior to 1980, optics research could generally be categorized as foundational with numerous basic device and proof-of-concept demonstrations that would potentially provide the framework for future optical system demonstrations. In the early 1980s, the focus of the community tended toward all-optical techniques. During this period, the community also began to collectively work on system-level demonstrations in an effort to gain acceptance and recognition in a predominantly electronics-based society. In 1984, B. K. Jenkins and colleagues at the University of Southern California demonstrated an optical master/slave flip flop using holograms and imaging techniques [1]. In 1985, D. Psaltis and co-workers at the California Institute of Technology demonstrated an optical implementation of an artificial neural network [2]. In the late 1980s, the first digital optical computer, demonstrated by A. Huang and co-workers at AT&T Bell Laboratories [3], showed the potential for an optical architecture to implement the functionality necessary for computing functions. However, this was far from a fully programmable optical computer. Before long, it was recognized that all-optical processing was synonymous with special-purpose processing and therefore was limited in its applications. By the mid-1980s, the term *photonics* had been coined, reflecting the growing ties between electronics and optics. Continued advances in materials science and semiconductor growth techniques led to novel device structures that could be engineered at the atomic level. These new quantum structures provided optical-to-electronic conversion at the quantum level, which improved device efficiency and provided the potential for direct integration with classical semiconductor circuits. Since then, the field has grown considerably, primarily as the result of a number of successful demonstrations of high-speed, high-throughput systems along with the commercialization of several other systems.

Recently, the area of optical information processing has progressed to yet another level. Systems that incorporate smart pixel-based processing using hybrid technologies [4] are being employed to take advantage of the parallelism and throughput of optics and the general-purpose processing ability of very large-scale integration (VLSI) techniques. From about 1988 to the present, AT&T's Bell Laboratories has demonstrated several generations of optical switching fabrics. $System_6$ demonstrated a switching fabric capable of supporting 4096 optical inputs and 256 optical outputs, operating at a maximum rate of 450 Mbps [5]. AT&T's most recent system demonstration, $System_7$ produced an interconnection fabric with 512 optical inputs and 512 optical outputs with individual channels tested above 600 Mbps [6]. This most recent switching fabric demonstrated channel data rates on the order of a gigabit per second (10^9 bps) with nearly 1000 total system connections, yielding an aggregate throughput approaching 10^{12} bps. This smart pixel technology is also being successfully applied to applications such as optical neural networks [7], image processing techniques [8], and telecommunications applications [9,10,11]. The fundamental question at this point is, what are the prospects for further breakthroughs in this technology and what role will smart pixel technology play in future optoelectronic and data communication applications?

To address this question, we will begin by considering some of the early semiconductor device development, which led to and continues to dominate a significant part of smart pixel technology.

10.2. Multiple Quantum Well Devices

Nonlinear operations are fundamental to any processing, switching, or logic operations. Much of the early work in this area was motivated by the idea that perhaps optical devices could avoid some of the inherent speed limitations exhibited by electronic devices. During the early 1980s, this search for nonlinear optical devices coincided with a maturing of sophisticated compound semiconductor growth techniques and advances in material sciences. With improvements in growth techniques, researchers were able to engineer semiconductor devices at the atomic level, one atomic layer at a time, and were able to create novel optoelectronic devices such as the multiple quantum well device.

Multiple quantum well (MQW) modulators consist of alternating thin layers of two semiconductor materials, the most studied to date being gallium arsenide (GaAs) and aluminum gallium arsenide (AlGaAs). The thin crystal layers are typically grown using advanced growth techniques such as molecular beam epitaxy (MBE) or metal-organic chemical vapor deposition (MOCVD), which have the ability to grow these layers with atomic precision. The operation of these devices can best be understood by considering the effect of the layered semiconductor structure on the electrons and holes within the material. Using GaAs and AlGaAs as the material example, the bandgap energy of GaAs (1.424 eV) is lower than that of AlGaAs (1.773 eV) and as a result, the electrons and holes in the semiconductor material see a minimum energy in the GaAs "well" material and the AlGaAs material on either side therefore acts as a "barrier." The semiconductor layers are so thin that the electron–hole pairs behave as "particles-in-a-box," often described in elementary quantum mechanics. The resulting quantum confinement causes discrete energy levels of the electron–hole pairs. One important consequence of this energy discretization is that very strong exciton absorption peaks appear at the edges of these steps, even at room temperature.

10.2.1. THE QUANTUM CONFINED STARK EFFECT

When an electric field is applied perpendicular to the layers, the electrons and holes move to lower energies, and the optical transition energy decreases, resulting in a shift in the wavelength of the absorption peak. This can be understood with the aid of the energy band diagrams shown in Fig. 10.1.

Figure 10.1(a) shows the energy band diagram under conditions of no applied electric field. Here, the optical transition energy is represented by the length of the arrow. When an electric field is applied, the band diagram tilts as shown in Fig. 10.1(b), both the electrons and holes move to lower energies, and the optical transition energy decreases. Because photon energy is inversely proportional to wavelength, this decrease in optical energy corresponds to an increase or red-shift in the wavelength of the absorption peaks. This shifting of the absorption peaks with applied field is the underlying principle of the quantum confined Stark effect (QCSE) [12,13]. This is shown in the responsivity characteristics of Fig. 10.2.

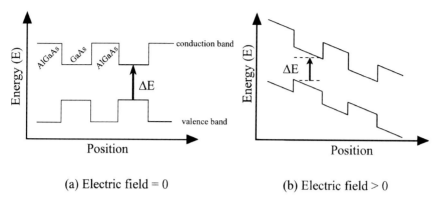

Fig. 10.1 Band diagram for GaAs/AlGaAs MQW structure. (a) No applied electric field, and (b) with applied electric field.

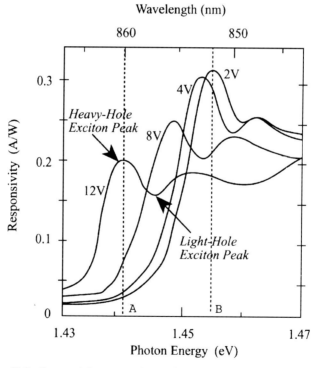

Fig. 10.2 Responsivity vs wavelength for specific applied electric fields.

10. New Devices for Optoelectronics: Smart Pixels

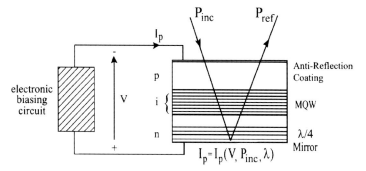

Fig. 10.3 Schematic diagram of a reflection-type MQW modulator.

10.2.2. THE SELF ELECTROOPTIC EFFECT DEVICE

Because a change in optical absorption at a specific wavelength can be affected by a change in the applied electric field, QCSE or electroabsorptive modulators can be produced. Here, optical intensity modulation is achieved using an external electric field applied to the modulator. In most instances to date, the structure used in these modulators is a PIN diode, where the quantum well layers are placed within the intrinsic region of the diode. The structure for a specific reflection-type MQW modulator called the self electrooptic effect device (SEED) [14] is shown in Fig. 10.3. Here P_{inc} is the incident optical power, P_{ref} is the reflected or output optical power, I_p is the series current in the external electric circuit, which by Kirchoff's current law is the photocurrent generated by the MQW modulator, and V is the applied electric field across the MQW structure. A quarter-wave dielectric stack provides a high reflectivity mirror at the rear of the device.

An important advantage of this type of structure is that the *p-i-n* diode is operated in a reverse bias configuration and therefore very low operating energies can be achieved while maintaining large electric fields across the quantum wells.

This type of electroabsorption modulator has enjoyed a great deal of success and widespread use. Switching speeds of 33 psec have been experimentally demonstrated [15] although this does not appear to represent any fundamental limit. Input optical energies on the order of 1 pJ for a 10 μm × 10 μm device are typical of current device technology. Arrays as large as 64 × 32 have been used for optical logic and memory operations [16]. There are, however, some less-than-desirable characteristics of these types of electroabsorption modulators. Contrast ratios for SEED-type modulators are typically low, on the order of 4:1. Also, since the operation of these modulators relies on the shift of narrow exciton absorption peaks, optimum performance requires an accurate and stable monochromatic source.

10.2.3. OTHER ELECTROABSORPTION MODULATORS

Another important class of electroabsorption modulator is the asymmetric Fabry-Perot modulator. Here, a partially reflecting mirror is grown on the front of the reflection-type electroabsorption modulator creating a Fabry-Perot étalon around the MQW structure. The cavity produces an increase in the field strength at the Fabry-Perot wavelength, which produces larger shifts of the exciton absorption peaks than in a modulator without an étalon. This particular modulator structure has gained considerable attention because of its lower insertion loss, higher contrast ratios, and lower operating voltages [17,18,19]. Although contrast ratios in excess of 100:1 have been reported [20], the wavelength sensitivity in these electroabsorption modulators is even more pronounced than with SEED-type modulators.

10.2.4. FUNCTIONALITY

To first order, the MQW modulator can be modeled as a parallel plate capacitor while an analysis of the circuit results in a first-order differential equation. For a stable solution to exist, it can be shown that the change in responsivity with applied field must be positive, resulting in a mode of operation in which the absorption of the MQW modulator is directly proportional to the current in the electronic bias circuit. If a constant current source is used as the external bias circuit, optical level shifting [14] or noninterferometric optical subtraction [21] can be realized. For this mode of operation, the modulator is operated at a wavelength where the responsivity of the modulator increases with increasing applied voltage, shown by the dashed line at point A in Fig. 10.2. If, however, the modulator is operated at a wavelength such that the responsivity decreases with increasing applied voltage, such as point B in Fig. 10.2, the solution to the differential equation is unstable and the resulting functionality is bistability.

To date, the majority of the applications using electroabsorption modulators have used the bistable switching functionality of the modulator. Applications such as optical logic [22], optical interconnection [23], and analog-to-digital conversion [24] have been demonstrated with this functionality. By integrating two MQW modulators in series, a symmetric SEED (S-SEED) [22] is formed. This device provides improved bistable characteristics, freedom from critical biasing issues characteristic of bistable devices, and additional features such as time-sequential gain [16]. A smaller number of researchers have investigated applications of the analog functionality including image processing [25] and laser power stabilization [26]. A good review of the analog functionality of SEEDs can be found in [27].

10.3. Smart Pixel Technology

Although the advent of the MQW modulator provided an important component for optical switching and information processing systems, there remained a growing need for programmability and increased functionality. Smart pixels were the next step toward finding this flexibility.

Until this point, individual optical devices provided enhanced capabilities but generally only with regard to a specific point process. Cascading optical devices to provide increased complexity and functionality was difficult and impractical due to fan-out and losses in the optical path. The concept of the smart pixel was to integrate both electronic processing and individual optical devices on a common chip to take advantage of the complexity of electronic processing circuits and the speed of optical devices. Arrays of these smart pixels would then bring with them the advantage of parallelism that optics could provide.

10.3.1. APPROACHES TO SMART PIXELS

There are a number of different approaches to smart pixels, which generally differ in the way in which the electronic and optical devices are integrated. Monolithic integration, direct epitaxy, and hybrid integration are the three most common approaches in use today. While some consider direct epitaxy to be a subset of monolithic integration [28], the distinction between the fabrication of smart pixels by growth in a common semiconductor material and growth of compound semiconductors onto silicon presents sufficiently different challenges that we describe these here as different integration techniques.

10.3.1.1. Monolithic Integration

Monolithic integration is a technique that allows both the electronics and the optical devices to be integrated in a common semiconductor material in a single growth process or by utilizing a re-growth technique. The material of choice in most cases is a compound semiconductor material such as gallium arsenide (GaAs), indium gallium arsenide (InGaAs), or indium phosphide (InP). Potentially, this approach would produce higher-speed smart pixels and may ultimately enable the manufacture of large arrays at lower cost than a hybrid integration approach. However, the simultaneous fabrication of both the electronic and optical circuits on the same substrate generally results in an overall system with less than optimum performance.

There are generally two types of monolithically integrated smart pixels: ones that incorporate passive optical modulators and ones that incorporate active optical emitters.

10.3.1.1.1. Modulators

Probably the most studied monolithically integrated smart pixels to date are those that incorporate MQW-type optical modulators. Figure 10.4 shows a schematic of a SEED that has been monolithically integrated with a field effect transistor (FET) [29]. This particular device was one of the first attempts at smart pixel development using MQW modulators. In the experimental demonstration, a *p-i-n* photodetector and load resistor, a depletion mode GaAs-AlGaAs heterostructure FET (HFET) and self-biased HFET load, together with an output AlGaAs MQW optical modulator were monolithically integrated within a 50 μm × 50 μm area. Here, the FET provides electronic gain and also electronic control of the SEED modulator.

Since this first demonstration, FET-SEED-type experiments have demonstrated 32 × 16 switching fabrics in which each pixel contained 25 FETs and 17 *p-i-n* diodes operating at switching speeds as high as 155 Mbps [5].

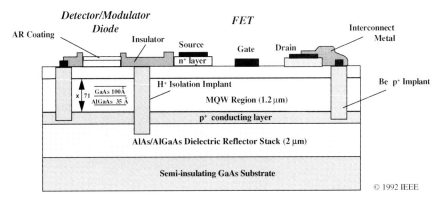

Fig. 10.4 Monolithic integration of Field Effect Transistor (FET) with a Self Electrooptic Effect Device (SEED).

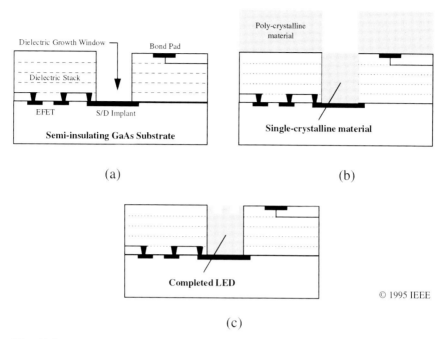

Fig. 10.5 Three stages of growing GaAs LEDs on GaAs VLSI circuitry. (a) Custom-designed GaAs VLSI chip with dielectric growth window and Source/Drain (S/D) implant, (b) Single-crystal material grows in the dielectric window and polycrystaline material deposits on the overglass and bond pads, (c) Planarizing polycrystalline etch, current confining mesa etch, Si_3N_4 chemical vapor deposition, and AuZn-Au-ohmic contact evaporation, completes the process.

10.3.1.1.2. Emitters

The second type of monolithically integrated smart pixel integrates active emitters such as light-emitting diodes (LEDs) or laser diodes with electronic circuitry. One such project integrates LEDs, optical FET (OPFET) photodetectors, metal-semiconductor-metal (MSM) photodetectors, and GaAs MESFET electronics in a common 0.6 μm GaAs process [30,31]. Figure 10.5 shows the three stages of integrating GaAs LEDs with GaAs VLSI circuitry using a re-growth technique. Here, GaAs-based heterostructures are epitaxially grown on fully metalized commercial VLSI GaAs MESFET integrated circuits.

LED-based structures are generally power-inefficient devices requiring high drive current and producing low optical output power. The broad angular distribution of the emitted light also introduces optical cross-talk in

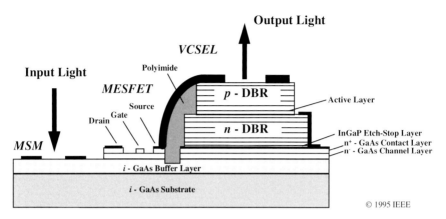

Fig. 10.6 Schematic cross-section showing the monolithic integration of a VCSEL, an MSM photodetector, and a MESFET.

densely packed arrays of devices. However, the advantage of using LED structures is that they are generally simpler to fabricate and are less susceptible to thermal variations than laser diodes.

Surface-emitting lasers have also been monolithically integrated to form smart pixels [32]. Most of this work to date has focused on a specific type of surface emitting laser called the vertical-cavity surface-emitting laser (VCSEL). Figure 10.6 shows one specific demonstration of the monolithic integration of a VCSEL with a metal-semiconductor-metal (MSM) photodetector and several metal semiconductor field-effect transistors (MESFETs) [33]. In this demonstration, the MSM photodetector and MESFET layers were deposited using MBE while the highly doped InGaP etch-stop layer, the GaAs buffer layer, and the VCSEL layers were deposited using MOCVD. The VCSEL layers contain a C-doped top distributed Bragg reflector (DBR), a Si-doped bottom DBR, and an active layer consisting of quantum wells. The total thickness of the VCSEL layer is approximately 8 μm. Here, input light is detected by the MSM and amplified by the MESFET, which then controls the current through the VCSEL, thus modulating the output light. Functionality of both NOR- and OR-type operation with optical gain and a 3 dB bandwidth of 220 MHz was demonstrated. The operating wavelength in this particular demonstration was 860 nm.

Some of the advantages of VCSELs are inherent single longitudinal mode operation and small divergence angle. However, historically, VCSELs have suffered from high series resistance and threshold voltage and nonuniformity in the VCSEL characteristics across arrays of devices. The high series resistance limits the optical output power, the temperature range for continuous wave (CW) operation, and the overall power efficiency of the laser. VCSELs with output optical power of 1 mW, threshold current of 2 mA, series resistance of 22 Ω, and power efficiency of 5% have been demonstrated [34]. The nonuniformity in the VCSEL characteristics is typically manifest in variations in output power and threshold and currently remains a significant problem for smart pixel arrays.

10.3.1.1.3. Observations

Although this integration approach promises improved performance, there are some limitations to its use. These limitations stem from two distinct issues: simultaneous optimization of both the optical devices and the electronic circuitry and a general lack of maturity of compound semiconductor technology. The optimization techniques used for the electronic circuitry which provides the increased functionality to the smart pixel are, in most cases, not the same as the optimization techniques required to produce high-quality optical devices. As a result, trade-offs must be made that ultimately reduce some performance metric associated with the overall smart pixel system. If a re-growth-type process is used, then additional processing complexity is also introduced. The second issue, technology maturity, is associated with the differences in the technological maturity of silicon processes compared to those of compound semiconductors. Mainstream silicon has enjoyed considerable success over the past several decades and, as a result, has benefited tremendously in terms of technology advancement and development. Compound semiconductors, however, have been used primarily for special-purpose, typically high-frequency applications and therefore the entire technology infrastructure is not as well developed. As a result, the tools necessary for designing, modeling, and fabricating compound semiconductor devices are not as well-developed as those used for silicon devices. Taken together, these two issues severely limit the usefulness of monolithic integration as a viable integration approach for smart pixel technology.

10.3.1.2. Direct Epitaxy

Direct epitaxy of compound semiconductors onto silicon is another approach to the integration of smart pixels. Here the compound semiconductor devices are directly grown onto the silicon crystal substrate. Because a significant amount of recent research has focused on the use of optical interconnects to alleviate some of the input and output bottlenecks associated with high-performance silicon integrated circuits, this integration technique appears to be a natural match. The principal limitation of this integration approach is a result of the large lattice mismatch between the silicon and compound semiconductor materials; approximately 4% for GaAs grown on silicon. In order to overcome this mismatch, a high-quality buffer layer is typically grown between the compound semiconductor and the silicon substrate. This buffer layer usually takes the form of a strained-layer superlattice, which provides an intermediate lattice constant between that of the compound semiconductor and the silicon substrate for crystalline bonding.

Figure 10.7 shows the structure of a GaAs-AlGaAs multiple quantum well reflection modulator grown directly on a silicon substrate using direct epitaxy techniques [35]. Here, a 1000 Å n^+ GaAs buffer layer is grown atop an n^+ silicon substrate.

After the entire growth process, etching is used to fabricate the optical devices. In this particular demonstration, the etching process produced cracks in the strained-layer superlattice, common in this approach. Although these cracks did not limit device performance, they did limit the overall usable device area.

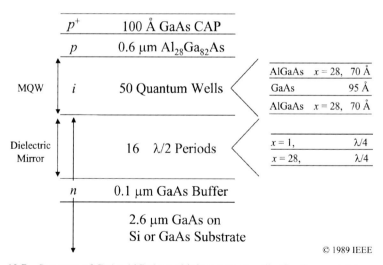

Fig. 10.7 Structure of GaAs-AlGaAs multiple quantum well reflection modulator grown on silicon substrate using direct epitaxy techniques.

To date, integration of optoelectronic devices with electronic circuitry using direct epitaxy has focused on growing passive optical devices such as MQW modulators on silicon. The difficulty associated with active emitter integration using this approach can be attributed to differences in the thermal expansion coefficients of the two dissimilar materials. For GaAs and silicon, this difference in thermal expansion coefficients is 50%. Active optical emitters that generate large amounts of heat are therefore problematic for this integration approach. Even with passive optical devices, the large differences in the thermal expansion coefficients can result in a potentially unstable bond between the two dissimilar materials. Catastrophic failure of these devices typically occurs at this interface as a result of stress, strain, and sheer of the crystalline material. Reduction of on-chip heat generation is therefore an important consideration in the design of smart pixel architectures using this integration technique.

Another problem associated with direct epitaxy is that the integrated circuit must be subjected to temperatures as high as 850°C during the compound semiconductor growth. This exposure to high temperatures can cause degradation to some or all of the transistors in the silicon circuitry. Also, the metalization of the integrated circuit must be performed after the growth of the compound semiconductor because aluminum, which is the standard metal used in silicon, degrades at such high temperatures.

10.3.1.3. Hybrid Integration

Hybrid integration is the third approach to developing smart pixels. Here the optical devices are grown separately from the silicon electronic circuitry. Then in a subsequent processing step, the optical devices are bonded to the silicon circuitry using a variety of bonding techniques. These include flip-chip bonding, epitaxial lift-off and subsequent contact bonding, and the creation of a physical cavity atop the silicon circuitry and flowing-in optical material such as liquid crystal material. Using this third approach, both the optical devices and the electronic circuits can be independently optimized resulting in an overall optimization of the smart pixel system. In contrast to the direct epitaxy approach, hybridization allows the silicon circuitry to be fully fabricated before attaching the optical devices and therefore, no high temperature exposure of the silicon is incurred in the process.

In the following sections, we will consider the three most popular hybrid integration techniques: flip-chip bonding, epitaxial lift-off, and liquid crystal on silicon.

10.3.1.3.1. Flip-Chip Bonding

In this integration approach, bonding pads are incorporated into both the compound semiconductor optical devices as well as the silicon circuitry and then, in a subsequent processing step, associated pads are brought into contact and a mechanical solder bond is effected. In some of the earlier approaches [36,37,38,39] to flip-chip bonding, gold-gold or gold-semiconductor bonds were employed. These types of bonds require a high-temperature anneal, which again can be detrimental to the silicon circuitry. In more recent flip-chip approaches [40,41], both the p and n contacts are attached to the silicon, eliminating the need for the high-temperature anneal. Because the optical devices are inverted during the flip-chip bonding process, the substrate must be either transparent to the optical radiation or removed during a subsequent substrate removal step to allow access to the optical devices. As with monolithic integration, both passive modulators and active emitters can be integrated using this approach.

10.3.1.3.2. Modulators

Figure 10.8 shows one particular example of hybrid integration in which a SEED MQW modulator is flip-chip bonded to complementary metal oxide semiconductor (CMOS) silicon circuitry [40]. In this approach, a three-step hybridization process is employed. First, the MQW modulator is fabricated

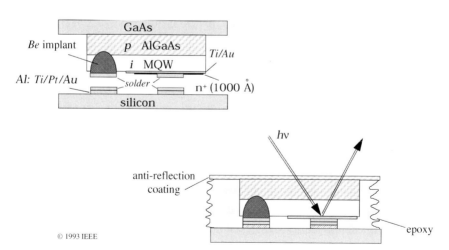

Fig. 10.8 Hybrid Integration: Complementary Metal Oxide Semiconductor (CMOS) Self Electrooptic Effect Device (SEED).

and gold and gold-germanium-gold contacts are deposited. Subsequently, indium contacts are deposited on the existing contacts, completing the MQW contact. The silicon circuitry is then patterned with aluminum contacts of the same size and spacing as those on the MQW modulators and finally, indium is deposited on the aluminum contacts. The modulator and silicon contacts are then aligned and heated to 200°C, causing the indium from each contact to melt and fuse. In the second step, epoxy is flowed between the samples, providing additional mechanical support. Finally, the GaAs substrate is removed using a chemical etch and an antireflection coating is applied to complete the hybridization.

CMOS-SEED technology has demonstrated tremendous potential for use in smart pixel applications. Early devices demonstrated superb heat sinking capability and ohmic contact, cycling from 30°C to 100°C with no degradation in device performance [40]. Later, research focused on device yield and demonstrated 32×32 arrays of MQW modulators flip-chip bonded to silicon CMOS circuitry with greater than 95% device yield [41]. In the first demonstration of this technology to smart pixels, an 8×8 array of CMOS-SEED switching nodes demonstrated 250 Mbps operation [42]. More recently, circuits with 4096 optical detectors, 256 optical modulators, 140,000 FETs operating beyond 400 Mbps have been demonstrated [43].

The majority of the previously mentioned smart pixels operate at 850 nm. Other similar hybrid smart pixel arrays have also been demonstrated operating at 1.6 μm [44]. We will discuss wavelength selection as a design consideration in Section 10.4. Another important point is that this technique is not limited exclusively to silicon CMOS but could be used just as effectively with other silicon circuit families such as emitter coupled logic (ECL), bipolar, and bipolar CMOS.

10.3.1.3.3. Emitters

Active emitters are the alternative to passive modulators when considering hybrid integration of smart pixels. These emitters come in the form of either LEDs or laser diodes. The majority of the work on active emitters has focused on VCSEL emitters for parallel processing and interconnect applications. Some of the advantages attributable to VCSELs are inherent single longitudinal mode operation, small divergence angle, low threshold current, and the capability of high bit-rate modulation. The out-of-plane geometry provided by these devices is particularly advantageous for smart pixel applications. One-dimensional arrays of 64×1 individually addressed VCSELs [45] and two-dimensional arrays of 7×20 VCSELs [46] have been demonstrated. In [47], an eight-element section of an 8×8 array of VCSELs has experimentally demonstrated 622 Mbps operation showing the potential for high-speed operation. VCSELs are finding applications in conventional duplex transceivers and single devices and small arrays are being used for intramachine bus applications. The use of VCSELs in smart pixel systems can also simplify the overall optical system because the laser source is integrated within the system.

While research focused on ultra-low threshold VCSELs continues, the majority of VCSELs currently available require threshold currents in the range of 100 μA to 1 mA with threshold voltages from 1.5 V to 2.0 V. When operating at high speeds, this bias power represents only a portion of the total operating power that results from the dynamc power dissipation of the VCSEL and the driver circuitry. When arrays of VCSELs are integrated on-chip, significant static power dissipation can result when the lasers are biased above threshold. One operational method to minimizing the power consumption of VCSELs is to operate them in a pulsed mode rather than continuous wave (CW) mode. Heat dissipation and nonuniformity of the output characteristics remain a concern with the application of these emitters to smart pixel applications. Also, since many VCSEL integration demonstrations to date have integrated contiguous arrays that include the substrate, the silicon circuitry underlying the VCSEL array is completely covered and not available for detector integration. As a result, detectors must be integrated with the VCSELs if detectors are to be included in the smart pixel architecture. These challenges continue to be active areas of research within the VCSEL community.

10.3.1.3.4. *Epitaxial Lift-Off*

A slightly different approach to hybrid integration of smart pixels is that of epitaxial lift-off [48,49]. Here thin compound semiconductor optical devices are grown separately from the silicon circuitry, removed from a sacrificial growth substrate using a technique called epitaxial lift-off (ELO) and then bonded to the CMOS silicon circuitry using either Van der Waals (contact) bonding or adhesion layers between the host and ELO epitaxial layer. A more recent approach enables the alignment and selective deposition of both individual and arrays of devices onto a host substrate and also allows the devices to be processed on both top and bottom of the epitaxial sample [50].

10. New Devices for Optoelectronics: Smart Pixels   369

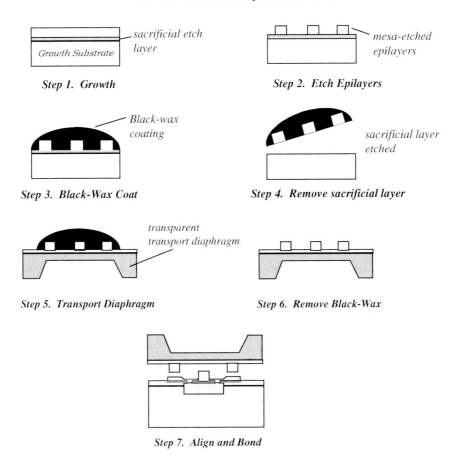

Fig. 10.9 Hybrid integration using the epitaxial lift-off technique.

This improved approach to the original ELO process, which uses a transparent polyimide transfer diaphragm, is illustrated in the seven-step process shown in Fig. 10.9 [51]. The as-grown material, with the epitaxial layers of interest atop the sacrificial etch layer, is shown as step 1. First, any contacts or coatings that are desired are applied to the as-grown material and individual devices are defined through standard mesa etching techniques. Next, the sample is coated with a black-wax handling layer by either melting or spray coating the wax onto the material. The sample is then immersed into a selective etch solution, which etches away the sacrificial layer and separates the epitaxial layers from the growth substrate. In steps 5 and 6, the thin film devices, embedded in the back wax, are contact bonded to a transparent diaphragm and the black wax is removed to reveal the thin-film devices.

To transfer and bond these thin-film devices to the host substrate, the transparent diaphragm is inverted so that the devices face the host substrate. The thin-film devices are visually aligned and then a pressure probe is applied to transfer individual, subarray, or the entire array of devices to the host substrate. High-quality thin-film devices have been successfully fabricated using this process. Thin-film devices as large as 2 cm × 4 cm and film thickness ranging from 200 Å [52] to 4.5 μm [53] have been reported. To date, thin-film devices made of AlGaAs, InGaAsP, CdTe, and CdS have been integrated with host substrates that include silicon, lithium niobate, and various types of polymers.

One of the advantages of this type of hybrid integration technique is that since the thin-film devices are inverted from the growth structure, both top and bottom of the epitaxial sample are available for processing. This is an important feature when considering three-dimensional interconnect applications such as in [53]. One of the limitations of this approach is that, to date, the process of transferring the thin-film devices to the host substrate has been labor intensive. However, recently, thin-film detector arrays as large as 64 × 64 have been demonstrated [54], opening the possibility of integrating large, homogeneous arrays of thin-film devices into smart pixel applications.

10.3.1.3.5. Liquid Crystal on Silicon

Techniques other than compound semiconductor-on-silicon are also being explored for smart pixel applications. Liquid crystal on silicon (LCOS) is another hybrid integration technique that has demonstrated tremendous potential for smart pixel applications. In this approach, a cover glass that has been coated with a transparent conducting material is placed on top of the silicon circuitry, separated either by spacers or the natural contour of the processed circuitry to create a physical cavity. The liquid crystal material is then flowed into the cavity and placed in direct contact with the silicon circuitry. An electric field is created between the glass electrode and metal pads, which are fabricated on the silicon chip, defining individual pixels at the locations of the metal pads. The metal pads also provide a high-reflectivity mirror from which the optical signal is reflected. Modulation of the electric field results in a modulation of the polarization, the intensity, or the phase of the illumination incident on the pixel.

Liquid crystal materials have previously found numerous applications in the area of projection display systems and are particularly attractive for smart pixels because of their large birefringence, low voltage operation, and their ease of fabrication for large arrays.

Liquid crystals most commonly used for optoelectronic applications have elongated molecules that are shaped like tiny rods. The directional alignment of these rods varies from completely random in the liquid phase to almost completely ordered in the solid phase. When a liquid crystal is heated, usually above 100°C, the molecules are randomly oriented and the material is isotropic. As the material cools, the molecules orient themselves, predominately in one direction. There are three types of liquid crystals: nematic, smectic, and cholesteric. In nematic liquid crystals, the molecules tend to be oriented parallel but their positions are random. In smectic liquid crystals, the molecules are again parallel, but their centers are stacked in parallel layers within which they have random positions. Cholesteric liquid crystals are a distorted form of the nematic liquid crystal in which the molecular orientation undergoes a helical rotation about a central axis.

In the early 1970s, Meyer [55] recognized that smectic liquid crystals exhibited ferroelectricity, a permanent dipole moment that can be switched by an externally applied electric field. As a result, the polarization vector of light propagating through a ferroelectric liquid crystal material can be altered by applying an electric field to the liquid crystal. Optoelectronic applications using these types of liquid crystals take advantage of the material's large birefringence, Δn. Typical values for Δn range between 0.1 and 0.2. With this large birefringence, a controllable half-wave plate can be fabricated. A π-phase shift between the ordinary and extraordinary waves of light propagating through the liquid crystal material results if the thickness of the material d is set at $d = \lambda_0/(2\Delta n)$, where λ_0 is the wavelength of light in a vacuum. For $\lambda_0 = 0.632$ μm and $\Delta n = 0.15, d = 2$ μm requiring a *very* thin layer of liquid crystal material to produce a half-wave plate. Binary phase modulation is accomplished by using an input polarizer oriented along the bisector of the two maximum switching states and an output polarizer oriented perpendicular to the input polarizer. Analog phase modulators can be made directly using nematic liquid crystals because the birefringence of these devices is directly voltage controlled. Today ferroelectric liquid crystals are used almost exclusively for LCOS smart pixels.

Figure 10.10 shows a cross-sectional schematic of the construction of one type of LCOS smart pixel. Here, an optically flat glass plate is coated with a thin transparent conducting electrode such as indium tin oxide (ITO) and is then placed on top of a conventional silicon integrated circuit. A cavity is then created between the glass and the processed silicon circuitry.

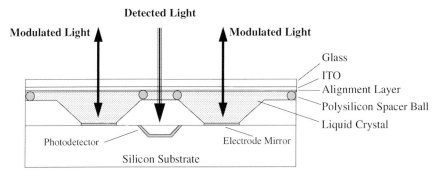

Fig. 10.10 Cross-sectional diagram of liquid crystal on silicon structure.

Either polyball spacers are epoxy-bonded to the processed silicon CMOS circuitry or the natural contour produced by the multi-layered silicon growth process forms the cavity. An alignment layer of obliquely evaporated silicon monoxide or rubbed polyvinyl alcohol is deposited onto the ITO. This alignment layer makes contact with the liquid crystal material and induces spatial alignment of the liquid crystal molecules. Metalized pad structures are fabricated during the silicon fabrication step and serve as highly reflective electrode mirrors. These metalized pads define the smart pixel modulators.

In operation, a polarized optical beam is imaged through the liquid crystal material onto the electrode mirror. When an electrical field is applied between the ITO and the electrode mirror, the state of the liquid crystal is modulated, thereby modulating the polarization of the optical beam as it reflects from the electrode mirror. In the case of a photodetector, the liquid crystal material is not dynamically modulated, therefore the incident beam is simply absorbed by the detector.

An important issue that must be considered when designing LCOS smart pixels is dc balancing of the external applied electric field. Ionic impurities in the liquid crystal material can cause a performance degradation if measures are not taken. Under the influence of an electric field, these ions drift and accumulate at the interface between the alignment layer and the liquid crystal material, causing a reduction of the total effective field. If the field is subsequently removed, these ions generate an electric field of the opposite polarity, which could cause the liquid crystal to switch to the opposite state. DC balancing is an active design measure in which this reverse field buildup is eliminated by driving the device with an ac signal, actively cycling the device between states to ensure the net field equals zero. This dc balancing requirement necessitates serious consideration during the design phase of any system application of LCOS smart pixels.

Smart pixel arrays that use LCOS integration have found applications in many different areas, including optoelectronic neural networks [56], image processing [57], interconnects [58], and competitive learning and adaptive holographic interconnections [59]. An excellent review of LCOS technology can be found in [60]. Additional research continues into the application of LCOS smart pixel technology [61,62].

10.3.1.3.6. Other Approaches to Hybrid Integration

A relatively new class of hybrid integration technique is known as wafer fusion [63, 64, 65, 66]. This approach is applied at the wafer level before the photonic devices are processed. This hybrid integration technique involves fusing two wafers with different lattice constants under high pressure and elevated temperature. To date, this technique has been applied to long-wavelength VCSELs and photodetectors primarily for telecommunication applications. This technique also holds promise for smart pixel fabrication. Another hybrid integration approach that is employed as a prefabrication technique is called polyimide bonding [67,68]. Here, multiple III-V substrates are bonded to a processed silicon wafer using a polyimide layer. Epitaxial layers are grown on the III-V chips before bonding and the processing of the photonic devices is completed after bonding to the processed silicon wafer.

10.3.2. PERFORMANCE METRICS

In order to characterize and compare the performance of different types of smart pixels, standard performance metrics must be developed. Numerous methodologies exist by which the performance of smart pixel technology can be characterized and evaluated. In this section, we consider two performance metrics that have been used by others [69] in the community: aggregate capacity of the smart pixel array and complexity of the individual smart pixel.

10.3.2.1. Aggregate Capacity

Aggregate capacity is a measure of the total information-handling capability of the smart pixel array. It combines the *individual* channel data rate with the total number of channels in the array to produce an aggregate

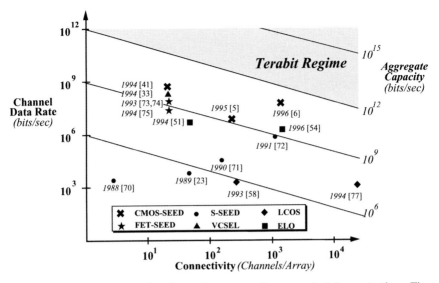

Fig. 10.11 Aggregate capacity of several representative smart pixel demonstrations. The year of the demonstration and the reference are also shown. Channel data rates are maximum channel data rates experimentally demonstrated.

information carrying capacity. Figure 10.11 shows a number of smart pixel demonstrations plotted as a function of channel data rate and array connectivity. This listing of demonstrations is not meant to be all-inclusive but instead attempts to represent the trend in smart pixel aggregate capacity. Notice also in Fig. 10.11 the terabit aggregate capacity regime — aggregate capacity in excess of 10^{12} bps.

10.3.2.2. Complexity

The underlying concept of smart pixels is to provide increased functionality and programmability to arrays of optical devices. One measure of this level of functionality and programmability deals with the number of transistors associated with each individual smart pixel. Higher transistor counts indicates more processing capability per smart pixel and therefore more functionality. Figure 10.12 shows a number of smart pixel demonstrations plotted as a function of individual channel data rate and number of transistors per smart pixel. Conspicuously absent from this plot are the S-SEED demonstrations previously plotted in Fig. 10.11. The reason for this purposeful omission is that the S-SEEDs do not meet the standard definition of smart pixels in that they have no additional transistors or drive electronics to provide increased flexibility, functionality, or programmability.

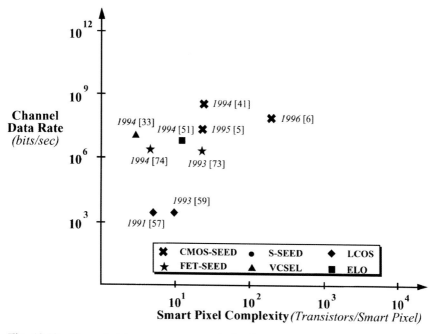

Fig. 10.12 Channel date rate versus complexity for several representative smart pixel demonstrations.

One can deduce the current level of smart pixel complexity as well as identify the smart pixel technology capable of supporting a specific application by first considering the number of transistors necessary for some simple operations. A simple CMOS inverter requires 2 transistors, and two-input CMOS NOR and NAND gates each require 4 transistors to provide this functionality. A two-input Exclusive-OR gate requires 8 transistors and an And-or-Invert (AOI) full-adder requires 24 transistors. A wide-range transconductance amplifier that produces a smooth sigmoidal function for thresholding requires 12 transistors.

10.4. Design Considerations

Smart pixel design considerations depend largely on the specific application and fabrication technology. Because smart pixels cover a wide range of applications and approaches, this section will not attempt to present an all-inclusive set of design considerations. Instead, some of the more universal constraints that apply to current technologies will be discussed.

Table 10.1 Silicon Complementary Metal Oxide Semiconductor (CMOS) Roadmap Technology Characteristics

Year of first DRAM shipment	1995	1998	2001	2004	2007	2010
Minimum feature size (µm)	0.35	0.25	0.18	0.13	0.10	0.07
Number of Chip I/Os						
Chip-to-package (pads) high-performance	900	1350	2000	2600	3600	4800
Number of Package Pins/Balls						
Microprocessor/controller	512	512	512	512	800	1024
ASIC (high-performance)	750	1100	1700	2200	3000	4000
Package cost (cents/pin)	1.4	1.3	1.1	1.0	0.9	0.8
Chip Frequency (MHz)						
On-chip clock, cost/performance	150	200	300	400	500	625
On-chip clock, high-performance	300	450	600	800	1000	1100
Chip-to-board speed, high-performance	150	200	250	300	375	475
Chip Size (mm^2)						
DRAM	190	280	420	640	960	1400
Microprocessor	250	300	360	430	520	620
ASIC	450	660	750	900	1100	1400
Maximum Number Wiring Levels (logic)						
On-chip	4–5	5	5–6	6	6–7	7–8

10.4.1. SILICON TECHNOLOGY ROADMAP

Because the majority of smart pixel technology relies heavily on silicon electronics, it is essential to consider projections for mainstream silicon technology when considering future capabilities for smart pixels. Selected excerpts from the 1994 Semiconductor Industry Association *National Technology Roadmap for Semiconductors* [78] are provided in Table 10.1. The reductions in feature size and improvements in clock speed and chip size will have a profound impact on the information-processing capability of silicon-based smart pixels. If we consider a pixel of constant physical area, as the process feature size decreases, the number of devices per pixel will nearly double every three years. This translates directly to increased smart pixel processing and functionality.

10.4.2. WAVELENGTH SELECTION

Wavelength selection is an important consideration in any optical system design. Applications often dictate a specific wavelength such as in long-haul optical fiber systems where dispersion-shifted fibers are used and the preferred wavelength is 1.55 μm. Similarly, applications using mainstream compound semiconductors, such as GaAs and AlGaAs, operate at wavelengths near 850 nm. Optical spot size is a key consideration in smart pixel system design. Because spot size is proportional to wavelength, shorter wavelength light will allow for smaller detectors and higher density smart pixels. As a result, shorter wavelengths may be more attractive for near-term smart pixel applications that are based on high-density arrays and free-space optical inputs. Pixel area can be decreased even further by utilizing GaN devices that operate around 400 nm, potentially reducing the diffraction-limited spot size by nearly a factor of four in comparison to 1.55 μm sources. In [79], optical-to-optical modulation using a MQW modulator and a *p-i-n* photodiode was experimentally demonstrated. Here, digital optical information modulated on a carrier of 632.8 nm was translated to a wavelength of 850 nm. This class of optical-to-optical conversion could also be applied to telecommunication wavelengths such as 1.3 μm or 1.55 μm and would allow the actual processing system to operate at 850 nm, providing higher density smart pixel arrays.

A major advantage of using longer wavelengths for three-dimensional memory and optical interconnect applications is that silicon is transparent to light above approximately 1.0 μm. Vertical interconnections using 1.3-μm light have been demonstrated by emitting light from the bottom wafer through an upper silicon wafer to detectors on the top surface [53]. In this demonstration, thin film InGaAsP/InP-based emitters and detectors were bonded to the respective silicon substrates using an ELO intergation process to allow operation at the longer wavelength.

Despite decreasing transistor feature size, the minimum dimensions of an individual smart pixel will be constrained by the size of the optical input and output devices. If the input is an array of light beams from an image, a diffractive optical element, or a spatial light modulator, each pixel must have the ability to detect and process the incoming information. The required photoreceptor will be much larger than the minimum feature size of the electronic circuits in order to accommodate the diffraction-limited spot size of the optical input. Currently, GaAs metal-semiconductor-metal (MSM) detectors and silicon photodetectors operate efficiently in the

700–900 nm range. Low and intermediate growth temperature MSM devices that detect in the 850-nm range are also capable of very fast response times.

Linewidth is another important specification closely related to wavelength selection. Most passive modulators to date are based on MQW technology and, as a result, rely on the shift of narrow exciton peaks. These modulators require narrow linewidths for optimum device performance, which requires a stable, single-frequency source. In constrast, ferroelectric liquid crystals in conjunction with VLSI offers the designer some flexibility in wavelength selection.

10.4.3. INDIVIDUAL DEVICE SPEED

A typical smart pixel architecture includes an input optical device and circuitry capable of receiving optical information, an electronic processing circuit, and output driver circuitry and optical output device. The input may be from an off-chip source or an on-chip emitter such as a VCSEL or LED. The output may be an emitter or modulator driven by the processing circuit. When considering operating speed, each element in the smart pixel system contributes to the overall performance. Many different types of devices are capable of converting the optical input into an electronic signal. Monolithic integration using compound semiconductors allows for MSM detectors at the front end of the smart pixels. These devices can be fabricated to operate at very high speeds with a wide range of efficiencies. Typically, however, process adjustments necessary to fabricate acceptable electronic devices will result in less-than-optimal detectors in a monolithic process. Specifically, conventional silicon CMOS fabrication processes are far from optimal for photodetector fabrication, resulting in CMOS detectors that are slow and inefficient in comparison with GaAs devices. However, the designer still must decide between slower, more highly efficient silicon phototransistors and faster, less efficient photodiodes. Liquid crystal on silicon systems are slowed by the requirement for these silicon photoreceptor devices as well as the switching speed of the liquid crystal material itself. Smart pixels using LCOS technology typically operate with microsecond switching speeds.

Bandwidth will often be limited by voltage or current amplifier circuits that serve as an interface between the detector and the processing circuit. Gain must usually be sacrificed to obtain higher speed amplifiers. Wider bandwidth amplifiers may be obtained by cascading a series of high-speed stages, each having lower gain. Similar constraints apply to the smart pixel output circuit and optical device. Tradeoffs between gain, speed, and power must be carefully balanced when selecting the output emitter and designing the necessary driver circuit. On-chip power consumption and heat removal are other important considerations in the selection of the output optical device. Active emitters produce more on-chip heat than passive modulators, which derive their optical energy from off-chip sources. Currently, off-chip heat removal techniques can be managed more easily at the system level.

Finally, the processing circuitry is unique to each smart pixel application. Bandwidth is a function of the number of devices necessary to satisfy the required functionality. As anticipated improvements in silicon and compound semiconductor processing become reality, the electronic processing circuits will be capable of progressively higher bandwidths, placing more emphasis on the limitations of the input and output circuits.

10.4.4. ARRAY SIZE

One of the fundamental reasons for the development of smart pixel technology is that large arrays of optical devices provide the mechanism to leverage the advantages associated with optics. To date, 1024×1024 arrays of SEED modulators and 256×256 arrays of LCOS devices have been demonstrated. Even larger arrays are necessary for high-capacity telecommunications switching fabrics and massively parallel image processing applications. As array sizes increase, smart pixels will experience the challenges associated with high-density circuits. These constraints include power dissipation, yield, alignment, and addressing. Monolithic integration and LCOS processes may offer better yields due to relaxed requirements for electrical continuity at the bonding points and spatial alignment of massive numbers of contacts. Although LCOS is not a monolithic process, alignment issues are minimized as a result of the fabrication of reflective electrodes as part of the normal silicon fabrication process [80]. Hybrid integration has the advantage of well-established, on-chip addressing techniques routinely used in CMOS for large device arrays.

A prime consideration in array size is the size of each pixel. The diffraction limited spot size of incident light determines the size of the photoreceptor within a pixel and the pitch of the array or pixel spacing. Reduction of optical cross-talk in high-density arrays may drive the system designer to consider using a VCSEL or laser input instead of LEDs. The large divergence angle of LEDs limits the potential role of this device in high-density smart pixel applications. Resonant cavity-enhanced (RCE) LEDs have improved linewidth and divergence [81], and can serve as an alternative to VCSELs when conditions permit.

10.4.5. SYSTEM INTEGRATION ISSUES

System integration issues are numerous and often specific to the selected architecture. Once the fabrication process has been determined, the designer must evaluate the type of optical input (VCSEL, laser diode, RCE LED, or LED) and the input medium (free space, fiber, light modulator, or diffractive/refractive element). The input photoreceptor is selected simultaneously with amplifier requirements while considering associated tradeoffs in noise, gain, speed, and overall bandwidth. Processing circuitry requires interfacing with both the input and output circuits through buffers, impedance matching amplifiers, or transconductance circuits. Output circuit considerations are similar to input requirements in terms of gain-speed, gain-bandwidth, and noise-resistance tradeoffs. A compatible output device and output driver are required for systems having optical outputs. The overall system must then be evaluated for speed/bandwidth, power consumption and dissipation, power supply, and current limits.

Finally, different approaches to packaging and testing are available to the designer of smart pixel systems. Pixel arrays may be packaged on a single integrated circuit chip using monolithic or hybrid processing, designed as individual elements for use on either a standard optical bench or a custom bench such as slotted plate technology, or integrated into a three-dimensional system using technologies such as multi-chip modules [82]. In the latter case, implementation alternatives include the use of diffractive, refractive, or reflective optical components in conjunction with microlens and/or macrolens hardware [83]. If the system architecture relies on monochromaticity, then the system designed is forced to use laser sources, either on- or off-chip. Because packaging and testing can have a major impact on system performance [84] and cost, these two requirements are normally a part of the original system design. Optomechanical considerations are also important to consider in the context of system design. As the array size of smart pixels continues to grow, the susceptibility of the system to vibration and misalignment increases. As a result, smart pixel system designers have recently focused on improvements to the optomechanical system in an attempt to minimize these effects [85].

10.4.6. VLSI CIRCUIT DESIGN CONSIDERATIONS

Hybrid integration on silicon takes advantage of the maturity of CMOS fabrication processes and VLSI technology in general. As a result of smaller feature sizes, complex smart pixel circuitry will require less physical area, but be susceptible to photon-induced charge transfer between devices. This phenomena occurs when optical energy, intended for a detector or modulator, falls instead on the surrounding electronic circuitry and creates charge carriers that are directly injected into the circuitry. Physical limitations on optical beam diameter require emphasis on controlling cross-talk within and between pixels. Metal shielding, guard rings, and other isolation procedures must be carefully considered and incorporated into the electronic circuit design.

With improvements in processing techniques also comes an increase in the number of metalization layers available to the silicon circuit designer. Hybrid, direct epitaxy, and LCOS processes, however, can be negatively affected by the addition of these metalization layers. Additional processing steps may be required to ensure adequate planarization of the silicon surface prior to the integration of the optical devices. This additional processing potentially adds to the complexity of bonding GaAs circuits to silicon or adding a liquid crystal layer to the chip surface. These additional processing steps could potentially add to the rising costs associated with re-tooling plants to meet expectations of the silicon roadmap.

10.4.7. SUMMARY

Table 10.2 summarizes some of the design considerations discussed in this section. This table in not meant to be all-inclusive, but instead, is meant to provide a quick-reference for system designers considering the use of smart pixel technology.

10.5. Applications

Applications of smart pixels to date have focused primarily on digital applications with a heavy emphasis toward addressing electronic interconnection limitations. These applications span from backplane interconnects, which support printed circuit board (PCB)-to-PCB communications, to high-speed, wide-bandwidth switching and routing for telecommunication applications. Some work has also been done in the area of signal and

Table 10.2 Design Considerations for Smart Pixel Technology

Smart pixel technology	Switching speed	Array size	System issues	Principal applications
Monolithic integration	Potentially the fastest	Limited	Reduced flexibility for optimization of electronic and optical technologies Process uniformity is critical for arrays	High-speed, single-channel and small array applications
Direct epitaxy	Medium	Limited	Limited lifetime and reliability resulting from lattice mismatch and thermal expansion coefficient differences Process uniformity is critical for arrays	Predominately research demonstrations, to-date
Hybrid integration				
Flip-chip bonding	Fast	Medium-Large		High-speed, high-density free-space interconnection applications
Epitaxial Lift-Off	Medium-Fast	Medium	Detectors and RCE LEDs, to-date	Through-chip and fiber optic interconnection applications
Liquid Crystal on Silicon	Slow	Large	Active DC balancing required to prevent liquid crystal degradation No active emitters	Image processing and displays

Passive modulators

	Switching Speed	Array Size		
MQW Modulators	Fast	Large	Precise, narrow linewidth optical sources required Low contrast ratio	Off-chip sources provide improved heat dissipation Differential signaling required to overcome limited contrast ratio
LCOS Modulators	Slow	Medium-Large	Active DC balancing required to prevent long-term liquid crystal degradation	

Active emitters

VCSELs		Small	Increased heat generation requires additional attention to heat removal techniques Device characteristics across arrays not yet uniform	Architectures cannot employ diffractive components
RCE LEDs		Medium	Incoherent	Short, through wafer 3-D interconnections and multi-mode fiber applications
LEDs		Medium	Incoherent	Architectures cannot employ diffractive components Low cost, multi-mode fiber applications

Switching Speed Slow: μsec; Fast: psec *Array Size*: Small: $< 8 \times 8$; Large: $> 256 \times 256$

image processing applications requiring massive parallelism such as neural networks and image halftoning. In this section, we introduce several representative applications of smart pixel technology, including backplane interconnects, high-speed switching and routing, digital image halftoning, and analog-to-digital (A/D) conversion.

10.5.1. OPTICAL INTERCONNECTIONS

When considering interconnection issues, one can broadly categorize the different potential architectures to be interconnected in a hierarchical fashion from the highest to lowest levels as: machine-to-machine, processor-to-processor, board-to-board, chip-to-chip, and finally intrachip interconnections. Optical techniques have already been successfully applied to machine-to-machine interconnections in the form of optical fiber communication links. Recently, considerable research has been directed toward bringing optics to the next lower levels of processor-to-processor, board-to-board, and chip-to-chip communications.

There exist several fundamental reasons for this interest in optics as an interconnect technology [86]. First, optical interconnects offer freedom from mutual coupling not afforded by electrical interconnects. This particular advantage becomes increasingly significant with increased bandwidth because the mutual coupling associated with electrical connections is proportional to the frequency of the signal propagating on the line. Increased routing flexibility is another advantage of optical interconnections. Electrical interconnect paths cannot cross and are constrained to reside close to a ground plane. In contrast, optical interconnections can be routed through one another and are not constrained to be physically located in the vicinity of a ground plane. Another potential advantage is freedom from capacitive loading effects. With electrical interconnects, this comes in the form of line-charging where, as the interconnect line becomes longer, an increased number of electrons is employed to charge the capacitance of the line. No such line-charging effects are present with optical interconnects. Finally, at the quantum level there exists the potential advantage of impedance transformation that matches the high impedances of small devices to the low impedances associated with electromagnetic field propagation [87]. This impedance matching results in more energy-efficient communications using optical interconnects for all except the shortest intrachip interconnects.

Smart pixel technology provides additional advantages in terms of interconnect applications. Much of the work on interconnects at the board-to-board and chip-to-chip level is focused on reducing communication bottlenecks associated with high-performance processing systems, which are predominately silicon-based. The majority of smart pixels rely on silicon technology for increased functionality and programmability, so optical interconnect applications of smart pixels provide a unique opportunity for producing compact systems with large aggregate input and output capacity. Another attractive feature is that most smart pixel interconnect architectures employ optical input and output that is normal to the surface of the chip, allowing two-dimensional arrays of interconnections to be formed and thereby increasing the potential throughput of the interconnect system.

10.5.1.1. Backplane Interconnections

Future digital systems such as asynchronous transfer mode (ATM) switching systems and massively parallel processing systems will require large PCB-to-PCB connectivity to support the large aggregate throughput demands being placed on such systems. Projections from the Semiconductor Industry Association indicate that by the year 2001, silicon CMOS feature size will be on the order of 0.18 μm, transistor density will be 13 million transistors per 360 mm^2, and on-chip clock rates will be approaching 600 MHz with off-chip rates in excess of 250 MHz. These requirements provide the motivation to use photonics-based approaches to backplane designs, which overcome issues such as ground-bounce, cross-talk, and reflections that result in signal distortion when using parallel electrical approaches at these frequencies.

One approach to solving the interconnection problem created by such high-performance systems is to exploit the temporal and spatial bandwidth of optical interconnections. At the system backplane, free-space optical technology can be employed, which supports a large number of digital optical communication channels, each of which is comprised of simple optical connections between arrays of smart pixel devices on successive PCBs. This type of optical backplane can potentially provide in excess of 10^3 high-performance connections while the electronic circuitry associated with each smart pixel can provide the intelligence necessary to provide reconfigurable interconnections.

One recent approach to building such a free-space optical backplane, which implements a *Hyperplane* [88] based ATM switching fabric, is shown in Fig. 10.13 [69]. Here, electronic PCBs connect to a rigid backplane, which contains both electrical and free-space optical interconnection channels. The *HyperChannel* is comprised of several smart pixel arrays which

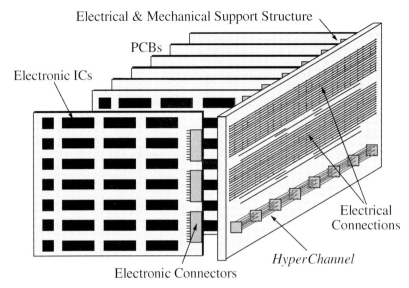

Fig. 10.13 Example of the application of smart pixel technology to backplane interconnects.

provide high-speed backplane interconnections to and from the individual PCBs.

The smart pixels in this application integrate SEED modulators with CMOS silicon circuitry using the hybrid, flip-chip bonding approach. Arrays of differential reflection-mode MQW modulators and detectors are used as a method of improving noise immunity and overcoming the low-contrast ratio limitation of the SEED modulators. The proximity of the optoelectronic devices with the CMOS circuitry allows decisions on the incoming optical data to be made local to the optical inputs and outputs and results in parallel whole-word processing and reduced requirements for driving long capacitive traces on the chip [89].

A photomicrograph of a buffered *Hyperplane* CMOS-SEED smart pixel is shown in Fig. 10.14. The core of this smart pixel array contains individual smart pixels organized into node channels. Control logic provides arbitration of data flow into and out of ATM cell queues, while each of three queues buffer an entire ATM cell. A queue-addressing block places segments of the ATM cell into the queue at the appropriate location. An output multiplexer controls which of the three queues writes output data to the electrical pinouts of the chip.

10. New Devices for Optoelectronics: Smart Pixels

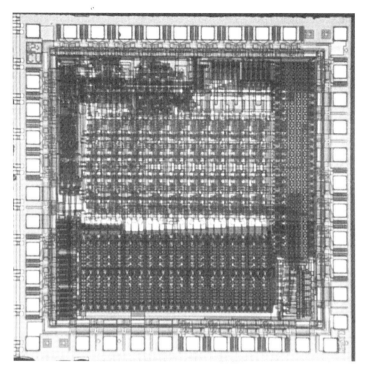

Fig. 10.14 Buffered *Hyperplane* CMOS-SEED Smart Pixel Array.

The physical size of this particular smart pixel chip is 2mm × 2mm. The electronic circuitry was designed using an 0.8 μm CMOS silicon process and provided a total transistor count in excess of 20,000 transistors with approximately 60 transistors per smart pixel. A 10 × 20 array of GaAs/AlGaAs SEED modulators was subsequently flip-chip bonded onto the processed silicon circuitry. The MQW modulator array provided the potential for up to 200 optical inputs or outputs, each of which had an optical window size of 18 μm × 18 μm.

The functionality of this smart pixel array to include packet injection, address recognition, packet extraction, and transparency was experimentally verified to 10 Mbps. A detailed system analysis predicts scalability to 2060 interconnections for a 1 cm × 1 cm chip area at a backplane clockrate of 125 MHz. This provides a maximum aggregate optical throughput of 4.5 Gbps which is sufficient to provide four users each with a 622 Mbps SONET STS-12 link [89].

The natural extension to this approach is to replace all of the electrical interconnections on the backplane with these high-speed *HyperChannels* and to add additional capability by incorporating fiber interfaces at other locations on the individual PCBs.

10.5.1.2. High-Speed Switching and Routing

As the demand for telecommunication switching capabilities continues to grow, the need for switching fabrics capable of switching large bandwidths of data becomes increasingly important. Examples of applications that are currently pushing the limits of electronic switching technology include ATM and video switching fabrics, pattern recognition and image processing systems, and high-performance computing systems. To meet the demands of these and future digital electronic systems, system designers are turning to optical interconnects to provide a potential solution to these massive interconnect applications.

Since about 1988, researchers at AT&T Bell Laboratories have been investigating the use of optical interconnects in high-speed switching applications. Figure 10.15 shows a representative block diagram of a switching fabric based on smart pixel technology [73].

AT&T Bell Laboratories has enjoyed tremendous success with various high-speed switching fabrics based on the SEED technology. Early

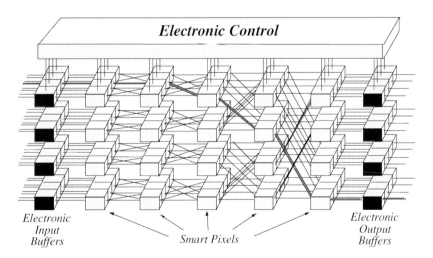

Fig. 10.15 Example of a high-speed switching fabric that incorporates electronic input and output buffers, electronic control, and smart pixels.

demonstrations incorporated individual SEEDs and S-SEEDs in basic switching and logic demonstrations. In 1993, their first smart pixel switching demonstration used monolithically integrated FET-SEED technology. The most recent advances use hybrid integration and the CMOS-SEED technology. Table 10.3 provides an overview of AT&T Bell Laboratories free-space photonic switching system demonstrations [90] to date and describes some of the technical details of each. These demonstrations have clearly led the community in the development of massively parallel interconnect and switching systems, custom optomechanical hardware, novel packaging and heat removal techniques, and other contributions critical to the successful evolution of a new technology such as smart pixels. These free-space photonic switching demonstrations proved that large arrays of SEED-based switching fabrics could be fabricated with high yield, excellent uniformity, and high-speed individual channel performance. Since the first edition of this chapter, demonstrations of these types have been limited. Instead, focus has shifted to improvements to transmitter and receiver circuit performance including higher speed operation, improved noise immunity of the receivers, and reductions in power consumption and circuit size.

In an attempt to provide higher efficiency interfaces between the optical devices and the electronic digital-processing elements, new optical receiver designs have been investigated. Clocked-sense-amplifier-based optical receivers are synchronous optical receivers that provide the potential for low-power, compact digital amplification. In [91], receivers of this type experimentally demonstrated operation in excess of 750 Mbps within a layout area of 44 μm × 22 μm with a bias-dependent estimated power dissipation of 1 to 2 mW. A two-beam transimpedance smart pixel optical receiver using a CMOS-SEED design demonstrated 1 Gbps performance within the same physical layout area [92]. Both of the receiver circuits used in these demonstrations were designed in a 0.8-μm CMOS foundry process. Recently, an integrated CMOS photoreceiver designed in a 0.35-μm production CMOS process demonstrated 1 Gbps operation with an average input power of -6 dBm and 1.5 mW of dissipated power [93]. Others have demonstrated similar 1 Gbps performance in a monolithically integrated silicon NMOS optical receiver [94]. These demonstrations clearly show the potential for high-speed operation of integrated receiver circuitry in smart pixel designs.

Other groups have also investigated novel optoelectronic routing architectures [95], approaches for optoelectronic interconnections of smart pixel arrays [96, 97], and other smart pixel-based switching architectures.

Table 10.3 AT&T Bell Laboratories Free-Space Photonic Switching System Demonstrations. †System$_7$ figures are Goals. To Date, the System$_7$ Smart Pixel has Demonstrated 625 Mbps Operation with 1024 Differential Optical Inputs and 1024 Differential Optical Outputs

System parameters	System$_1$ (1988)	System$_2$ (1989)	System$_3$ (1990)	System$_4$ (1991)	System$_5$ (1993)	System$_6$ (1995)	System$_7$† (1996)
Channels	2	32	64	1024	32	256	1024
Bit Rate	10 Kbps	30 Kbps	100 Kbps	1 Mbps	155 Mbps	208 Mbps	622 Mbps
Technology	S-SEED	S-SEED	S-SEED	S-SEED	FET-SEED	CMOS-SEED	CMOS-SEED
Chip/System I/O	12/20	256/1024	1024/3072	10,240/61,440	192/960	4352/65,536	4096/1,048,576
Hardware:							
Optics	Catalog	Catalog	Catalog	Custom	Custom	Custom	Custom
Mechanics	Catalog	Hybrid	Custom	Custom	Custom	Custom	Custom
System Area	32 ft^2	16 ft^2	6 ft^2	0.78 ft^2	1.16 ft^2	0.29 ft^2	0.18 ft^2
Reference	70	23	71	72	73, 74	5	6

Other groups are investigating the implementation of digital signal-processing algorithms in smart pixel technology. Several groups are applying these techniques to image processing applications [98, 99], optoelectronic Field Programmable Gate Arrays [100,101], advanced imaging systems [102,103] and artificial retinal prosthesis [104].

10.5.2. SIGNAL PROCESSING

A significantly smaller subset of the smart pixel community is focusing on analog or mixed signal applications of smart pixels for signal or image processing. In the following sections, we introduce two specific applications of smart pixel technology to a mixed technology implementation of error diffusion coding: digital image halftoning and oversampled analog-to-digital (A/D) conversion.

10.5.2.1. Digital Image Halftoning

Digital image halftoning is an important class of A/D conversion within the context of image processing. Halftoning can be thought of as an image compression technique whereby a continuous-tone, gray-scale image is printed or displayed using only *binary-valued* pixels. Error diffusion is one method of achieving digital halftoning in which the error associated with a nonlinear quantization process is diffused within a local region and subsequent filtering methods employed in an effort to improve some performance metric such as signal-to-noise ratio. Classical error diffusion [105, 106] is a one-dimensional, serial technique in which the algorithm raster scans the image from upper-left to lower-right and, as a result, introduces visual artifacts directly attributable to the halftoning algorithm itself.

Artificial neural networks and their application to image halftoning have received considerable attention lately [107,108]. This popularity lies in their ability to minimize a particular metric associated with a highly nonlinear system of equations. Specifically, the problem of creating a halftoned image can be cast in terms of a nonlinear quadratic optimization problem where the performance metric to be minimized is the difference between the original and the halftoned images.

Figure 10.16 shows the error diffusion architecture and an electronic implementation of a four-neuron error diffusion-type neural network. Here the individual neurons are represented as amplifiers (standard and inverting)

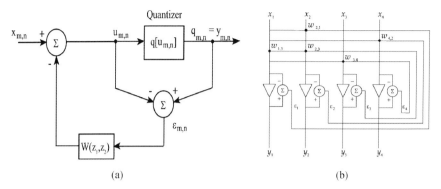

Fig. 10.16 (a) Two-dimensional error diffusion architecture, (b) Four-neuron electronic implementation.

and the synapses by the physical connections between the input and output of the amplifiers. Resistors are typically used to make these connections.

The energy function of this error diffusion neural network can be described by

$$E(x, y) = \mathbf{y}^T \mathbf{A} \mathbf{y} - 2\mathbf{y}^T \mathbf{A} \mathbf{x} + \mathbf{x}^T \mathbf{A} \mathbf{x} = [\mathbf{B}(\underbrace{\mathbf{y} - \mathbf{x}}_{\text{error}})]^T [\mathbf{B}(\underbrace{\mathbf{y} - \mathbf{x}}_{\text{error}})],$$

where $\mathbf{y} \in \{-1, 1\}$ is a vector of quantized states with one element per pixel, $\mathbf{A} = (\mathbf{I} + \mathbf{W})^{-1}$, $\mathbf{W}$ is derived from the original error diffusion filter weights through the relationship $W(i, j) = -w[(j-i) \text{ div } N, (j-i) \text{ mod } N]$ and $\mathbf{A} = \mathbf{B}^T \mathbf{B}$. From this equation it is clear that as the neural network converges and the energy function is minimized, so too is the error between the output halftoned image and the input gray-scale image.

Figure 10.17 (a) and (b) show 348 × 348 halftoned images of the Cadet Chapel at West Point using pixel-by-pixel thresholding and this error diffusion neural network, respectively. Recently, a smart pixel approach to digital image halftoning has been investigated [7, 109, 110]. The smart pixel implementation provides the advantage that all pixel quantization decisions are computed in parallel and therefore the error diffusion process becomes two-dimensional and symmetric. Visual artifacts attributable to the halftoning algorithm are eliminated and overall halftoned image quality is significantly improved. Also, the inherent parallelism associated with optical processing reduces the computational requirements while decreasing the total convergence time of the halftoning process. Over the past several years, research efforts have been focused on designing, characterizing, and

(a)   (b)

Fig. 10.17 Halftoned images of the Cadet Chapel at West Point. (a) using pixel-by-pixel thresholding, and (b) using an error diffusion neural network.

incrementally improving a smart pixel implementation of the error diffusion neural network based on a specific hybrid integration technique. In this research, SEED MQW modulators are flip-chip bonded to CMOS VLSI silicon circuitry to create the CMOS-SEED smart pixel.

10.5.2.1.1. 5 × 5 CMOS-SEED Smart Pixel Array

The following summarizes results of two consecutive generations of CMOS-SEED smart pixel arrays. The first generation CMOS-SEED array consisted of a 10 × 10 array of SEED MQW modulators integrated with 0.8-μm CMOS silicon circuitry while the second generation integrated an identical SEED array with 0.5-μm CMOS silicon circuitry. The focus of both of these designs was to demonstrate the usefulness of smart pixel technology to this specific analog neural application. Although optical weighting and interconnections are clearly the preferred method of achieving the necessary neural interconnections, neither of these designs incorporated this approach. Instead, electronic weighting and interconnections were incorporated in the design with analysis focused on optimizing the specific performance of the analog neural circuitry and the receiver and driver circuitry for the optical input and output signals, respectively. In a parallel effort, diffractive optical weighting and interconnections have been designed and experimentally characterized for a future smart pixel architecture [111,112].

10.5.2.1.2. First-Generation CMOS-SEED

The functionality necessary to implement the error diffusion neural network consists of a one-bit quantizer, two differencing nodes, and the interconnection and weighting of the error diffusion filter. Figure 10.18 shows the first-generation circuitry for a single neuron of the error diffusion neural network using the CMOS-SEED smart pixel technology.

All state variables in each circuit are represented as currents. Beginning in the upper left of the circuit and then proceeding in a clockwise direction, the input optical signal incident on the SEED is continuous in intensity and represents the individual analog pixel intensity. The input SEED at each neuron converts the optical signal to a photocurrent and subsequently, current mirrors are used to buffer the input. The width-to-length ratio of the metal oxide semiconductor field effect transistors (MOSFETs) determine the current gain used to amplify the photocurrent. The first circuit produces two output signals: $+I_u$, which represents the state variable $u(m, n)$ as the input to the quantizer, and $-I_u$, which represents the state variable $-u(m, n)$ as the input to the feedback differencing node. The function of the quantizer is to provide a smooth, continuous thresholding function for the neuron producing the output signal I_{out}, which corresponds to the state variable $y(m,n)$. This second electronic circuit is a modified wide-range transconductance amplifier that produces a hyperbolic tangent sigmoidal function when operated in the sub-threshold regime. Additional transistors are included in the modification to provide input current to output current functionality. The third circuit takes as its input I_{out}, the state variable $y(m,n)$, produces a replica of the original signal, and drives an output optical SEED. In this case, the output optical signal is a binary quantity represented as the presence or absence of light. Here, the photoemission process is electroluminescence, which results from the SEED being forward-biased. The last circuit at the bottom of the schematic implements the error weighting and distribution function of the error diffusion filter. The weighting is implemented by again scaling the width-to-length ratio of the MOSFETs to achieve the desired weighting coefficients. The neuron-to-neuron interconnections are accomplished using the four metalization layers of the 0.8-μm silicon CMOS process. The difficulty encountered using this weighting and interconnect approach was that we were forced to constrain the size of the diffusion kernel to a 5 × 5 array because of the physical extent of the circuitry necessary to implement this functionality. Even with this reduced filter kernel, the weighting and interconnect circuitry consumed over 75%

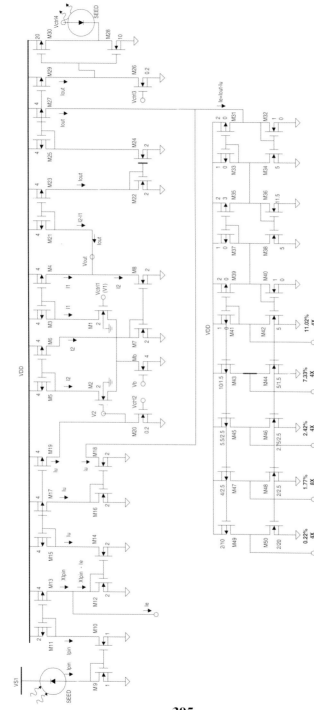

Fig. 10.18 Circuit diagram for the first-generation smart pixel implementation of the error diffusion neural network based on a CMOS-SEED-type smart pixel architecture.

Fig. 10.19 Photomicrograph of a smart pixel implementation of a 5 × 5 error diffusion neural network for digital image halftoning.

of the total silicon area. As a result, a smaller filter with a 5 × 5 region of support was designed and these coefficients were used in the smart pixel architecture. The five unique coefficients in this filter are shown at the bottom of the error diffusion circuitry along with the number of additional replications necessary to implement the 5 × 5 filter.

Figure 10.19 shows a photomicrograph of a 5 × 5 error diffusion neural network that was fabricated using CMOS-SEED smart pixels.

The CMOS circuitry was produced using an 0.8-μm silicon process and the SEED modulators were subsequently flip-chip bonded to silicon circuitry using the same integration process described in Section 10.3.1.3.1. This smart pixel implementation resulted in a system with approximately 90 transistors per smart pixel and a total transistor count of almost 1800. A total of 50 optical input/output channels is provided in this implementation.

Figure 10.20 shows a photomicrograph of a single neuron of the 5 × 5 neural network. The rectangular features are the MQW modulators while the silicon circuits are visible between and beneath the modulators. The MQW modulators are approximately 70 μm × 30 μm and have optical windows that are 18 μm × 18 μm.

10. New Devices for Optoelectronics: Smart Pixels 397

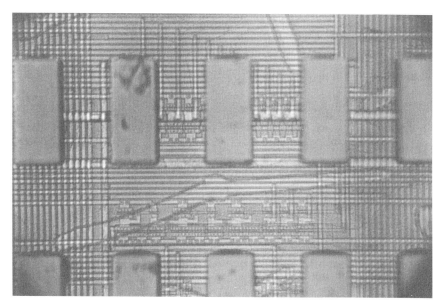

Fig. 10.20 Photomicrograph of a single neuron of the 5 × 5 error diffusion neural network.

Experimental testing of this first-generation smart pixel array confirmed electrical functionality of the quantizer operation as well as the replication features of the current mirrors within the architecture. Electroluminescence of the SEED modulators has also been verified and characterized. These experimental results demonstrated the importance of transistor matching within the wide-range transconductance amplifier and the error weighting circuitry. The results from this first-generation design and characterization were incorporated into the second-generation design in an effort to improve individual circuit operation and, as a result, improve overall network performance.

10.5.2.1.3. Second-Generation CMOS-SEED

In the design of the second-generation CMOS-SEED smart pixel, we concentrated on improving the performance of the neural circuitry. Specifically, we improved the error weighting circuitry and the quantizer performance. We also attempted to accurately model the operation of each of the functional elements as well as the total nonlinear network performance using PSpice simulation tools. The same 5 × 5 array size and functionality of one-bit quantization, subtraction, neuron-to-neuron weighting and interconnection, and optical input and output were retained. The electronic circuitry for the error diffusion neural network was implemented this time in 0.5-μm silicon CMOS technology. Figure 10.21 shows the circuitry associated with a single neuron of this second-generation error diffusion neural network.

The one-bit quantizer is again implemented using a modified wide-range transconductance amplifier [113] operated in the sub-threshold regime. The slope of this sigmoidal function was carefully designed by matching MOSFET transistors to meet the convergence criteria and the nonlinear dynamics of the error diffusion neural network. The error weighting circuitry at the bottom represents only the largest weight (11.02%) with interconnects to its four local neighbors ($I_{outA} - I_{outD}$). Again, all state variables are represented as currents. The error weighting and distribution circuitry for the 5×5 array was again implemented in silicon circuitry, matching individual and stage-to-stage MOSFET transistors. Bi-directional error currents were implemented to provide the circuitry and the network with fully symmetric performance. The specific improvements to the second-generation circuitry include improving stage-to-stage isolation by tying the drain and source of selected transistors together, as seen with transistors M19 and M20 in the schematic. This configuration places a constraint on the common node, which helped isolate adjoining stages. This was particularly important to accurate circuit operation because the design of the wide-range transconductance amplifier required small feature size transistors compared to the input buffer and error weighting circuitry. In the circuit layout, we also changed several of the transistors in the wide-range transconductance amplifier to ensure that small width-to-length ratios were used in an effort to minimize channel length modulation effects. The SEED driver circuit was redesigned to provide additional isolation between the large currents used to forward bias the SEED and the balance of the sensitive neural circuitry. Finally, the transistor characteristics were accurately matched throughout the network using extracted transistor data from a previous 0.5-μm foundry run. The central neuron of this new smart pixel array consists of approximately 160 transistors while the complete 5×5 array accounts for over 3600 transistors, a nearly two-fold increase in the total transistor count over our first-generation design. The central neuron is interconnected to the surrounding 24 neurons in the 5×5 array using the same fixed interconnect and weighting scheme as in the first-generation realization. PSpice simulations of the complete 5×5 nonlinear dynamical neural network were performed using transistor parameters extracted from a previous 0.5 μm

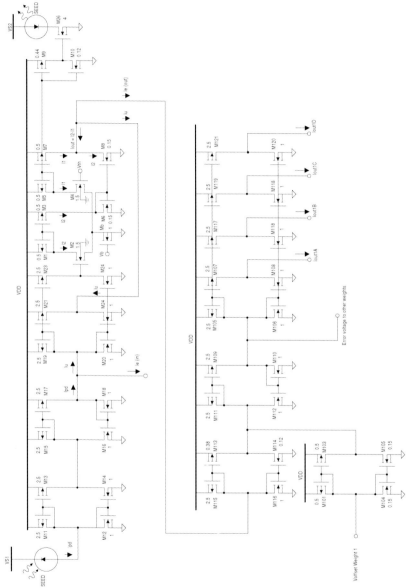

Fig. 10.21 Circuit diagram of a single neuron and a single error weight of the 5 × 5 error diffusion neural network based on a CMOS-SEED-type smart pixel architecture.

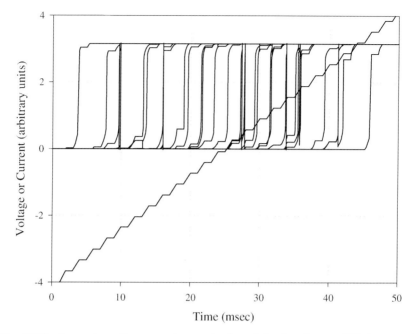

Fig. 10.22 Dynamic performance characterization of the 5 × 5 error diffusion neural network to a stepped linear input.

MOSIS foundry. Figure 10.22 shows the PSpice simulation for the performance of the complete 5 × 5 array to a linear input signal.

As the input signal is linearly increased over the input dynamic range, the number of neurons in the *on*-state increases from 0 to 25, representing the analog input level at any given time interval as the ratio of the number of neurons in the *on*-state to those in the *off*-state. The non-uniform rate of increase in the number of neurons in the *on*-state is predictable and is a result of the artificially small network size.

Figure 10.23 shows an expanded segment of Fig. 10.22, which provides an analysis of the transient behavior of the network.

These simulations predict individual neuron switching speeds of less than $1\mu s$, demonstrating the capability for real-time digital image halftoning. Individual component functionality was experimentally characterized and dynamic operation of the full 5 × 5 neural array was also experimentally characterized. Dynamic stimulus of single and multiple neurons was conducted and demonstrated correct error diffusion and network operation. Figure 10.24 shows CCD images of the operational 5 × 5 CMOS-SEED smart pixel array. Figure 10.24(a) shows the fully-functioning array while

10. New Devices for Optoelectronics: Smart Pixels   401

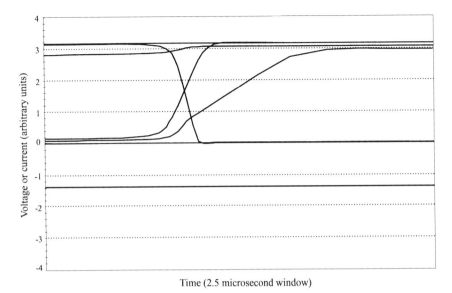

Fig. 10.23 Transient analysis of 5 × 5 network performance.

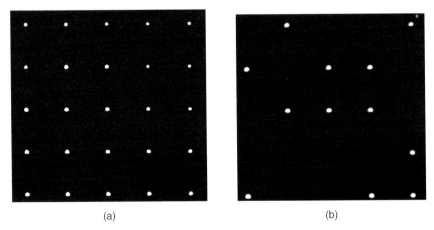

(a)   (b)

Fig. 10.24 CCD images of the 5 × 5 CMOS-SEED array (a) fully-operational array and (b) under 50% gray scale illumination.

Fig. 10.24(b) shows the 5 × 5 CMOS-SEED neural array under 50% grayscale input. Here 50% of the SEED modulators are shown to be in the *on*-state.

Both the PSpice simulations and the experimental results demonstrate that this approach to a smart pixel implementation of the error diffusion neural network provides sufficient accuracy for the digital halftoning application. The individual neuron switching speeds also demonstrate the capability for this smart pixel hardware implementation to provide real-time halftoning of video images.

Partitioning of large images for parallel computation and compatibility with smart pixel hardware has also been an active area of research recently [114,115].

10.5.2.2. Analog-to-Digital Conversion

Because the majority of signals encountered in nature are continuous in both time and amplitude, the A/D interface is generally considered to be the most critical part of any overall signal acquisition and processing system. Because of the difficulty in achieving high-resolution and high-speed A/D converters, this A/D interface has been and continues to be a barrier to the realization of high-speed, high-throughput systems.

Recently, there has been renewed interest in new and innovative approaches to A/D conversion, with a significant emphasis on photonic approaches that provide high-speed clocking, broadband sampling, reduced mutual interference of signals, and compatibility with existing photonic-based systems. A more complete description of photonic A/D conversion advantages and approaches can be found in [116]. Most of the efforts involving smart pixel technology have focused on the use of oversampling A/D converters such as $\Delta\Sigma$- or $\Sigma\Delta$-modulators and error diffusion modulators. Here we will provide a brief overview of these applications of smart pixel technology.

Within the electronic A/D converter community, temporal oversampling and spectral noise shaping have become common practice in high-fidelity audio applications. Oversampling $\Delta\Sigma$- or $\Sigma\Delta$-modulators [117,118] are routinely used to provide resolution in excess of 16-bits at audio-frequency bandwidths. Here, a low-resolution quantizer is embedded in a feedback architecture in an effort to reduce the quantization noise through spectral noise shaping. A large error associated with a single sample is diffused over many subsequent samples and then linear filtering techniques are applied to remove the spectrally shaped noise, thereby improving the overall signal-to-noise ratio (SNR) of the converter. In the temporal oversampling approach, the output data sequence is subsequently processed by a digital postprocessor, which consists of a low-pass filter and decimation circuitry. The final output is a high-resolution multi-bit digital word, which accurately approximates the analog input sample. Although not normally

10. New Devices for Optoelectronics: Smart Pixels

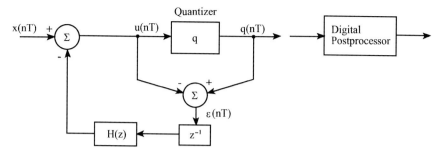

Fig. 10.25 Block diagram of an oversampled A/D converter consisting of a recursive error diffusion modulator and a digital postprocessor.

used in high-speed applications, oversampling converters have been shown to be tolerant to nonidealities in components, imperfections in sampling, and even broadband noise.

Figure 10.25 shows the block diagram for one realization of an oversampled A/D converter using a recursive error diffusion modulator and a digital postprocessor.

In this architecture, the error introduced by the quantizer is physically computed and temporally redistributed to subsequent samples according to the feedback filter H(z) in an effort to influence future quantization decisions. An analysis of this process in the frequency domain shows that the quantization error is spectrally shaped and redistributed to frequencies above the Nyquist frequency of the original sampled signal and then subsequently removed by low-pass filtering and decimation provided by the digital postprocessor.

The improvement in the performance of an oversampled A/D converter over that of a conventional Nyquist-rate converter can be seen by examining the SQNR. The maximum SQNR for both a Nyquist rate A/D converter and an oversampled A/D converter can be quantitatively described by

$$SQNR_{max}(b) = 3 \cdot 2^{2b-1} \quad \text{and} \quad SQNR_{max}(M, N) = \frac{3}{2} \cdot \left[\frac{2N+1}{\pi^{2N}}\right] \cdot M^{2N+1},$$

respectively. Here, b is the number of bits resolution, N represents the order of the modulator defined as the order of the feedback filter, and M is the oversampling ratio that is the ratio of the sampling frequency to the Nyquist frequency of the sampled signal. Second- and third-order modulators are common practice in the audio industry, providing in excess of 16 bits of resolution at audio frequency bandwidths.

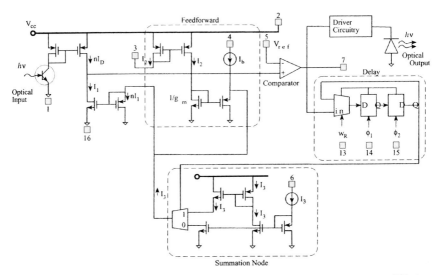

Fig. 10.26 Circuit diagram for a smart pixel implementation of a first-order error diffusion modulator based on an ELO-type smart pixel architecture.

In an effort to extend the advantages of oversampled A/D conversion beyond audio frequency bandwidths, optoelectronic implementations have been investigated [24]. In this demonstration, discrete MQW modulators were used to provide the functionality necessary to implement a first-order oversampled modulator. Recently, extensions to this development have included smart pixel technology. Figure 10.26 shows the circuitry necessary to implement a first-order error diffusion modulator using an ELO-type smart pixel technology.

Here, the intensity of the optical input represents the continuous amplitude signal to be A/D converted. The numbers represent electrical pin-outs on the final integrated circuit chip. A silicon photodetector converts the light intensity to a photocurrent $x(nT)$, which is then buffered and amplified by the first set of current mirrors. The feedforward section provides a scaled replica of the state variable $u(nT)$ for use in the quantizer differencing node and as the input to the comparator. The comparator provides the one-bit quantization necessary for A/D conversion, producing the output state variable $y(nT)$. The driver circuitry provides the necessary current gain to drive the active emitter, which in this demonstration is a resonant cavity-enhanced (RCE) LED. The delay circuit is driven by a two-phase, non-overlapping clock and provides a one-bit delay for the first-order modulator. The summation node circuitry provides voltage-to-current conversion for the feedback state variable $y(nT)$.

The RCE LEDs [119] provide several distinct advantages over conventional LEDs. The RCE LED is fabricated by placing the gain structure of a conventional LED into a Fabry-Perot cavity. The resonant cavity enhances the LED output spectrum, reducing the linewidth and providing increased directionality in the output. RCE LED's therefore offer a compromise between conventional LEDs and VCSELs. These active emitters can also be fabricated as thin-film structures and are therefore ideal candidates for ELO-type hybrid integration in smart pixel applications. Recently, RCE thin-film AlGaAs/GaAs/AlGaAs LEDs with metal mirrors experimentally demonstrated turn-on voltages of 1.3 volts, linewidths of 10.4 nm, dispersion half-angles of 23.7°, and stable output over more than 1700 hours of operation [119].

Figure 10.27 shows a photomicrograph of a smart pixel implementation of a first-order error diffusion modulator. The electronic circuitry was fabricated using a 2.0-μm CMOS process and was integrated with thin-film RCE LEDs using the hybrid ELO approach described previously in Section 10.3.1.3.2. Two RCE LEDs and associated drivers are prominently visible in the center of the picture.

The concept of optoelectronic $\Delta\Sigma$-modulation has also been investigated by others for applications in smart focal plane arrays and enhanced

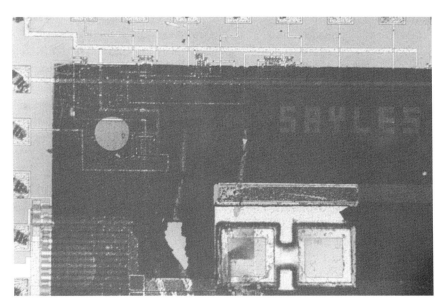

Fig. 10.27 Photomicrograph of an ELO-type smart pixel implementation of a first-order error diffusion modulator.

imaging arrays [120,121]. In these demonstations, 2-D arrays of $\Delta\Sigma$-modulators are used to provide front-end A/D conversion in imaging applications.

10.5.2.2.1. Distributed Photonic A/D Conversion

Recently, the concept of error diffusion, which is a fully equivalent, alternative form of $\Delta\Sigma$-modulation, has been extended to a 2-D symmetric error diffusion neural network for application to photonic A/D conversion [122,123,124]. This new approach to photonic A/D conversion leverages the 2-D nature of an optical architecture to extend the concept of spectral noise shaping to include both temporal and spatial error diffusion. While the concept of spatial oversampling for spectral noise shaping has been previously considered [125], this is the first to provide a methodology for interconnect weight design and to extend this concept to 2-D and photonics applications. This approach uses a mode-locked laser to generate the optical sampling pulses, an electrooptically driven interferometer to modulate the electronic analog signal onto the optical pulses, and a 2-D smart pixel hardware implementation of a distributed error diffusion neural network. In this approach, the input signal is first sampled at a rate higher than that required by the Nyquist criterion and then presented spatially as the input to a 2-D error diffusion neural network consisting of $M \times N$ neurons, each representing a pixel in the image space. The neural network processes the input image, producing an $M \times N$ pixel binary output image, referred to as a halftoned image. The data contained in the resulting 2-D halftoned image can be thought of as the 2-D corollary to the output data sequence from the $\Delta\Sigma$-modulator. In the ideal case, with a constant grayscale input image in which each pixel value is g, the sample average across the output binary image equals g. Upon convergence, the neural network minimizes an energy function representing the frequency-weighted squared error between the input analog image and the output halftoned image. Decimation and low-pass filtering techniques, common to temporal oversampling A/D converters, subsequently process the $M \times N$ pixel output binary image using high-speed digital electronic circuitry to obtain the desired high-resolution Nyquist-rate A/D conversion. By employing a 2-D smart pixel neural approach, each pixel constitutes a simple oversampling modulator that is interconnected to all other pixels in the array, thereby producing a distributed A/D architecture. Each quantizer within the network is embedded in a fully connected, distributed mesh feedback loop, which spectrally shapes the overall quantization noise. This spectral noise shaping diffuses individual quantizer errors across the array, thereby improving overall SNR performance and significantly reducing the effects of component mismatch typically associated with parallel or channelized A/D approaches.

In the 2-D error diffusion neural network, the effect of quantizer threshold mismatch can be analytically characterized as a modification to the neural network energy function

$$E = (\mathbf{y} - \mathbf{x})^T \mathbf{A} (\mathbf{y} - \mathbf{x}) + \varepsilon^T \mathbf{A}^{-1} \varepsilon - 2\varepsilon^T (\mathbf{y} - \mathbf{x}),$$

where $\mathbf{x}$ is a vector of input samples, $\mathbf{y}$ is a vector of quantized output values, $\mathbf{A}$ is a matrix describing the neural interconnect weighting and distribution, and ε is a vector containing the threshold mismatch of the array. The first term is the original energy function of the error diffusion neural network while the last two terms are contributions from threshold mismatch. By design $\mathbf{A}^{-1} \leq 1$ and $\mathbf{A}^{-1}$ has high-pass spectral characteristics that spectrally shape contributions from the mismatch to higher frequencies. In an effort to assess the potential performance of this new approach to distributed A/D conversion, a simulation model of the error diffusion neural network was developed and the 2-D output halftone data sequence was evaluated using classical spectral estimation techniques. Several different 2-D error diffusion filters were tested to verify the relationship between the spectral characteristics of the filter and the resulting noise shaping of the output halftone. Threshold and interconnect weight mismatch were modeled in a Monte Carlo style simulation. Threshold mismatch of $\pm 12\%$ and interconnect weight mismatch of $\pm 8\%$ for a 7×7 interconnect weight matrix resulted in less than 0.5 dB reduction in SNR.

Figure 10.28 shows a 3-D perspective plot of the power spectrum of the quantized output data from the error diffusion network using a 2-D input sinusoidal image described by $x = 0.5 + 0.5 \sin(2\pi f_1/m_i) \sin(2\pi f_2/n_i)$, $i = 1, 2, 3 \ldots N$ where $N = 256$ and $f_1 = f_2 = 2$. The inset shows the frequency response of the interconnect weights that constitute the error diffusion filter. The noise shaping of the output power spectrum corresponds, as predicted, to the frequency response of the error diffusion filter. Approximately 23 dB of low-frequency noise suppression is achieved with this filter, resulting in an SNR of 51.4 dB. Other, more aggressive filters have also been designed, which resulted in excess of 40 dB of low-frequency noise suppression resulting in an SNR in excess of 70 dB.

The smart pixel hardware implementation for this distributed approach to A/D conversion is identical to that described previously for the digital image halftoning application. The difference between the two hardware

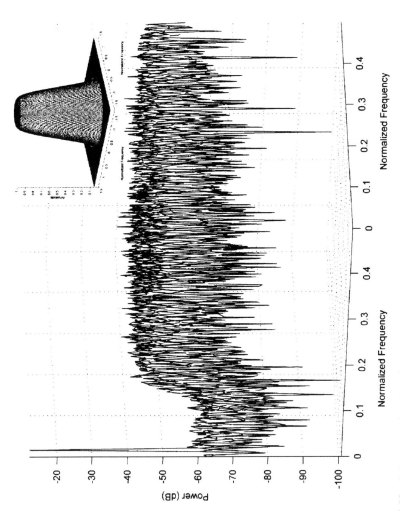

Fig. 10.28 3-D perspective plot of the power spectrum of the fully connected distributed mesh feedback architecture.

realizations is that this A/D application requires digital postprocessor functionality. Because this is a deterministic mapping, matrix-vector techniques can be applied as either hidden-layers in the error diffusion neural network or in an FPGA implementation that will integrate directly with the smart pixel chip.

10.6. Future Trends and Directions

Although smart pixels are fairly young in terms of technological maturity, they have received tremendous focus and have emerged as a potential solution to many interconnection and massively parallel processing challenges. Individual smart pixel device development has been replaced by system demonstrations and architectural development. At a recent conference dedicated to smart pixel technology [126], a majority of the papers were dedicated to issues such as optical power budget, efficient architectural designs, optimization of smart pixel receivers, and system demonstrations. Each of these topics relates to engineering issues reflecting a maturing of the basic technology. As both silicon and compound semiconductor processing techniques mature, so too will smart pixel devices. Larger arrays of optical devices will provide increased aggregate capacity while smaller feature sizes will lead to higher density electronic circuits which, in turn, will provide increased complexity. Systems that provide aggregate capacity in excess of 10^{12} bps can now be envisioned. Video-on-demand is one application that could immediately use this capability. The various integration techniques available with smart pixels provide the potential to integrate multi-spectral detectors on a common processing substrate. Large-scale and multi-spectral imagery applications in both the defense and commercial sectors could leverage some of the features of smart pixel technology.

Another potential area of application includes human perception and visualization. Portable smart pixel displays that project images the size of the human retina could provide processed information better suited to the reception and processing of the human brain. An interesting extension to the smart pixels discussed here is that of programmable smart pixels. Some recent work on field-programmable smart pixels [127,128] demonstrated the capability to dynamically program the electronic circuitry and therefore change the functionality of the smart pixel system. Although restricted to digital electronic circuitry, this approach to smart pixels holds promise for quickly reprogrammable interconnections for circuit switching and telecommunication applications.

Another technology area closely related to and which could have a significant impact on future smart pixel applications is that of microelectromechanical (MEM) device technology. Here, miromachining and nanofabrication techniques are used to create submicron mechanical devices in silicon and compound semiconductor materials. Optical applications of MEMs which would complement and extend the applications of smart pixel technology include mechanically tunable VCSELs [129], optical scanning mirrors [130], and tunable optical filters [131]. Other uses include microsensors and microactuators that could potentially be integrated on a common substrate with the smart pixels providing the capability for mechanical and optical sensing, processing, and mechanical actuating from the same integrated device.

The integration of electronics with optics has produced hybrid devices fully capable of leveraging the advantages of each individual technology and providing total system performance well beyond the capability of either individual technology. This technology could provide solutions to many of the electrical interconnection problems currently facing many high-speed optoelectronic data communication, telecommunication, and signal and image-processing applications.

References

1. Jenkins, B. K., A. A. Sawchuk, T. C. Strand, R. Forchheimer, and B. H. Soffer. 1984. "Sequential optical logic implementation." *Appl. Opt.* 23: 3455–3464.
2. Abu-Mostafa, Y. S., and D. Psaltis. 1987. "Optical neural computers." *Scientific American*, pp. 88–95.
3. Huang, A. 1984. "Architectural considerations involved in the design of an optical digital computer." *Proc. IEEE* 72: 780–786.
4. Hinton, H. S. 1988. "Architectural considerations for photonic switching networks." *IEEE J. Select Areas Commun.* 6: 1209–1226.
5. Hinterlong, S. J., A. L. Lentine, D. J. Reiley, J. M. Sasian, R. L. Morrison, R. A. Novotny, M. G. Beckman, D. B. Buchholz, T. J. Cloonan, and G. W. Richards. 1996. "An ATM switching system demonstration using a 40 Gb/s throughput smart pixel optoelectronic VLSI chip." *Proc. 1996 IEEE/LEOS Summer Topical Meeting on Smart Pixels*, Keystone, Colo., pp. 47–48.
6. Lentine, A. L., K. W. Goossen, J. A. Walker, J. E. Cunningham, W. Y. Jan, T. K. Woodward, A. V. Krishnamoorthy, B. J. Tseng, S. P. Hui, R. E. Leibenguth, L. M. F. Chirovsky, R. A. Novotny, D. B. Buchholz, and R. L. Morrison. 1996. "Optoelectronic switching chip with greater than 1 terabit

per second potential optical I/O bandwidth." *Proc. 1996 IEEE/LEOS Summer Topical Meeting on Smart Pixels*, Keystone, Colo., Postdeadline Paper 001.
7. Sayles, A. H., B. L. Shoop, E. K. Ressler. 1996. "A novel smart pixel network for signal processing applications." *Proc. 1996 IEEE/LEOS Summer Topical Meeting on Smart Pixels*, Keystone, Colo., pp. 86–87.
8. Shoop, B. L., and E. K. Ressler. 1994. "Optimal error diffusion for digital halftoning using an optical neural network." *Proc. First International Conf. on Image Processing*, (Institute of Electrical and Electronics Engineers, Austin, Tex.), pp. 1036–1040.
9. McCormick, F. B., F. A. P. Tooley, T. J. Cloonan, J. L. Brubaker, A. L. Lentine, R. L. Morrison, S. J. Hinterlong, M. J. Herron, S. L. Walker, and J. M. Sasian. 1991. "S-SEED-based photonic switching network demonstration." *OSA Proceedings on Photonics in Switching*, H. S. Hinton and J. W. Goodman (Eds.), pp. 48–55.
10. Cloonan, T. J., G. W. Richards, A. L. Lentine, F. B. McCormick, and J. R. Erickson. 1992. "Free-space photonic switching architectures based on extended generalized shuffle networks." *Appl. Opt.* 31: 7471–7492.
11. Woodward, T. K., A. L. Lentine, L. M. F. Chirovsky, M. W. Focht, J. M. Freund, G. D. Guth, R. E. Leibenguth, and L. E. Smith. 1993. "GaAs/AlGaAs FET-SEED receiver/transmitters." *OSA Proceedings on Photonics in Switching*, J. W. Goodman and R. C. Alferness (Eds.), pp. 81–84.
12. Miller, D. A. B., D. S. Chemla, T. C. Damen, A. C. Gossard, W. Wiegmann, T. H. Wood, and C. A. Burrus. 1984. "Band-edge electroabsorption in quantum well structures: the quantum-confined stark effect." *Phys. Rev. Lett.* 53: 2173–2177.
13. Miller, D. A. B. 1984. "Optical bistability and differential gain resulting from absorption increasing with excitation." *J. Opt. Soc. Am. B* 1: 857–864.
14. Miller, D. A. B., D. S Chemla, T. C. Damen, T. H. Wood, C. A. Burrus, A. C. Gossard, and W. Wiegmann. 1985. "The quantum well self-electro-optic effect device: optoelectronic bistability and oscillation, and self-linearized modulation," *IEEE J. Quantum Electron.* QE-21: 1462–1476.
15. Boyd, G. D., A. M. Fox, D. A. B. Miller, L. M. F. Chirovsky, L. A. D'Asaro, J. M. Kuo, R. F. Kopf, and A. L. Lentine. 1990. "33 ps optical switching of symmetric self electro-optic effect devices." *Appl. Phys. Lett.* 57: 1843–1845.
16. Lentine, A. L., F. B. McCormick, R. A. Novotny, L. M. F. Chirovsky, L. A. D'Asaro, R. F. Kopf, J. M. Kuo, and G. D. Boyd. 1990. "A 2-kbit array of symmetric self-electrooptic effect devices." *IEEE Photonics Technol. Lett.* 2: 51–53.
17. Whitehead, M., A. Rivers, G. Parry, J. S. Roberts, and C. Button. 1989. "Low-voltage multiple quantum well modulators with on:off ratios > 100:1." *IEEE Photonics Technol. Lett.* 25: 984–986.

18. Yan, R. H., R. J. Simes, and L. A. Coldren. 1989. "Electroabsorptive fabry-perot reflection modulators with asymmetric mirrors." *IEEE Photonics Technol. Lett.* 1: 273–275.
19. Pezeshki, B., D. Thomas, and J. S. Harris, Jr. 1990. "Optimization of modulation ratio and insertion loss in reflective electroabsorption modulators." *Appl. Phys. Lett.* 57: 1491–1493.
20. Law, K.-K., R. H. Yan, L. A. Coldren, and J. L. Merz. 1990. "Self-electrooptic device based on superlattice asymmetric fabry-perot modulator with an on:off ratio >100." *Appl. Phys. Lett.* 57: 1345–1347.
21. Shoop, B. L., B. Pezeshki, J. W. Goodman, and J. S. Harris, Jr. 1992. "Non-interferometric optical subtraction using reflection-electroabsorption modulators." *Optics Lett.* 17: 58–60.
22. Lentine, A. L., H. S. Hinton, D. A. B. Miller, J. E. Henry, J. E. Cunningham, and L. M. F. Chirovsky. 1989. "Symmetric self-electrooptic effect device: optical set-reset latch, differential logic gate, and differential modulator/detector." *IEEE J. Quantum Electron.* 25: 1928–1936.
23. Cloonan, T. J., M. J. Herron, F. A. P. Tooley, G. W. Richards, F. B. McCormick, E. Kerbis, J. L. Brubaker, and A. L. Lentine. 1990. "An all optical implementation of a 3D crossover network." *IEEE Photonics Technol. Lett.* 2: 438–440.
24. Shoop, B. L., and J. W. Goodman. 1992. "Optical oversampled analog-to-digital conversion." *Appl. Opt.* 31: 5654–5660.
25. DeSouza, E. A., L. Carraresi, G. D. Boyd, and D. A. B. Miller. 1994. "Self-linearized analog differential self-electro-optic-effect device," *Appl. Opt.* 33: 1492–1497.
26. Shoop, B. L., B. Pezeshki, J. W. Goodman, and J. S. Harris, Jr. 1992. "Laser power stabilization using a quantum well modulator." *IEEE Photonics Technol. Lett.* 4: 136–139.
27. Miller, D. A. B. 1993. "Novel analog self-electrooptic-effect devices." *IEEE J. Quantum Electron.* 29: 678–698.
28. Krishnamoorthy, A., and K. W. Goosen. 1998. "Optoelectronic-VLSI: photonics integrated with VLSI circuits." *IEEE J. Select. Topics Quantum Electron.* 4: 899–912.
29. Woodward, T. K., L. M. F. Chirovsky, A. L. Lentine, L. A. D'Asaro, E. J. Laskowski, M. Focht, G. Guth, S. S. Pei, F. Ren, G. J. Przybylek, L. E. Smith, R. E. Leibenguth, M. T. Asom, R. F. Kopf, J. M. Kuo, and M. D. Feuer. 1992. "Operation of a fully-integrated GaAs-Al_xGa_{1-x}As FET-SEED: a basic optically addressed integrated circuit." *IEEE Photonics Technol. Lett.* 4: 614–617.
30. Grot, A. C., D. Psaltis, K. V. Shenoy, and C. G. Fonstad, Jr. 1994. "Integration of LED's and GaAs circuits by MBE regrowth." *IEEE Photonics Technol. Lett.* 6: 819–821.

31. Shenoy, K. V., C. G. Fonstad, Jr., A. C. Grot, and D. Psaltis. 1995. "Monolithic optoelectronic circuit design and fabrication by epitaxial growth on commercial VLSI GaAs MESFETs." *IEEE Photonics Technol. Lett.* 7: 508–510.
32. Cheng, J., P. Zhou, S. Z. Sun, S. Hersee, D. R. Myers, J. Zolper, and G. A. Vawter. 1993. "Surface-emitting laser-based smart pixels for two-dimensional optical logic and reconfigurable optical interconnections." *IEEE J. Quantum Electron*. QE-2: 741–756.
33. Matsuo, S., T. Nakahara, Y. Kohama, Y. Ohiso, S. Fukushima, and T. Kurokawa. 1995. "Monolithically integrated photonic switching device using an MSM PD, MESFET's, and a VCSEL." *IEEE Photonics Technol. Lett.* 7: 1165–1167.
34. Zhou, P., J. Cheng, C. F. Schaus, S. Z. Sun, K. Zheng, E. Armour, W. Hsin, D. R. Myers, and G. A. Vawter. 1991. "Low series resistance high efficiency GaAs/AlGaAs vertical-cavity surface-emitting lasers with continuously-graded mirrors grown by MOCVD." *IEEE Photonics Technol. Lett.* 3: 591–593.
35. Goossen, K. W., G. D. Boyd, J. E. Cunningham, W. Y. Jan, D. A. B. Miller, D. S. Chemla, and R. M. Lum. 1989. "GaAs-AlGaAs multiquantum well reflection modulators grown on GaAs and silicon substrates." *IEEE Photonics Technol. Lett.* 1: 304–306.
36. Weiland, J., H. Melchior, M. Q. Kearley, C. Morris, A. J. Moseley, M. G. Goodwin, and R. C. Goodfellow. 1991. "Optical receiver array in silicon bipolar technology with self-aligned, low parasitic III/V detectors for DC-1 Gbit/s parallel links." *Electron. Lett.* 27: 2211–2113.
37. Camperi-Ginestet, C., M. Hargis, N. Jokerst, and M. Allen. 1991. "Alignable epitaxial liftoff of GaAs materials with selective deposition using polyimide diaphragms." *IEEE Photonics Technol. Lett.* 3: 1123–1125.
38. Yoffe, G. W., and J. M. Dell. 1991. "Multiple-quantum-well reflection modulator using a lift-off GaAs-AlGaAs film bonded to gold on silicon." *Electron. Lett.* 27: 557–559.
39. Yanagisawa, M., H. Terui, K. Shuto, T. Miya, and M. Kobayashi. 1992. "Film-level hybrid integration of AlGaAs laser diode with glass waveguide on silicon substrate." *IEEE Photonics Technol. Lett.* 4: 21–23.
40. Goossen, K. W., J. E. Cunningham, and W. Y. Jan. 1993. "GaAs 850 nm modulators solder-bonded to silicon." *IEEE Photonics Technol. Lett.* 5: 776–778.
41. Goossen, K. W., J. A. Walker, L. A. D'Asaro, S. P. Hui, B. Tseng, R. Leibenguth, D. Kossives, D. D. Bacon, D. Dahringer, L. M. F. Chirovsky, A. L. Lentine, and D. A. B. Miller. 1995. "GaAs MQW modulators integrated with silicon CMOS." *IEEE Photonics Technol. Lett.* 7: 360–362.
42. Lentine, A. L., K. W. Goossen, J. A. Walker, L. M. F. Chirovsky, L. A. D'Asaro, S. P. Hui, B. T. Tseng, R. E. Leibenguth, D. P. Kossives, D. W.

Dahringer, and D. A. B. Miller. 1995. "8 × 8 array of optoelectronic switching nodes comprised of flip-chip-solder-bonded MQW modulators on silicon CMOS circuitry." *Photonics In Switching*, 1995 OSA Technical Digest Series, 12: 13–15.

43. Lentine, A. L., K. W. Goossen, J. A. Walker, L. M. F. Chirovsky, L. A. D'Asaro, S. P. Hui, B. T. Tseng, R. E. Leibenguth, J. E. Cunningham, D. W. Dahringer, D. P. Kossives, D. D. Bacon, R. L. Morrison, R. A. Novotny, and D. B. Buchholz. 1995. "High speed optoelectronic VLSI switching chip with greater than 4000 optical I/O based on flip chip bonding of MQW modulators and detectors to silicon CMOS." *1995 Conf. on Lasers and Electro-Optics*, Postdeadline Paper.

44. Mosely, A. J., M. Q. Kearley, R. C. Morris, J. Urquhart, M. J. Goodwin, and G. Harris. 1991. "8 × 8 flipchip assembled In GaAs detector arrays for optical interconnect." *Electron. Lett.* 27: 1566–1567.

45. Morgan, R. A., K. C. Robinson, L. M. F. Chirovsky, M. W. Focht, G. D. Guth, R. E. Leibenguth, K. G. Glogovsky, G. J. Przybylek, and L. E. Smith. 1991. "Uniform 64 × 1 arrays of individually-addressed vertical cavity top surface emitting lasers." *Electron. Lett.* 27: 1400–1401.

46. Chang-Hasnain, C. J., J. P. Harbison, C.-E. Zah, M. W. Maeda, L. T. Florez, N. G. Stoffel, and T.-P. Lee. 1991. "Multiple wavelength tunable surface-emitting laser arrays." *IEEE J. Quantum Electron.* 27: 1368–1376.

47. Banwell, T. C., A. C. Von Lehmen, and R. R. Cordell. 1993. "VCSE laser transmitters for parallel data links." *IEEE J. Quantum Electron.* 29: 635–644.

48. Yablonovitch, E., T. Gmitter, J. P. Harbison, and R. Bhat. 1987. "Extreme selectivity in the liftoff of epitaxial GaAs films." *Appl. Phys. Lett.* 51: 2222–2224.

49. Yablonovitch, E., E. Kapon, T. J. Gmitter, C. P. Yun, and R. Bhat. 1989. "Double heterostructure GaAs/AlGaAs thin film diode lasers on glass substrates." *IEEE Photonics Technol. Lett.* 1: 41–42.

50. Camperi-Ginestet, C., M. Hargis, N. Jokerst, and M. Allen. 1991. "Alignable epitaxial liftoff of GaAs materials with selective deposition using polyimide diaphragms." *IEEE Photonics Technol. Lett.* 3: 1123–1126.

51. Jokerst, N. M. 1994. "Parallel processing: into the next dimension." *Optics and Photonics News* 5: 8–14.

52. Yablonovitch, E., D. M. Hwang, T. J. Gmitter, L. T. Florez, and J. P. Harbison. 1990. "Van der Waals bonding of GaAs epitaxial liftoff films onto arbitrary substrates." *Appl. Phys. Lett.* 56: 2419–2421.

53. Calhoun, K. H., C. B. Camperi-Ginestet, and N. M. Jokerst. 1993. "Vertical optical communication through stacked silicon wafers using hybrid monolithic thin film In GaAsP emitters and detectors." *IEEE Photonics Technol. Lett.* 5: 254–257.

54. Wills, D. S. 1996. "Smart pixel architectures for image processing." *Proc. of 1996 IEEE/LEOS Summer Topical Meeting on Smart Pixels*, Keystone, Colo., pp. 93–94.
55. Meyer, R. B., L. Liebert, J. Strzelecki, and P. Keller. 1975. "Ferroelectric liquid crystals." *J. de Phys. Lett.* 36: L69–71.
56. Mao, C. C., and K. M. Johnson. 1993. "Optoelectronic array that computes error and weight modification for a bipolar optical neural network." *Appl. Opt.* 32: 1290–1296.
57. Jared, D. A., and K. M. Johnson. 1991. "Optically addressed thresholding very-large-scale-integration/liquid-crystal spatial light modulators." *Opt. Lett.* 16: 767–769.
58. Kranzdorf, M., K. M. Johnson, J. Bigner, and L. Zhang. 1989. "An optical connectionist machine with polarization-based bipolar weights." *Opt. Eng.* 28: 844–848.
59. Wagner, K., and T. M. Slagle. 1993. "Optical competitive learning with VLSI/liquid-crystal winner-take-all modulators." *Appl. Opt.* 32: 1408–1435.
60. Johnson, K. M., D. J. McKnight, and I. Underwood. 1993. "Smart spatial light modulators using liquid crystals on silicon." *IEEE J. Quantum Electron.* QE-2: 699–714.
61. Wilkinson, T. D., N. New, and W. A. Crossland. 2000. "Optical comparator based on FLC LCOS technology." *Proc. 2000 IEEE/LEOS Summer Topical Meeting on Electronic-Enhanced Optics*, Aventura, Fla., pp. 31–32.
62. Underwood, I., D. G. Vass, M. I. Newsam, J. M. Oton, X. Quintana, L. Chan, N. Flannigan, G. Swedenkrans, and M. Rampin. 2000. "Antiferroelectric liquid crystal on silicon." *Proc. 2000 IEEE/LEOS Summer Topical Meeting on Electronic-Enhanced Optics*, Aventura, Fla., pp. 21–22.
63. Nakahara, T., H. Tsuda, K. Tateno, N. Ishihara, and C. Amano. 2000. "High-sensitivity 1-Gb/s CMOS receiver integrated with a III-V photodiode by wafer-bonding." *Proc. 2000 IEEE/LEOS Summer Topical Meeting on Electronic-Enhanced Optics*, Aventura, Fla., pp. 17–18.
64. Wada, H., T. Takamori, and T. Kamijoh. 1997. "Room temperature photopumped operation of 1.58 μm vertical-cavity lasers fabricated on Si substrates using wafer-bonding." *IEEE Photon. Technol. Lett.* 8: 181–183.
65. Margalit, N. M., D. I. Babic, K. Streubel, R. P. Mirin, R. L. Naone, J. E. Bowers, and E. L. Hu. 1996. "Submilliamp long-wavelength vertical cavity lasers." *Electron. Lett.* 32: 1675–1677.
66. Zhu, Z. H., F. E. Ejeckam, Y. Qian, J. Zhang, G. L. Christenson, and Y. H. Lo. 1997. "Wafer-bonding technology and its applications in optoelectronic devices and materials." *IEEE J. Select Topics Quantum Electron* 3: 927–936.
67. Matsuo, S., K. Tateno, T. Nakahara, H. Tsuda, and T. Kurokawa. 1997. "Use of polyimide bonding for hybrid integration of a vertical cavity surface emitting laser on a silicon substrate." *Electron. Lett.* 33: 1148–1149.

68. Nakahara, T., H. Tsuda, K. Tateno, S. Matsuo, and T. Kurakawa. 1999. "Hybrid integration of smart pixels by using polyimide bonding: demonstration of a GaAs p-i-n photodiode/CMOS/receiver." *IEEE J. Select. Topics Quantum Electron.* 5: 209–216.
69. Hinton, H. S., 1995. "Progress and directions of smart pixel based systems." presented at the *AT&T/ARPA CO-OP Hybrid SEED Workshop*, Fairfax, Va., July 18–21.
70. Kerbis, E., T. J. Kloonan, and F. B. McCormick. 1990. "An all-optical realization of a 2 × 1 free-space switching node." *IEEE Photonics Technol. Lett.* 2: 600–602.
71. McCormick, F. B., F. A. P. Tooley, T. J. Cloonan, J. L. Brubaker, A. L. Lentine, R. L. Morrison, S. J. Hinterlong, M. J. Herron, S. L. Walker, and J. M. Sasian. 1991. "S-SEED-based photonic switching network demonstration." *OSA Proceedings on Photonics in Switching*, H. S. Hinton and J. W. Goodman (Eds), pp. 48–55.
72. McCormick, F. B., T. J. Cloonan, F. A. P. Tooley, A. L. Lentine, J. M. Sasian, R. L. Morrison, R. L. Morrison, S. L. Walker, R. J. Crisci, R. A. Novotny, S. J. Hinterlong, H. S. Hinton, and E. Kerbis. 1993. "Six-stage digital free-space optical switching network using symmetric self-electro-optic-effect devices." *Appl. Opt.* 32: 5153–5171.
73. McCormick, F. B., T. J. Cloonan, A. L. Lentine, J. M. Sasian, R. L. Morrison, M. G. Beckman, S. L. Walker, M. J. Wojcik, S. J. Hinterlong, R. J. Crisci, R. A. Novotny, and H. S. Hinton. 1993. "5-stage embedded-control EGS network using FET-SEED smart pixel arrays." *OSA Proc. on Photonics in Switching*, J. W. Goodman and R. C. Alferness (Eds.), pp. 81–84.
74. McCormick, F. B., T. J. Cloonan, A. L. Lentine, J. M. Sasian, R. L. Morrison, M. G. Beckman, S. L. Walker, M. J. Wojcik, S. J. Hinterlong, R. J. Crisci, R. A. Novotny, and H. S. Hinton. 1994. "Five-stage free-space optical switching network with field-effect transistor self-electro-optic-effect-device smart-pixel arrays." *Appl. Opt.* 33: 1601–1618.
75. Plant, D. V., A. Z. Shang, M. R. Otazo, B. Robertson, and H. S. Hinton. 1994. "Design and characterization of FET-SEED smart pixel transceiver arrays for optical backplanes." in *Proc. 1994 IEEE/LEOS Summer Topical Meeting on Smart Pixels*, Lake Tahoe, Nev., pp. 26–27.
76. Lentine, A. L., K. W. Goossen, J. A. Walker, L. M. F. Chirovsky, L. A. D'Asaro, S. P. Hui, B. T. Tseng, R. E. Leibenguth, D. P. Kossives, D. W. Dahringer, D. D. Bacon, T. K. Woodward, and D. A. B. Miller. 1995. "700Mb/s operation of optoelectronic switching nodes comprised of flip-chip-bonded GaAs/AlGaAs MQW modulators and detectors on silicon CMOS circuitry." *Tech. Dig Conf. on Lasers and Electro-Optics*, Postdeadline Paper CPD11.

77. McKnight, D. J., K. M. Johnson, and R. A. Serati. 1994. "A 256 × 256 liquid-crystal-on-silicon spatial light modulator." *Appl. Opt.* 33: 2775–2784.
78. *The National Technology Roadmap for Semiconductors.* 1994. Semiconductor Industry Association. SEMATECH, Incorporated.
79. Fisher, J. J., and B. L. Shoop. 1995. "Optical-to-optical modulator based on a multiple quantum well device." *Proc. of Ninth Nat. Conf. on Undergraduate Research* II: 614–618.
80. Kazlas, P. T., D. J. McKnight, and K. M. Johnson. 1996. "Integrated assembly of smart pixel arrays and fabrication of associated micro-optics." *Proc. of 1996 IEEE/LEOS Summer Topical Meeting on Smart Pixels*, Keystone, Colo., pp. 51–52.
81. Schubert, E., Y. Wang, A. Cho, L. Tu, and G. Zydzik. 1992. "Resonant cavity light emitting diode." *Appl. Phys. Lett.* 51: 921–923.
82. Twyford, E., J. Chen, N. M. Jokerst, and N. Hartman. 1995. "Optical MCM interconnect using thin film device integration." *OSA Annual Meeting*, Portland, Ore.
83. McCormick, F. B., 1996. "Smart pixel optics and packaging." *Proc. 1996 IEEE/LEOS Summer Topical Meeting on Smart Pixels*, Keystone, Colo., pp. 45–46.
84. Kabal, D. N., G. C. Boisset, D. R. Rolston, and D. V. Plant. 1996. "Packaging of two-dimensional smart pixel arrays." *Proc. 1996 IEEE/LEOS Summer Topical Meeting on Smart Pixels*, Keystone, Colo., pp. 53–54.
85. Derstine, M. W., S. Wakelin, F. B. McCormick, and F. A. P. Tooley, 1994. "A gentle introduction to optomechanics for free space optical systems." Available via anonymous ftp from the /pub/optomech directory of ftp.optivision.com, 18 pages.
86. Goodman, J. W. 1989. "Optics as an interconnect technology." In *Optical Processing and Computing*. H. H. Arsenault, T. Szoplik, and B. Macukow (Eds.). Academic Press, pp. 1–32.
87. Miller, D. A. B. 1989. "Optics for low-energy communication inside digital processors: quantum detectors, sources, and modulators as efficient impedance converters." *Optics Lett.* 14: 146–148.
88. Hinton, H. S., and T. H. Szymanski, "Intelligent optical backplanes." *Proc. Conf. on Massively Parallel Processing with Opt. Interconnect.*, San Antonio, Tex.
89. Devenport, K. E., H. S. Hinton, and D. J. Goodwill. 1996. "A *hyperplane* smart pixel array for packet based switching." *Proc. 1996 IEEE/LEOS Summer Topical Meeting on Smart Pixels*, Keystone, Colo., pp. 32–33.
90. Lentine, A. L., 1993. "Advances in SEED based free space switching systems." *OSA Proceedings on Photonics in Switching*, J. W. Goodman and R. C. Alferness (Eds.), pp. 81–84.

91. Woodward, T. K., A. V. Krishnamoorthy, K. W. Goossen, J. A. Walker, J. E. Cunningham, W. Y. Jan, L. M. F. Chirovsky, S. P. Hui, B. Tseng, D. Kossives, D. Dahringer, D. Bacon, and R. E. Leibenguth. 1996. "Clocked-sense-amplifier-based smart-pixel optical receivers." *IEEE Photonics Technol. Lett.* 8: 1067–1069.
92. Woodward, T. K., A. V. Krishnamoorthy, A. L. Lentine, K. W. Goossen, J. A. Walker, J. E. Cunningham, W. Y. Jan, L. A. D'Asaro, L. M. F. Chirovsky, S. P. Hui, B. Tseng, D. Kossives, D. Dahringer, and R. E. Leibenguth. 1996. "1-Gb/s two-beam transimpedance smart-pixel optical receivers made from hybrid GaAs MQW modulators bonded to 0.8-μm silicon CMOS." *IEEE Photonics Technol. Lett.* 8: 422–424.
93. Woodward, T. K., and A. V. Krishnamoorthy. 1999. "1-Gb/s integrated optical detectors and receivers in commercial CMOS technologies." *IEEE J. Select. Topics Quantum Electron.* 5: 146–156.
94. Schow, C. L., J. D. Schaub, R. Li, J. Qi, and J. C. Campbell. 1998. "A 1-Gb/s monolithically integrated silicon NMOS optical receiver." *IEEE J. Select. Topics Quantum Electron.* 4: 1035–1039.
95. Raksapatcharawong, M., and T. M. Pinkston. 1999. "Design issues for core-based optoelectronic chips: a case study for the WARRP network router." *IEEE J. Select. Topics Quantum Electron.* 5: 330–338.
96. Azadeh, M., R. B. Darling, and W. R. Babbitt. 1999. "Characteristics of optoelectronic feedback for smart pixels with smart illumination." *IEEE J. Select. Topics Quantum Electron.* 5: 172–177.
97. Azadeh, M., R. B. Darling, and W. R. Babbitt. 2000. "A model for optoelectronically interconnected smart pixel arrays." *J. Lightwave Tech.* 18: 1437–1444.
98. Kuznia, C. B., J.-M. Wu, C.-H. Chen, B. Hoanca, L. Cheng, A. G. Weber, and A. A. Sawchuck. 1999. "Two-dimensional parallel pipeline smart pixel array cellular logic (SPARCL) processors – chip design and system implementation." *IEEE J. Select. Topics Quantum Electron.* 5: 376–386.
99. Sawchuck, A. A., and C. B. Kuznia. 2000. "Optoelectronic interconnections for high throughput networks and signal processing." *Proc. 2000 IEEE/LEOS Summer Topical Meeting on Electronic-Enhanced Optics*, Aventura, Fla., pp. 35–36.
100. Campenhout, J. V., H. V. Marck, J. Depreitere, and J. Dambre. 1999. "Optoelectronic FPGA's." *IEEE J. Select. Topics Quantum Electron.* 5: 306–315.
101. Bockstaele, R., M. Brunfaut, J. Depreitere, W. Meeus, J. M. Campenhout, H. Melchior, R. Annen, P. Zenklusen, L. Vanwassenhove, J. Hall, A. Neyer, B. Wittmann, P. Heremans, J. V. Koetsem, R. King, H. Thienpont, and R. Baets. 2000. "Latency study of an area I/O enhanced FPGA with 256 optical channels per CMOS IC." *Proc. 2000 IEEE/LEOS Summer Topical Meeting on Electronic-Enhanced Optics*, Aventura, Fla., pp. 27–28.

102. Cathy, W. T., E. R. Dowski, G. E. Johnson, R. H. Cormack, and H. B. Wach. 2000. "Optics coupled with electronics enables imaging systems with previously impossible performance." *Proc. 2000 IEEE/LEOS Summer Topical Meeting on Electronic-Enhanced Optics*, Aventura, Fla., pp. 63–64.
103. Lee, K., S. Seo, S. Huang, Y. Joo, W. A. Doolittle, S. Fike, N. M. Jokerst, M. Brooke, and A. Brown. 2000. "Design of a smart pixel multispectral imaging array using 3D stacked thin film detectors on Si CMOS circuits." *Proc. 2000 IEEE/LEOS Summer Topical Meeting on Electronic-Enhanced Optics*, Aventura, Fla., pp. 57–58.
104. Liu, W., and M. S. Humayun. 2000. "Artificial retinal prosthesis to restore vision for the blind." *Proc. 2000 IEEE/LEOS Summer Topical Meeting on Electronic-Enhanced Optics*, Aventura, Fla., pp. 26–62.
105. Floyd, R., and L. Steinberg. 1975. "An adaptive algorithm for spatial gray scale." *SID 75 Digest* 36: 35–36.
106. Jarvis, J. F., C. N. Judice, and W. J. Ninke. 1976. "A survey of techniques for the display of continuous-tone pictures on bilevel displays." *Computer Graphics and Image Processing* 5: 13–40.
107. Anastassiou, D. 1989. "Error diffusion coding for A/D conversion." *IEEE Trans. Circuits Syst.* 36: 1175–1186.
108. Crounse, K. R., T. Roska, and L. O. Chua. 1993. "Image halftoning with cellular neural networks." *IEEE Trans. Circuits Syst.* 40: 267–283.
109. Shoop, B. L., R. W. Sadowski, G. P. Dudevoir, E. K. Ressler, A. H. Sayles, D. A. Hall, and D. M. Litynski. 1998. "Smart pixel technology and an application to two-dimensional analog-to-digital conversion." *Opt. Eng.* 37: 3175–3186.
110. Shoop, B. L., D. A. Hall, D. M. Litynski, P. K. Das, and C. DeCusatis. 1998. "Applications of smart pixel technology to image compression for optical data compression." *OSA Annual Meeting Technical Digest*, Baltimore, Md.
111. Kilby, G. R., B. L. Shoop, J. N. Mait, T. D. Wagner, and E. K. Ressler. 1998. "Experimental characterization of a diffractive optical filter for use in an optoelectronic analog-to-digital converter." *Optics Comm.* 157: 1–6.
112. Shoop, B. L., T. D. Wagner, J. N. Mait, G. R. Kilby, and E. K. Ressler. 1999. "Design and analysis of a diffractive optical filter for use in an optoelectronic error diffusion neural network." *Appl. Opt.* 38: 3077–3088.
113. Mead, C. 1998. *Analog VLSI and Neural Systems*, Chapter 5, Addison Wesley.
114. Wagner, T. D., D. A. Nash, J. R. S. Blair, E. K. Ressler, and B. L. Shoop. 2000. "A partitioning scheme for optoelectronic neural networks." *Proc. 2000 IEEE/LEOS Summer Topical Meeting on Electronic-Enhanced Optics*, Aventura, Fla., pp. 69–70.

115. Nash, D. A., J. R. S. Blair, T. D. Wagner, E. K. Ressler, and B. L. Shoop. 2001. "Evaluating the fidelity of a partitioned digital image halftoning algorithm." *Proc. SPIE* 4388.
116. Shoop, B. L. 2001. *Photonic Analog-to-Digital Conversion.* Springer Verlag.
117. Inose, H., and Y. Yasuda, 1963. "A unity bit coding method by negative feedback." *Proc. IEEE* 51: 1524–1535.
118. Candy, J. C. 1974. "A use of limit cycle oscillations to obtain robust analog-to-digital converters." *IEEE Trans. Commun.* 22: 298–305.
119. Wilkinson, S. T., N. M. Jokerst, and R. P. Leavitt. 1995. "Resonant-cavity-enhanced thin-film AlGaAs/GaAs/AlGaAs LED's with metal mirrors." *Appl. Opt.* 34: 8298–8302.
120. Joo, Y., J. Park, M. Thomas, K. S. Chung, M. A. Brooke, N. M. Jokerst, and D. S. Wills. 1999. "Smart CMOS focal plane arrays: a Si CMOS detector array and sigma-delta analog-to-digital converter imaging system." *IEEE J. Select Topics of Quantum Electron.* 5: 296–305.
121. Brooke, M. A. 2000. "Enhanced imaging arrays using a sigma delta ADC in Si CMOS for each array pixel." *Proc. 2000 IEEE/LEOS Summer Topical Meeting on Electronic-Enhanced Optics*, Aventura, Fla., pp. 11–12.
122. Shoop, B. L., P. K. Das, G. P. Dudevoir, T. D. Wagner, R. W. Sadowski, E. K. Ressle. 2001. "A highly-parallel mismatch tolerant photonic A/D converter." *Proc. Conf. on Lasers and Electro-Optics*, Baltimore, Md.
123. Shoop, B. L., P. K. Das, E. K. Ressler, and T, J. Talty. 2000. "High-resolution photonic A/D conversion using oversampling techniques," *Proc. Conf. on Lasers and Electro-Optics.* San Francisco, Calif., pp. 491–492.
124. Shoop, B. L., P. K. Das, E. K. Ressler, R. W. Sadowski, G. P. Dudevoir, and A. H. Sayles. 2000. "A fully-connected, distributed mesh feedback architecture for photonic A/D conversion." *Proc. 2000 IEEE/LEOS Summer Topical Meeting on Electronic-Enhanced Optics*, Aventura, Fla., pp. 65–66.
125. Adams, R. W. 1997. "Spectral noise-shaping in integrate-and-fire neural networks." *Proc. Int. Conf. on Neural Networks*, pp. 953–958.
126. Hinton, H. S., and N. M. Jokerst (Conf. Chairs). *Digest of IEEE/LEOS 1996 Summer Topical Meeting on Smart Pixels*, Keystone, Colo.
127. Szymanski, T. H., and H. S. Hinton. 1994. "Architecture of a field programmable smart pixel array." *Proc. Int. Conf. Optical Computing*, Edinburgh, Scotland, pp. 497–500.
128. Sherif, S. S., T. H. Szymanski, and H. S. Hinton. 1996. "Design and implementation of a field programmable smart pixel array." *Proc. 1996 IEEE/LEOS Summer Topical Meeting on Smart Pixels*, Keystone, Colo., pp. 78–79.
129. Harris, J. S., Jr., M. C. Larson, and A. R. Massengale. 1996. "Broad-range continuous wavelength tuning in microelectromechanical vertical-cavity surface-emitting lasers." *Proc. 1996 IEEE/LEOS Summer Topical Meeting on Optical MEMs and Their Applications*, Keystone, Colo., pp. 31–32.

130. Goto, H. 1996. "Si micromachined 2D optical scanning mirror and its application to scanning sensors." *Proc. 1996 IEEE/LEOS Summer Topical Meeting on Optical MEMs and Their Applications*, Keystone, Colo., pp. 17–18.
131. Arch, D., T. Ohnstein, D. Zook, and H. Guckel. 1996. "A MEMS-based tunable infrared filter for spectroscopy." *Proc. 1996 IEEE/LEOS Summer Topical Meeting on Optical MEMs and Their Applications*, Keystone, Colo., pp. 21–22.

Chapter 11 | Emerging Technology for Fiber Optic Data Communication

Chung-Sheng Li

IBM Thomas J. Watson Research Center, Hawthorne,
New York 10532

Due to the explosive growth in the demand for network bandwidth, devices and subsystems that can support gigabit and multigigabit throughput have become increasingly important. In this chapter, we review several emerging technologies for fiber optical data communication. In particular, we focus on the technologies for wavelength division multiplexing (WDM), as it is the most important technique in advancing the communication bandwidth for the next-generation broadband networks.

11.1. Introduction

Due to the explosive growth of the number of the Internet users and the world wide web sites, there has been a significant increase in the demand for the network bandwidth. It has been estimated by the Internet Geography project at UC Berkeley that there were a total of more than 29 million Internet domains by the end of March 2001. These web sites have spawned many academic and commercial applications such as digital library, distance learning, electronic commerce (B2C and B2B), cybermall, on-demand streaming video/audio, and Napster-like music/file-sharing applications. Consequently, the network congestion is aggravated at both the backbone and regional levels.

As of May 2001, most of the interconnections between Internet routers at the backbone level operated by AT&T, Worldcom, and Sprint have been implemented in OC-48 (2.4 Gbps). As the internet traffic doubles every 90 to 180 days, the network infrastructure will have serious difficulty keeping up with the growth of the traffic even though there was a temporary excess in capacity during 2001. Needless to say, future Internet applications such as collaborative computing, peer-to-peer music/video/file sharing, digital library and distance learning will create even more pressure on the transmission and switching capabilities of the system as more bandwidth-hungry information (graphics, images and video) will be distributed electronically. Therefore, it is natural to expect that the backbone of the future Internet will be based on faster data rate such as OC-192 or beyond and employ some form of optical switching to alleviate the bandwidth problem. As a matter of fact, many carriers have already been in the middle of upgrading these trunk-level capacity to OC-192c (10 Gbps) or even OC-768 (40 Gbps).

Asynchronous Transfer Mode (ATM) over Synchronous Optical Network (SONET) and Multi-Protocol Label Switching (MPLS) have already been adopted as the primary transport mechanism for carrying broadband traffic for the future. Currently, the broadband traffic is carried on single-mode fiber between major switching hubs for data rates up to OC-48 (2.5 Gb/s). However, the speed of each fiber cannot be increased indefinitely. When the bandwidth required is more than can be supported by a single OC-48 connection, additional multiplexing techniques have to be incorporated in order to advance the link capacity. Currently, as the technology for OC-192 becomes mature, significant research has been devoted for developing OC-768 (40 Gbps) transmission: *Electronic Time Division Multiplexing (ETDM)* in which four OC-192 channels are time multiplexed together into a single OC-768 channel electronically. In contrast, *Optical Time Division Multiplexing (OTDM)* multiplexes four OC-192 channels using optical means.

In this chapter, we will survey a number of promising technologies for fiber optic data communications. The goal is to investigate the potentials and limitations of each technology. The organization of the rest of this chapter is as follows: Section 11.2 describes the architecture of all-optical networks including both broadcast-and-select networks and wavelength routed networks. The device aspects of tunable transmitters and tunable receivers for WDM networks are discussed in Section 11.3 and 11.4, respectively. Section 11.5 describes the optical amplifiers, which is by far the most important technology for increasing the distance between data regeneration so far. Wavelength (de)multiplexer technologies are described in Section 11.6, while wavelength router technologies are discussed in

Section 11.7. Section 11.8 discusses the wavelength converters, and this chapter is briefly summarized in Section 11.9.

11.2. Architecture of All-Optical Network

11.2.1. BROADCAST-AND-SELECT NETWORKS

A broadcast-and-select network consists of nodes interconnected to each other via a star coupler, as shown in Fig. 11.1. An optical fiber link, called the star carries signals from each node to the star. The star combines the signals from all the nodes and distributes the resulting optical signal equally among all its outputs. Another optical fiber link carries the combined signal from an output of the star to each node. Examples of such networks are Lambdanet [1] and Rainbow [2].

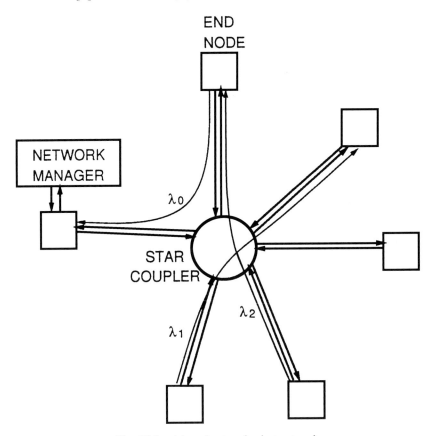

Fig. 11.1 A broadcast-and-select network.

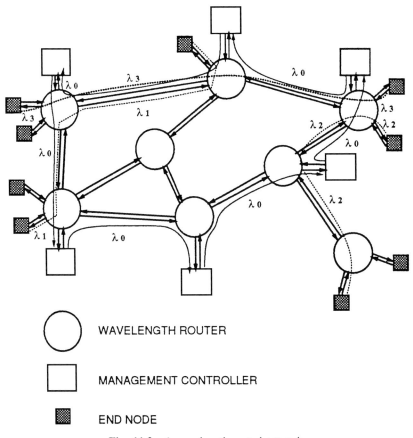

Fig. 11.2 A wavelength-routed network.

11.2.2. WAVELENGTH-ROUTED NETWORKS

A wavelength-routed network is shown in Fig. 11.2. The network consists of *static* or *reconfigurable* wavelength routers interconnected by fiber links. Static routers provide a fixed, non-reconfigurable routing pattern. A reconfigurable router on the other hand allows the routing pattern to be changed dynamically. These routers provide static or reconfigurable *lightpaths* between end-nodes. A lightpath is a connection consisting of a path in the network between the two nodes and a wavelength assigned on the path. End-nodes are attached to the wavelength routers. One or more controllers that perform the network management functions are attached to the end node(s).

11.2.3. A BRIEF WDM/WDMA HISTORY

The first field test of a WDM system by the British Telecom in Europe dated back to 1991. This test was based on a 5-node, 3-wavelength OC-12 (622 Mb/s) ring around London with a total distance of 89 km. The ESPRIT program, funded by the European government since 1991, is a consortium funding multiple programs including OLIVES on optical interconnects and several WDM-related efforts. The RACE project, which is a joint university corporate program, has also included demonstrations of multiwavelength transport network (MWTN). Since 1995, ACTS (Advanced Communications Technologies and Services), which includes a total of 13 projects, started building a trans-European Information Infrastructure based on ATM and developing Metropolitan Optical Network (METON). In Japan, NTT is building a 16-channel photonic transport network with over 320 Gb/s throughput. In the United States, ARPA/DARPA has funded a series of WDM/WDMA activities between 1991 and 1996:

- AON (All-Optical Network) consortium, which includes AT&T, DEC, and MIT, focused on developing architectures and technologies that can support point-to-point or point-to-multipoint high-speed circuit-switched multigigabits-per-second digital or analog sessions.
- ONTC (Optical Network Technology Consortium) which includes Bellcore, Columbia University, focused on scalable multiwavelength multihop optical network.
- MONET (Multiple wavelength Optical Network), which is a consortium including Bell Labs, Bellcore, and three regional Bells, is chartered to develop WDM testbeds and come up with commercial applications for the technology.

Since 1995, there are many commercially available WDM and DWDM systems. These systems are described in more detailed in Chapter 5: Optical Wavelength Division Multiplexing for Data Communication Networks.

11.3. Tunable Transmitter

The tunable transmitter is used to select the correct wavelength for data transmission. As opposed to a fixed-tuned transmitter, the wavelength of a tunable transmitter can be selected by an externally controlled electrical signal.

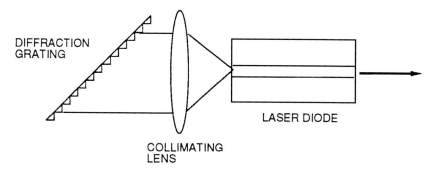

Fig. 11.3 Structure of an external cavity laser.

Currently, wavelength tuning can be achieved by one of following mechanisms:

- External cavity tunable lasers: A typical external cavity tunable laser, as shown in Fig. 11.3, includes a frequency selective component in conjunction with a Fabry–Perot laser diode with one facet coated with antireflection coating. The frequency selective component can be a diffraction grating or any tunable filter whose transmission or reflection characteristics can be controlled externally. This structure usually enjoys wide tuning range but suffers slow tuning time when a mechanical tunable structure is employed. Alternatively, an electrooptic tunable can be employed to provide faster tuning time but narrower tuning range.
- Two-section tunable Distributed Bragg Reflector (DBR) tunable laser diodes: In a two-section device, as shown in Fig. 11.4, separate electrodes carry separate injection current: One is for the active area while the other one is for controlling the index seen by the Bragg mirror. This type of device usually has a small continuous tuning range. For example, the tuning range of the device reported in [3] is limited to $\simeq$5.8 nm (720 GHz at 1.55 μm).
- Three-section DBR tunable Laser Diodes: The major drawback of the two-section DBR device is the big gap in the available tuning range. This problem can be solved by adding a *phase-shift* section. With this additional section, the phase of the wave incident on the Bragg mirror section can be varied and matched, thus avoiding the gap of the tuning range.

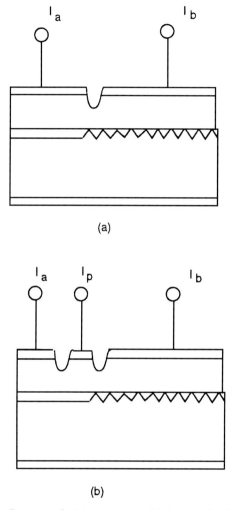

Fig. 11.4 Structure of a (a) two-section (b) three-section laser diode.

- One- or two-dimensional laser diode array [4], or a multichannel grating cavity laser [5]: allowing only a few signaling channels. In another extreme, the wavelengths covering the range of interests can be reached by individual lasers in a one-dimensional or a two-dimensional laser array [4] with each lasing element emitting at a different wavelength. Single-dimensional laser array, with each lasing element emitting at single transverse and longitudinal mode, can be fabricated. One possible scheme of single-mode operation can be achieved by means of short cavity (with high reflectivity coatings) where the neighboring cavity modes from the lasing modes are far from the peak gain. Different emission wavelengths can be achieved by tailoring the cavity length of each array element. For the two-dimensional laser array such as the vertical surface-emitting laser array, each array element emits at a different wavelength by tailoring the length of the cavity [4]. It is possible to turn on more than one laser at any given time in both of these approaches. Heat dissipation, however, might limit the number of lasers that can be turned on.

The coupling of the laser emission from the one-dimensional or two-dimensional laser into a single fiber or waveguide can be achieved with gratings or computer-generated holographic coupler. Coupling of four wavelengths from a vertical surface-emitting laser array has been demonstrated recently. Holographic coupling has also been demonstrated to couple over 100 wavelengths with 2% of coupling efficiency.

Due to the cross-talk and the limited bandwidth of the electronic switch, an external modulator might be required to modulate the laser beam for higher bit rate. The light signals can be modulated by either using a directional coupler type modulator, a Mach–Zehnder type modulator [7], or a quantum well modulator [8]. The operation of these devices is required to be wavelength independent over the entire tuning range of the tunable transmitter.

11.4. Tunable Receiver

The tunable receiver is used to select the correct wavelength for data reception. As opposed to a fixed-tuned receiver the wavelength of a tunable receiver can be selected by an externally controlled electrical signal.

Ideally, each tunable receiver needs a tuning range that covers the entire transmission bandwidth with high resolution and can be tuned from any channel to any other channel within a short period of time. Tunable receiver structures that have been investigated include:

- Single-Cavity Fabry–Perot Interferometer:
  The simplest form of a tunable filter is a tunable Fabry–Perot interferometer, which consists of a movable mirror to form a tunable resonant cavity.

The electric field at the output side of the FP filter in the frequency domain is given by

$$S_{out}(f) = H_{FP}(f) S_{in}(f) \tag{11.1}$$

where H_{FP} is the frequency domain transfer function given by

$$H_{FP}(f) = \frac{T}{1 - R e^{j2\pi \frac{f-f_c}{FSR}}} \tag{11.2}$$

where R are the power transmission coefficient and the power reflection coefficient of the filter, respectively. The parameter f_c is the center frequency of the filter. The parameter FSR is the *free spectral range* at which the transmission peaks are repeated and can be defined as $FSR = c/2\mu L$, where L is the FP cavity length and μ is the refractive index of the medium bounded by the FP cavity mirror. The 3 dB transmission bandwidth FWHM (*full width at half maximum*) of the FP filter is related to FSR and F by

$$FWHM = \frac{FSR}{F}, \tag{11.3}$$

where the reflectivity *finesse* F of the FP filter is defined as

$$F = \frac{\pi \sqrt{R}}{1 - R} \tag{11.4}$$

Based on this principle, both fiber Fabry–Perot (as shown in Fig. 11.5) and liquid crystal Fabry–Perot tunable filters have been

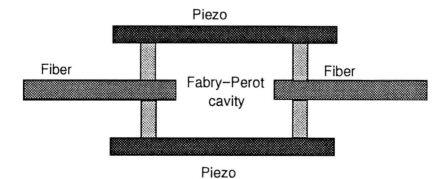

Fig. 11.5 Structure of a fiber Fabry–Perot tunable filter.

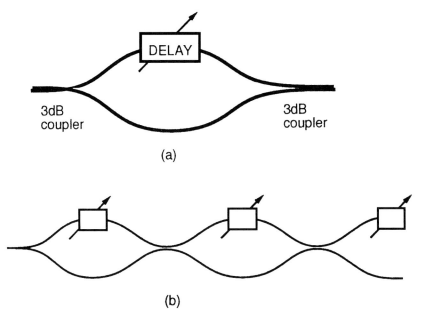

Fig. 11.6 (a) Single-stage (b) Multi-stage Mach–Zehnder tunable filter.

realized. The tuning time of these devices is usually on the order of milliseconds because of the use of electromechanical devices.
- Cascaded Multiple Fabry–Perot Filters [36]: The resolution of Fabry–Perot filters can be increased by cascading multiple Fabry–Perot filters by using either vernier or coarse-fine principle.
- Cascaded Mach–Zehnder tunable filter [37]: The structure of a Mach–Zehnder interferometer is shown in Fig. 11.6(a). The light is first split by a 3 dB coupler at the input, then goes through two branches with a phase shift difference and is then combined by another 3 dB coupler. The path length difference between two arms causes constructive and destructive interference depending on the input wavelength, resulting in a wavelength selective device. To tune each Mach–Zehnder, it is only necessary to vary the differential path length by $\lambda/2$. Successive Mach–Zehnder filters can be cascaded together, as shown in Fig. 11.6(b). In order to tune to a specific channel, filters with different periodic ranges will have to be centered at the same location. This can be accomplished by tuning the differential path length of the individual filter.

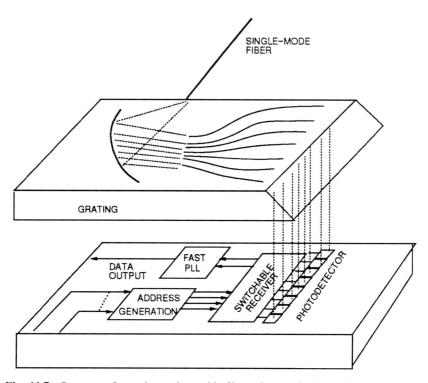

Fig. 11.7 Structure of optoelectronic tunable filter using grating demultiplexer and photodetector array.

- Acoustooptic tunable filter [37]: The structure of an acoustooptic tunable filter based on surface-acoustic-wave (SAW) principle is shown in Fig. 11.8. The incoming beam goes through a polarization splitter, separating the horizontally polarized beam from the vertically polarized beam. Both of these beams travel down the waveguide with a grating established by the surface acoustic wave generated by a transducer. The resonant structure established by the grating rotates the polarization of the selected wavelength while leaving the other wavelength unchanged. Another polarization beam splitter at the output collects the signals with rotated polarization to output 1 while the rest is passed to output 2.
- Switchable grating [5]: The monolithic grating spectrometer is a planar waveguide device where the grating and the input/output waveguide channels are integrated as shown in Fig. 11.7. A polarization independent, 78 channel (channel separation of 1 nm)

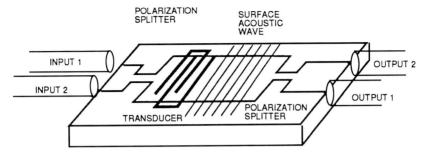

Fig. 11.8 Structure of a grating demultiplexer.

device has recently been demonstrated with cross-talk ≤ -20 dB [5]. For a 256-channel system, the detector array can be grouped into 64 groups with 4 detector array elements in each bar from which a preamplifier is connected to [9]. The power for the MSM detectors is provided by a 2-bit control as shown. The outputs from the preamplifiers are controlled by gates through a 3-bit control such that every 8 outputs from the gates are fed into a postamplifier. The outputs from the postamplifiers are further controlled by another 3-bit controller. In this way, any channel from the 256 channels can be selected. The switching speed could be very fast, mostly due to the power-up time required by the MSM detectors (the total capacitance that needs to be driven is ~ 100 fF $\times$ 64).

11.5. Optical Amplifier

In order to achieve all-optical metropolitan/wide area networks (MAN/WAN), optical amplification is required to compensate for various losses such as fiber attenuation, coupling and splitting loss in the star couplers, as well as coupling loss in the wavelength routers. Both rare-earth-ion-doped fiber amplifiers [10, 11] and semiconductor-laser amplifiers can be used to provide amplification of the optical signals.

11.5.1. SEMICONDUCTOR OPTICAL AMPLIFIER

A semiconductor amplifier is basically a laser diode that operates below lasing threshold. Two basic types of semiconductor amplifier can be distinguished: Fabry–Perot amplifiers (FPA) and traveling wave amplifiers (TWA). In FPA structures, two cleaved facets act as partial reflective mirrors that form a Fabry-Perot cavity. The natural reflectivity for air-semiconductor facets is 32%, but can be modified through a wide range by using antireflection coating or high-reflection coating. FPA is less desirable in many applications because of the nonuniform gain across the spectrum. TWA has the same structure as FPA except that antireflection coating is applied to both facets to minimize internal feedback.

The maximum available signal gain of both the FPA and the TWA is limited by gain saturation. The TWA gain G_s as a function of the input power $P_{in} = P_{out}/G_s$ is given by the following equation:

$$G_s = 1 + \frac{P_{sat}}{P_{in}} \ln \frac{G_o}{G_s} \qquad (11.5)$$

where G_o is the maximum amplifier gain, corresponding to the single pass gain in the absence of input light. It is easy to observe that G_s monotonically decreases to 1 as the input signal power increases, resulting in the gain saturation effect.

Cross-talk occurs when multiple optical signals or channels are amplified simultaneously. Under this circumstance, the signal gain for one channel is affected by the intensity levels of other channels as a result of the gain saturation. This effect depends on the carrier lifetime, which is on the order of 1 ns. Therefore, cross-talk among different channels is most pronounced when the data rate is comparable to the reciprocal of the carrier lifetime.

Another limit to the amplifier gain is due to the amplifier spontaneous emission (ASE). The amplification of spontaneous emission is triggered by the spontaneous recombination of electrons and holes in the amplifier medium. This noise beats with the signal at the photodetector, causing signal-spontaneous beat noise and spontaneous-spontaneous beat noise.

11.5.2. *DOPED-FIBER AMPLIFIER*

When optical fibers are doped with rare-earth ions such as erbium, neodymium, or praseodymium, the loss spectrum of the fiber can be drastically modified. During the absorption process, the photons from the optical pump at wavelength λ_p are absorbed by the outer orbital electrons of the rare-earth ions and these electrons are raised to higher energy levels. The de-excitation of these high-energy levels to the ground state might occur either radiatively or nonradiatively. If there is an intermediate level, additional de-excitation can be stimulated by the signal photon, providing that the bandgap between

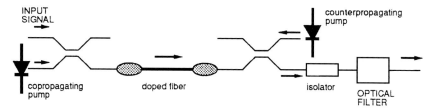

Fig. 11.9 A typical doped-fiber amplifier system with either copropagation pump or counterpropagation pump.

the intermediate state and the ground state corresponds to the energy of the signal photons. The result would be an amplification of the optical signal at wavelength λ_s.

The main difference between doped-fiber amplifier and semiconductor amplifiers is that the amplifier gain of the doped-fiber amplifier is provided by means of optical pumping as opposed to electrical pumping.

Figure 11.9 shows a typical fiber amplifier system. Currently, the most popular doped-fiber amplifiers are based on erbium doping. Similar to semiconductor amplifier, the gain of erbium-doped fiber amplifier also saturates. However, the cross-talk effect is much reduced thanks to the long fluorescence lifetime.

11.5.3. GAIN EQUALIZATION

The gain spectra of these optical amplifiers are non-flat over the fiber transmission windows at 1.3 and 1.55 μm, resulting in non-uniform amplification of the signals. Together with the near-far effect resulting from optical signals originating from various nodes at locations separated by large distances, there exists a wide dynamic range among various signals arriving at the receivers. The best dynamic range of lightwave receivers with high sensitivity reported thus far is limited to less than $\sim$20 dB at 2.4 Gbps [12] and less than $\sim$30 dB at 1 Gbps [13]. In addition, the signal with high average optical power saturates the gain of the optical amplifiers placed along the path of propagation. This limits the available gain for the remaining wavelength channels. Thus, signal power equalization among different wavelength channels is required.

Most of the existing studies on gain equalization have been focused on either statically or dynamically equalizing the non flat gain spectra of the optical amplifiers, but without addressing the near-far effect. For static gain equalization, schemes including grating embedded in the Er^{3+}

fiber amplifier [10], cooling the amplifiers to low temperatures [14], or a notch filter [15, 16, 17] were proposed previously to flatten the gain spectra. An algorithm is proposed to adjust the optical signal power at different transmitters to achieve equalization [18]. In [19], gain equalization is achieved by placing a set of attenuators in the arms of the back-to-back grating multiplexers to compensate for nonflat gain spectra of the fiber amplifier. For dynamic gain equalization, a two-stage fiber amplifier with offset gain peaks was proposed in [20] to equalize the optical signal power among different WDM channels by adjusting the pump power. This scheme, however, has a very limited equalized bandwidth of ∼2.5 nm. Dynamic gain equalization can also be achieved through controlling the transmission spectra of tunable optical filters. Using this scheme, a 3-stage (for 29 WDM channels) [21] and 6-stage (for 100 WDM channels) [22] Er^{3+}-doped-fiber amplifier system with equalized gain spectra were demonstrated using a multi-stage Mach–Zehnder Interferometric filter. Acoustooptic tunable filter has also been used to equalize gain spectra for a very wide transmission window [23]. The combination of these schemes can, in principle, solve the near-far problem in the networks.

11.6. Wavelength Multiplexer/Demultiplexer

Wavelength multiplexers and demultiplexers are the essential components for constructing any wavelength routers. They can also be used for building tunable receivers and transmitters as described in the previous sections.

Two types of wavelength multiplexers/demultiplexers are most widely used: grating demultiplexers and phase arrays.

Figure 11.10 shows an etched-grating demultiplexer with N output waveguides and its cross-sectional view, respectively. The reflective grating

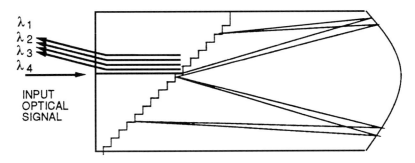

Fig. 11.10 Structure of a grating demultiplexer.

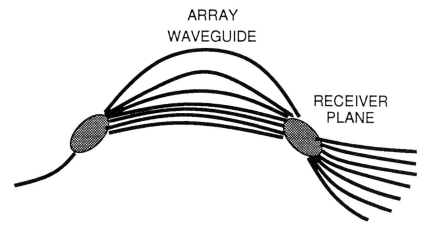

Fig. 11.11 Structure of phase array wavelength demultiplexer.

uses the Rowland Circle configuration in which the grating lies along a circle while the focal line lies along a circle of half the diameter.

Phase array wavelength multiplexer/demultiplexers have been shown to be the superior WDM demultiplexers for systems with a small number of channels. A phase array demultiplexer consists of a dispersive waveguide array connected to input and output waveguides through two radiative couplers as shown in Fig. 11.11. Light from an input waveguide diverging in the first star coupler is collected by the array waveguides, which are designed in such a way that the optical path length difference between adjacent waveguides equals an integer multiple of the central design wavelength of the demultiplexer. This results in the phase and intensity distribution of the collected light being reproduced at the start of the second star coupler, causing the light to converge and focus on the receiver plane. Due to the path length difference, the reproduced phase front will tilt with varying wavelength, thus sweeping the focal spot across different output waveguides.

11.7. Wavelength Router

Wavelength routing for all-optical networks using WDMA has received increasing attention recently [24, 25, 26, 27]. In a wavelength-routing network, wavelength-selective elements are used to route different wavelengths to their corresponding destinations. Compared to a network using only star couplers, a network with wavelength routing capability can avoid

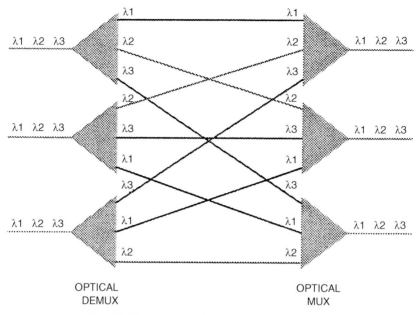

Fig. 11.12 Structure of a static wavelength router.

the splitting loss incurred by the broadcasting nature of a star coupler [28]. Furthermore, the same wavelength can be used simultaneously on different links of the same network and reduce the total number of required wavelengths [24].

The routing mechanism in a wavelength router can either be static, in which the wavelengths are routed using a fixed configuration [29], or dynamic, in which the wavelength paths can be reconfigured [30]. The common feature of these multi-port devices is that different wavelengths from each individual input port are spatially resolved and permuted before they are recombined with wavelengths from other input ports. These wavelength routers, however, have imperfections and nonideal filtering characteristics which give rise to signal distortion and cross-talk.

Figure 11.12 shows the structure of a static wavelength router that consists of K optical demultiplexers and multiplexers. Each input fiber to an optical demultiplexer is assumed to contain up to M different wavelengths where $M \leq K$. However, we only consider the case where $M \leq K$. The optical demultiplexer spatially separates the incoming wavelengths into M paths. Each of these paths is then combined at an optical multiplexer with the outputs from the other $M - 1$ optical demultiplexers.

The wavelength routing configuration in Fig. 11.12 is fixed permanently. The optical data at wavelength λ_j entering the ith demultiplexer exit at the $[(j-i) \bmod M]^{\text{th}}$ output of that demultiplexer. That output is connected to the ith input of the $[(j-i) \bmod M]^{\text{th}}$ multiplexer.

Because of the imperfections and nonideal filtering characteristics of the optical multiplexers and demultiplexers, cross-talk occurs in the wavelength routers. On the demultiplexer side, each output contains both the signals from the desired wavelength and that from the other $M-1$ cross-talk wavelengths. From reciprocity, both the desired wavelength and the cross-talk signals exit at the output on the multiplexer side. Thus, each wavelength at every multiplexer contains $M-1$ cross-talk signals originating from all demultiplexers.

Cross-talk phenomena in wavelength routers have previously been studied [31, 32, 33, 34]. It was shown in [31] that the maximum allowable cross-talk in each grating (grating as optical demultiplexers and multiplexers in the wavelength router) is -15 dB in an all-optical network with moderate size (say 20 wavelengths and 10 routers in cascade). The results are based on using a 1-dB power penalty criterion and only considering the power addition effect of the cross-talk. Cross-talk can also arise from beating between the data signal and the leakage signal (from imperfect filtering) at the same output channel. The beating of these uncorrelated signals converts the phase noise of the laser sources into the amplitude noise and corrupts the received signals [35] when the linewidths of the laser sources are smaller than the electrical bandwidth of the receiver. Coherent beating, in which the data signal beats with itself, can occur as a result of the beatings among the signals from multiple paths or loops caused by the leakage in the wavelength routers in the system. It was shown in [33] that the component cross-talk has to be less than -20, -30, and -40 dB in order to achieve satisfactory performance for a system consisting of a single, ten and hundred leakage sources, respectively.

Figure 11.13 shows the structure of a dynamic wavelength-routing device. This device consists of a total of N optical demultiplexers and N optical multiplexers. Each of the input fibers to an optical demultiplexer contains M different wavelengths. The optical demultiplexer spatially separates the incoming wavelengths into M paths. Each of these paths passes through a photonic switch before they are combined with the outputs from the other $M-1$ optical switches. When the cross-talk of the optical multiplexer/demultiplexer/switch is considered, each wavelength channel at each input optical demux can reach any of the output optical mux via M different paths.

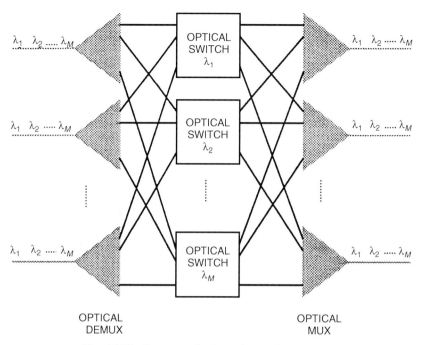

Fig. 11.13 Structure of a dynamic wavelength router.

11.8. Wavelength Converter

The network capacity of WDM networks is determined by the number of independent lightpaths. One way to increase the number of nodes that can be supported by network is to use wavelength router to enable spatial reuse of the wavelengths, as described in the previous section. The second method is to convert signals from one wavelength to another. Wavelength conversion also allows distributing the network control and management into smaller subnetworks and allows flexible wavelength assignments within the subnetworks.

There are three basic mechanisms for wavelength conversion:

1. Optoelectronic Conversion: The most straightforward mechanism for wavelength conversion is to convert each individual wavelength to electronical signals, and then retransmit by lasers at the appropriate wavelength. A nonblocking crosspoint switch can be embedded within the O/E and E/O conversion such that any wavelength can be converted to any other wavelength (as shown in

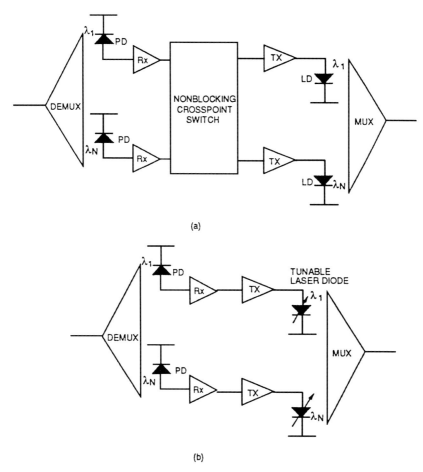

Fig. 11.14 (a) Structure of an optoelectronic wavelength converter using electronic crosspoint switch. (b) Structure of an optoelectronic wavelength converter using tunable laser.

Fig. 11.14(a)). Alternatively, a tunable laser can be used instead of a fixed tuned laser to achieve the same wavelength conversion capability (as shown in Fig. 11.14(b)). This mechanism only requires mature technology. The protocol transparency is completely lost if full data regeneration (which includes retiming, reshaping, and reclocking) is performed within the wavelength converter. On the other hand, limited transparency can be achieved by incorporating only analog amplification in the conversion process. In this case, other information associated with the signals including phase, frequency, and analog amplitude is still lost.

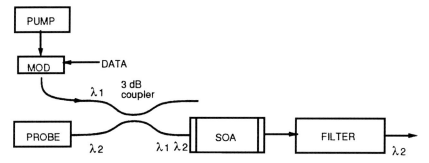

Fig. 11.15 Structure of an optical gating wavelength converter using semiconductor optical amplifier.

2. Optical Gating Wavelength Conversion: This type of wavelength converter, as shown in Fig. 11.15, accepts an input signal at wavelength λ_1 which contains the information and a cw probe signal at wavelength λ_2. The probe signal, which is at the target wavelength, is then modulated by the input signal through one of the following mechanisms:
   - Saturable absorber: In this mechanism, the input signal saturates the absorption and allows the probe beam to transmit. Due to carrier recombinations, the bandwidth is usually limited to less than 1 GHz.
   - Cross-gain modulation: The gain of a semiconductor optical amplifier saturates as the optical level increases. Therefore, it is possible to modulate the amplifier gain with an input signal, and encode the gain modulation on a separate cw probe signal.
   - Cross-phase modulation: Optical signals traveling through semiconductor optical amplifiers undergo a relatively large phase modulation compared to the gain modulation. The cross-phase modulation effect is utilized in an interferometer configuration such as in a Mach–Zehnder interferometer. The interferometric nature of the device converts this phase modulation to an amplitude modulation in the probe signal. The interferometer can operate in two different modes, a noninverting mode where an increase in input signal power causes a decrease in probe power, and an inverting mode where an increase in input signal power causes a decrease in probe power.

3. Wave-Mixing Wavelength Conversion: Wavelength mixing, such as three-wave and four-wave mixing, arises from nonlinear optical response when more than one wave is present. The phase and frequency of the generated waves are linear combinations of the interacting waves. This is the only method that preserves both phase and frequency information of the input signals, and is thus completely transparent. It is also the only method that can simultaneously convert multiple frequencies. Furthermore, it has the potential to accommodate signals at extremely high bit rates. Wave-mixing mechanisms can occur in either passive waveguides or semiconductor optical amplifiers.

11.9. Summary

In this chapter, we have surveyed a number of promising technologies for fiber-optic data communication systems. In particular, we have focused on technologies that can support gigabit all-optical WDM networks. Although most of these technologies are still far from being mature, they nevertheless hold the promise of dramatically improving the network capacity of existing fiber optical networks.

References

1. Goodman, M. S., H. Kobrinski, M. Vecchi, R. M. Bulley, and J. M. Gimlett. Aug. 1990. "The LAMBDANET multiwavelength network: architecture, applications and demonstrations." *IEEE J. Selected Areas in Commun.* **8**(6): 995–1004.
2. Janniello, F. J., R. Ramaswami, and D. G. Steinberg. May/June 1993. "A prototype circuit-switched multi-wavelength optical metropolitan-area network." *IEEE/OSA J. Lightwave Tech.*, 11: 777–782.
3. Murata, S., I. Mito, and K. Kobayashi. April 1987. "Over 720GHz (5.8nm) frequency tuning by a 1.5 μm DBR laser with phase and bragg wavelength control regions." *Electronic Letters* **23**(8): 403–405.
4. Maeda, M. W., C. J. Chang-Hasnain, J. S. Patel, H. A. Johnson, J. A. Walker, and Chinlon Lin. 1991. "Two dimensional multiwavelength surface emitting laser array in a four-channel wavelength-division-multiplexed system experiment." *OFC 91 Digest*, p. 73.
5. Kirkby, P. A. 1990. "Multichannel grating demultiplexer receivers for high density wavelength systems." *IEEE Journal of Lightwave Technology*, 8: 204–211.

6. Nyairo, K. O., C. J. Armistead, and P. A. Kirkby. 1991. "Crosstalk compensated WDM signal generation using a multichannel grating cavity laser." *ECOC Digest,* p. 689.
7. Alferness, R. C. Aug. 1982. "Waveguide electrooptic modulators." *IEEE Transactions on Microwave Theory and Techniques* **30**(8): 1121–1137.
8. Miller, D. A. B., D. S. Chemla, T. C. Damen, T. H. Wood, C. A. Burrus, Jr., A. C. Gossard, and W. Wiegmann. Sept. 1985. "The quantum well self-electrooptic effect device: optoelectronic bistability and oscillation and self-linearized modulation." *IEEE Journal of Quantum Electronics* **21**(9): 1462–1476.
9. Chang, G. K., W. P. Hong, R. Bhat, C. K. Nguyen, J. L. Gimlett, C. Lin, and J. R. Hayes. 1991. "Novel electronically switched multichannel receiver for wavelength division multiplexed systems." *Proc. OFC 91,* p. 6.
10. Tachibana, M., R. I. Laming, P. R. Morkel, and D. N. Payne. 1990. "Gain-shaped erbium-doped fiber amplifier (EDFA) with broad spectral bandwidth." *Topical Meeting on Optical Amplifier Application,* p. MD1.
11. Desurvire, E., C. R. Giles, J. L. Zyskind, J. R. Simpson, P. C. Becker, and N. A. Olsson. 1990. "Recent advances in erbium-doped fiber amplifiers at 1.5 μm." *Proc. Optical Fiber Communication Conference,* San Francisco, Calif.
12. Blaser, M., and H. Melchior. Nov. 1992. "High performance monolithically integrated $In_{0.53}Ga_{0.47}As/InP$ PIN/JFET optical receiver front end with adaptive feedback control." *IEEE Photonics Technology Letter* 4(11).
13. Mikamura, H. Y., Oyabu, S. Inano, E. Tsumura, and T. Suzuki. 1991. "GaAs IC chip set for compact optical module of giga bit rates." *Proceedings IECON'91 1991 International Conference on Industrial Electronics, Control and Instrumentation.*
14. Goldstein, E. L., V. da Silva, L. Eskildsen, M. Andrejco, and Y. Silberberg. Feb. 1993. "Inhomogeneously broadened fiber-amplifier cascade for wavelength-multiplexed systems." *Proc. OFC'93.*
15. Tachibana, M., R. I. Laming, P. R. Morkel, and D. N. Payne. Feb. 1991. "Erbium-doped fiber amplifier with flattened gain spectra." *IEEE Photonic Technology Letters* **3**(2): 118–120.
16. Wilinson, M., A. Bebbington, S. A. Cassidy, and P. Mckee. 1992. "D-fiber filter for erbium gain flattening." *Electronic Letters* 28: 131.
17. Willner, A. E., and S.-M. Hwang. Sept. 1993. "Passive equalization of nonuniform EDFA gain by optical filtering for megameter transmission of 20 WDM channels through a cascade of EDFA's." *IEEE Photonics Technology Letters* **5**(9): 1023–1026.
18. Chraplyvy, A. R., J. A. Nagel, R. W. Tkach. Aug. 1992. "Equalization in amplified WDM lightwave transmission systems." *IEEE Photonic Technology Letters* **4**(8): 920–922.

19. Elrefaie, A. F., E. L. Goldstein, S. Zaidi, and N. Jackman. Sept. 1993. "Fiber-amplifier cascades with gain equalization in multiwavelength unidirectional inter-office ring network." *IEEE Photonics Technology Letters* **5**(9): 1026–1031.
20. Giles, C. R., and D. J. Giovanni. Dec. 1990. "Dynamic gain equalization in two-stage fiber amplifiers." *IEEE Photonic Technology Letters* **2**(12): 866–868.
21. Inoue, K., T. Kominato, and H. Toba. 1991. "Tunable gain equalization using a Mach–Zehnder optical filter in multistage fiber amplifiers." *IEEE Photonics Technology Letters* **3**(8): 718–720.
22. Toba, H., K. Takemoto, T. Nakanishi, and J. Nakano. Feb. 1993. "A 100-channel optical FDM six-stage in-line amplifier system employing tunable gain equalizer." *IEEE Photonic Technology Letters,* pp. 248–250.
23. Su, S. F., R. Olshansky, G. Joyce, D. A. Smith, and J. E. Baran. 1992. "Use of acoustooptic tunable filters as equalizers in WDM lightwave systems." *OFC Proceeding,* pp. 203–204.
24. Brackett, C. A. 1993. "The principle of scalability and modularity in multi-wavelength optical networks." *Proc. OFC: Access Network,* p. 44.
25. Alexander, S. B., et al. 1993. "A precompetitive consortium on wide-band all-optical network." *IEEE Journal of Lightwave Technologies* **11**(5/6): 714–735.
26. Chlamtac, I., A. Ganz, G. Karmi. July 1992. "Lightpath communications: an approach to high-bandwidth optical WAN's." *IEEE Transactions on Communications* **40**(7): 1171–1182.
27. Hill, G. R. 1988. "A wavelength routing approach to optical communication networks." *Proc. INFOCOM,* pp. 354–362.
28. Ramaswami, R. Feb. 1993. "Multiwavelength lightwave networks for commputer communication." *IEEE Communications Magazine* **31**(2): 78–88.
29. Zirngibl, M., C. H. Joyner, B. Glance. 1994. "Digitally tunable channel dropping filter/equalizer based on waveguide grating router and optical amplifier integration." *IEEE Photonic Technology Letters* **6**(4): 513–515.
30. d'Alessandro, A., D. A. Smith, J. E. Baran. March 1994. "Multichannel operation of an integrated acousto-optic wavelength routing switch for WDM systems." *IEEE Photonic Technology Letters* **6**(3): 390–393.
31. Li, C.-S., and F. Tong, C. J. Georgiou. 1993. "Crosstalk penalty in an all-optical network using static wavelength routers." *Proc. LEOS Annual Meeting.*
32. Li, C.-S., and F. Tong. 1994. "Crosstalk penalty in an all-optical network using dynamic wavelength routers." *Proc. OFC'94.*
33. Goldstein, E. L., L. Eskildsen, A. F. Elrefaie. May 1994. "Performance implications of component crosstalk in transparent lightwave networks." *IEEE Photonics Technology Letters* **6**(5): 657–660.

34. Goldstein, E. L., and L. Eskildsen. Jan. 1995. "Scaling limitations in transparent optical networks due to low-level crosstalk." *IEEE Photonic Technology Letters* **7**(1): 93–94.
35. Gimlett, J., and N. K. Cheung. 1989. "Effects of phase-to-intensity noise conversion by multiple reflections on gigabit-per-second DFB laser transmission systems." *IEEE Journal of Lightwave Technologies* **7**(6): 888–895.
36. Hamdy, W. M., and P. A. Humblet. 1993. "Sensitivity analysis of direct detection optical FDMA networks with OOK modulation." *IEEE Journal of Lightwave Technologies* **11**(5/6): 783–794.
37. DeCusatis, C., and P. Das. 1991. *Acousto-Optic Signal Processing Fundamentals and Applications*. Boston: Artech House.

Chapter 12 | Manufacturing Challenges

Eric Maass

Motorola, Incorporated, Tempe, Arizona 85284

12.1. Customer Requirements — Trends

The manufacturing challenges faced by optoelectronics components and systems are the direct effect of the technology challenges posed, which are in turn the indirect result of customer needs and expectations. These needs include performance (principally data rate and distance), reliability, service, and price.

12.1.1. PERFORMANCE

With the increasing speeds of hardware, such as microprocessors, and the increasing expectations of the users to include realistic video and user friendliness in the software or information communicated comes a need for increasing data rates on the interconnects from, to, and within the system. Some have called this the need to "feed the beast" with increasing supplies of its favorite "food" — data.

    The need for higher data rates is driven by the increasing data rates required by applications that incorporate video or multimedia in general and the global sharing of information, including video conferencing, in which information in graphs, figures, spreadsheets, or other forms needs to be shared in real time. This trend for higher data rates is not expected to end — as the higher data rates are achieved, increasing expectations for better resolution, more interactiveness, and new capabilities not previously exploited because of system bottlenecks will drive new requirements for higher data rates. The trade-off in how to achieve the data rates will be based on the cost advantages and limitations of various approaches.

12.1.2. RELIABILITY

Historically, optoelectronics applications have required high reliability primarily due to the early adoption of optoelectronics for telecommunications applications. Any expectation that data communication applications will involve lower reliability requirements and expectations will likely be short-lived. There are several reasons for this, including the rising expectations of customers in general and business customers in particular. However, the overriding trend will be the merging of data communications, telecommunications, and entertainment.

With the possible use of optoelectronics close to the home, the need for reliability can relate to "lifetime" uses, much as the telephone is now. Although it would be nice to believe that products will be driven toward increased reliability by the need for customer satisfaction and altruism, it may well prove that the avoidance of deep-pocket lawsuits will be an even greater driving force.

12.1.3. DELIVERY

The trend toward higher expectations will also include service in general, with on-time delivery as a subset. Customers increasingly expect promises to be kept. Companies that meet their delivery promises are rewarded with further business.

A possible strategy for achieving high expectations for service is illustrated in Fig. 12.1.

12.1.4. PRICE

Finally "the bottom line is the bottom line." Assuming that the component or system meets the performance and reliability requirements, the customer will generally purchase the least expensive product.

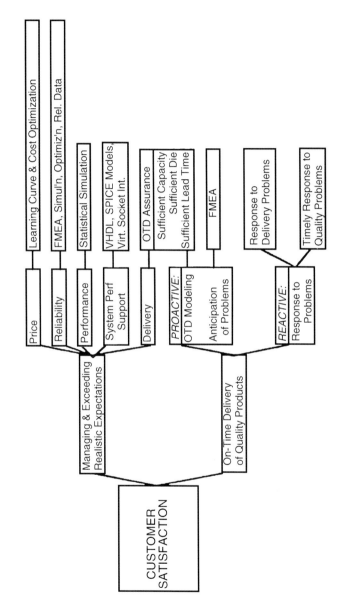

Fig. 12.1 Six σ service expectations.

12.2. Manufacturing Requirements — Trends

The performance, reliability, delivery, and price expectations of the customers drive technology and manufacturing challenges. The manufacturing challenges involve anticipating and developing manufacturing strategies to deal with these customer requirements.

12.2.1. PERFORMANCE

The data communication customers' rising expectations for performance — here defined as a high data rate, paired with a low error rate — will need to be met by combinations of system parameters, as illustrated in Figs. 4.2a and 4.2b in Vol. 1 Chapter 4. Most of these system parameters involve tight requirements and tradeoffs, which pose challenges to the manufacturing sites.

Some of the key performance parameters from Figs. 4.2a and 4.2b are listed in Table 12.1, in which the parameters are separated into the separate components of the block diagram, as shown in Fig. 12.2, reproduced from Fig. 4.1 of Vol. 1 Chapter 4.

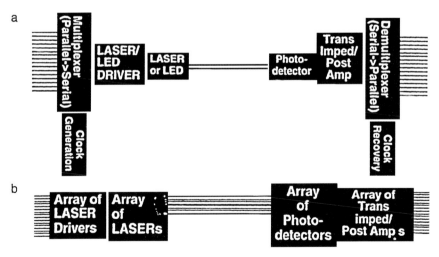

Fig. 12.2 Block diagrams of serial and parallel optoelectronic data communication systems.

12.2.1.1. Performance: Statistical Process Control, Reduction of Variability, Test, and Evaluation

In Vol. 1 Fig. 4.2, the customer requirements are related to the performance parameters. Each of these performance parameters needs to be placed under statistical process control (SPC). SPC involves:

Control charts
  Appropriate control limits
  Corrective actions based on out-of-control condition

Preventative actions
  Anticipation of potential problems
  Poka Yoke/mistake proofing
  Feedback and control mechanisms within automated equipment

Reduction of variability

Control charting involves determining the natural variation of the performance parameter and developing control limits based on this natural variation. The control charts detect if the process varies beyond the natural variation so that appropriate corrective actions can be executed. These corrective actions avoid sending marginal product to the customer and correct the problem that caused the "out of control" condition. If the natural variation of some performance parameters exceeds the limits that are acceptable to the customers (specification limits), then design and process development efforts are required to reduce the variability.

Although control charting is an effective feedback system, it has been argued that control charting of the performance parameters is "after the fact": The product that is measured is an output of the process that needs to be controlled. It has been argued that it would be much preferred to control the inputs to the process that needs to be controlled rather than the output. (S. Shingo, *Zero quality control: Source inspection and the Poka-Yoke system*). Poka Yoke, or mistake proofing, involves anticipating and preventing the possibility of an out-of-control situation by methods such as

Checks and double checks prior to processing
Assistance to the operator
Source inspections

Table 12.1 Key Component Performance Parameters Related to System Performance Requirements

	Transmit/ Receive IC	Lasers	Packaging	Fiber	Photodetectors	PLL
High data rate	F_{max}, F_t Bipolar: collector/ base capacitance MOS: left V_t	Capacitance Resistance Threshold current	Parasitics Lead inductance Lead capacitance	Attenuation	Capacitance Resistance Recombination?	High Q oscillation Filter w/low parasitics Substrate Isoin Matching/ differential Sharp edges
Long distance	Noise figure Parasitics Adaptive threshold	Wavelength match w/fiber	Opto coupling	Loss Graded index		
Clean Data (Eye Opening)	Sharp edges Current to laser when off I/O levels		Parasitics			High Q oscillation Filter w/low parasitics Substrate Isoin Matching/ differential Sharp edges

Low cost	Process cost/ complexity			
Reliability	Dielectric integration	Crystal defects	OEI coupling	Material (plastic?)
	Interface charges/ traps	Point defects	Laser thermal	
	Electromigration/ metal integration		Heat sink	
	Contact integration			
	Corrosion			
	ESD			
	Latchup			

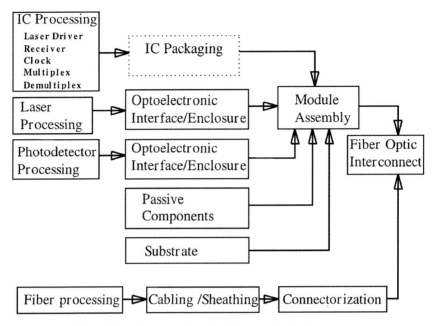

Fig. 12.3 Overview of optical link manufacturing process.

Source inspections detect nascent problems on the inputs and process motions or steps of the processing equipment and correct them, warn the operator, or shut down the processing before the product is affected.

The overall manufacturing process for an optoelectronic system is illustrated in Fig. 12.3. Within the integrated circuit (IC) processing areas, the controls for the ICs would be consistent with the requirements for other high-frequency integrated circuits, which would presumably be manufactured on the same processes. Nonetheless, for the sake of completeness, some of those requirements are listed below:

Complementary Metal Oxide Semiconductor
  Effective gate length, including photolithography and etch at the gate polysilicon definition step and the critical dimension measurements after develop and after etch
  Threshold voltages, both n and p channels
  Gate oxide thickness, in that it affects the transconductance, the threshold voltages, and reliability
  Capacitances, such as from gate to drain and drain to substrate/well

Isolation, especially radio frequency (RF) isolation, for the phase-locked loop (PLL) at higher frequencies

Bipolar

Current gain (beta) and the correlated $V\beta$

Capacitances, especially collector-base capacitance base resistance

Isolation, especially RF isolation, for the PLL at higher frequencies.

For all integrated circuits, the metal lines pose performance issues, contributing parasitic capacitances, inductances, and perhaps even act as undesired antennas. Note that a quarter wavelength at 1 GHz is about the length of some of the longer metal lines on larger ICs.

Within the assembly areas, the controls on die bonding, wire bonding, and molding operations would be required to prevent shorts and opens. For optoelectronics, tighter controls are required for all the parasitic inductances and capacitances associated with the packaging, especially the leads. The lead lengths should be minimized: Ideally, leads should be eliminated altogether.

Within the optoelectronic assembly area, the controls include the standard controls implemented for standard integrated circuits and new controls to meet the tolerances for good coupling between the light-emitting device and the packaging or between the photodetector and the packaging. This may require active alignment, which could use a feedback mechanism involving sensing the light output from the light-emitting device through the package to ensure that the alignment requirements are met. Although an active alignment system can be placed under controls that detect and prevent errors (Poka Yoke or mistake-proofing mechanisms), it can be an expensive system to implement.

Alternatively, passive alignment systems use the individual mechanical tolerances, determined by the manufacturing capabilities inherent in each of the components and piece parts as well as the assembly process. These tolerances determine where the light-emitting and light-detecting devices can be placed in the package and ensure that these tolerances are met during the assembly process without requiring activation during the assembly process. Passive alignment, if feasible, would be expected to have cost and cycle time benefits compared to active alignment of the light-emitting and photodetector devices during the assembly/alignment process.

Within the module assembly area, controls are required in the manufacture of the substrate (whether printed circuit or multichip module (MCM) substrate) and during the final assembly. The substrate manufacturing will generally require controls on the drilling (placement and diameter of holes), lamination (epoxy curing), photolithography, and etch processing. Final assembly will require controls on the insertion or placement of the components and on the solder joining process, including time, temperature, and ambient.

12.2.1.2. Meeting Customer Requirements: Design for Reliability, and Reliability Testing

Data communication customers will increasingly require high levels of reliability, here defined as the need to have the parts continue to meet the performance requirements over time. This requires that reliability be designed in and built in to every component, as illustrated in Fig. 12.4.

Although designing and building in reliability is critical, customer satisfaction will require that testing of the integrated circuits also ensures reliability. Because most reliability testing, historically, has involved destructive testing, which can only be performed on a sample of the parts, this reliability testing will necessarily involve some innovative testing, such as the use of acceleration of known reliability degradation mechanisms. In integrated circuits, current is an acceleration factor for electromigration and contact integrity, voltage for dielectric integrity, and temperature for several mechanisms: low temperatures for hot carrier effects and high temperatures for corrosion. Some of the reliability degradation mechanisms

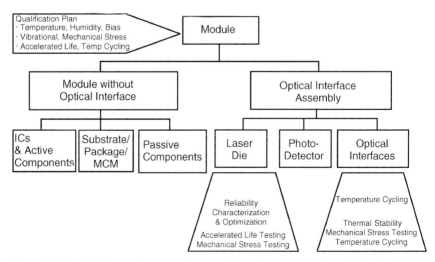

Fig. 12.4 Reliability requirements for optoelectronic systems, branched to individual component reliabilities.

Table 12.2 **Laser Diode Reliability Mechanisms, Related to Acceleration Factors and Possible Test Structures to Ensure Reliability Is Built In**

Laser diode mechanism	Acceleration factors	Test structures
Crystal defect migration Dark line/area propagation Point defect formation	Current, temperature Optical power, stress (packaging stress)	Standard laser
Contact integrity	Current, temperature	Laser, test pattern
Die bond integrity Corrosion Electromigration	 Temperature, humidity Current, temperature	Test patterns
Optical waveguiding Stress- or thermal-induced change in refractive index of III–V material	Stress, temperature (?)	Laser
Facet defect migration (not important for VCELs)	Optical power, current	Laser
Facet erosion		

for laser diodes are illustrated in Table 12.2 along with acceleration factors and some possible test structures for those reliability mechanisms.

12.2.1.3. Meeting Customer Requirements: Delivery, Optimizing On-Time Delivery, and Lead/Cycle Time

Figure 12.1 summarized customer service expectations. Service means many things, including the on-time delivery of quality products, providing the application support that the customers expect, and keeping the customer informed of progress in resolving issues that the customer expects to see resolved. The immediate expectation is that the customer will receive reliable, high-quality products delivered on time.

On-time delivery of optical links means meeting the customer's requirements on the quantity delivered and when it is delivered. A related issue is the lead time — how long of a delay the customers must experience in receiving product after placing an order. Often, the lead time is not an issue: the customer may place an order for product that is not needed immediately. At other times, the delay in receiving the optical links could result in lost sales for the customer.

The overall delivery issue, then, involves three issues: optimizing the percentage of the product that is delivered on time, minimizing the lead time, and minimizing the inventory. The latter, inventory, is primarily a financial trade-off. The customers would not mind if the supplier kept a great deal of excess inventory available, as safety stock, and responded immediately to orders placed by shipping from this safety stock — as long as the customers do not experience higher prices as a result. Unfortunately, holding product in inventory can be expensive and generally has a substantial impact on the bottom line.

The percentage on-time delivery can be optimized in the overall manufacturing flow by treating it as the product of several probabilities:

$$\Pr(\text{OTD}) = \Pr(\text{sufficent amount}) \times \Pr(\text{sufficient lead time}). \quad (12.1)$$

The probability that a sufficient amount of product, optoelectronic links, are produced to meet the customers' orders is obtained by determining the material that must be started, allowing for the yields of each of the components of the optoelectronic link, and ensuring that there is sufficient capacity available in the manufacturing line, described by

$$\Pr(\text{sufficient amount}) = \Pr(\text{sufficient capacity}) \times \Pr(\text{sufficient material started}) \times \Pr(\text{sufficient cumulative yield}). \quad (12.2)$$

The probability that sufficient capacity exists, Pr(sufficient capacity), is a matter of balancing the capacity development and expansion with the orders expected. Because anticipation of future orders is an inexact science, this can be a difficult issue, which is explored further under Section 12.2.2.1.

However, once the capacity is established and capacity expansions are planned and known, the key to Pr(sufficient capacity) is in order acceptance: Treating each order accepted as a promise that must be kept, for integrity reasons if not for customer satisfaction, allows the capacity to be matched to the orders. Accepting orders greater than maximum capacity, or accepting additional orders once all the capacity has been allocated, results in unpleasant and diversive trade-off issues on the level of "which customers are more important to us." From these perspectives, the probability that sufficient capacity exists is treated as a digital, 0 or 100% probability, that can reach 100% if sufficient integrity is maintained in the order-acceptance process.

The probability that sufficient material is started, Pr(sufficient material started), also encompasses the allowance for upsides in the demand from the customers. Generally, customers must project future sales when placing orders: This is an inexact science at best, and the customers will often underestimate or overestimate the sales. If the customers overestimate sales, the suppliers or the customer can keep the excess inventory, incurring a financial cost but the customer is generally satisfied. If the customers have underestimated sales, the customers are disappointing their customers and losing potential sales.

If the supplier provides sufficient product to meet the actual orders received but insufficient to meet the upside, the supplier may be on strong legal grounds, but the customer is still losing out. If the financial impact is not overwhelming, it might be preferable for the supplier to allow for some upside in the orders, allowing the potential to "delight the customer" by supplying the additional product in the happy case that the customers' sales exceed expectations: A delighted customer experiencing vigorous sales is a very favorable condition indeed!

Within the overall manufacturing flow, each of the steps has an associated yield — process, probe, assembly, and final test yields in the case of integrated circuit processing. Each of these yields follows a type of nonnormal distribution referred to as a beta distribution, which is a basic statistical distribution that is limited to the range of 0–100%. The beta distribution for MOS integrated circuits is illustrated in Fig. 12.5.

By determining the cumulative yield distribution associated with each of the components of the manufacturing flow illustrated in Fig. 12.3, the appropriate amount of material can be started through each of the processes to meet the delivery requirements in terms of quantity. A spreadsheet incorporating the statistical distributions of yield and cycle time can be used to determine the amount of material to be started and, when it

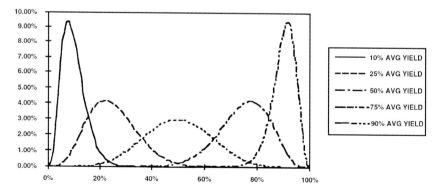

Fig. 12.5 Beta distributions for yields of MOS integrated circuits.

needs to be started, to ensure on-time delivery (perhaps with 95% confidence) (Fig. 12.6).

From Eq. (12.1), the final probability associated with meeting on-time delivery requirements is the probability that the parts will be delivered to the customer when promised, Pr(sufficient lead time).

Each of the steps in Fig. 12.3 has a cycle time associated with it. Cycle times tend to follow another nonnormal distribution, called a gamma distribution, which is obtained when there is a minimum but no maximum: The cycle time for IC processing, for example, has a minimum theoretical cycle time including all the sequential times that wafers need to spend in furnaces, photolithographic and etch equipment, and so on. However, the cycle time for individual lots can exceed the theoretical cycle time, such that there is no maximum—the distribution is bounded by minimum theoretical cycle time and infinity. An example of a cycle time gamma distribution is shown in Fig. 12.7.

The overall cycle time is a complex combination of the cycle times for each of the component processes illustrated in Fig. 12.3. Because the manufacturing cycle time is one of the trade-offs involved in the decision making to allow for the manufacturing challenges in optoelectronic systems, it may be best to generate a hypothetical example of the overall cycle time for a hypothetical optoelectronic link. Although the example used here is not a real example, it will illustrate the process of optimizing cycle time and on-time delivery relative to the inventory and other associated costs for various alternative manufacturing strategies.

The nested histograms of cycle times displayed in Fig. 12.8 each have the same horizontal axis—cycle time in days. Proceedings from left to right in the manufacturing flow, the cycle time distributions are for the labeled steps, in each case inclusive of the cycle times of the preceding steps of the flow. In other words, the cycle time distribution displayed as a histogram for connectorization would be the entire cycle time experienced at that step, including the cycle time for fiber processing and for cabling/sheathing.

Figure 12.9 illustrates the lead time and inventory trade-offs considering the overall fabrication and shipping of a complete link as a whole system. With no inventory, the total lead time is at its maximum, but the inventory cost is naturally at a minimum.

In this situation, the total lead time is also called the "visible horizon lead time." The visible horizon is the planning horizon—how far in advance the manufacturing company must plan starts. The manufacturer can quote a lead time substantially less than this total lead time to customers if the company is willing to start the processing early, in anticipation of orders.

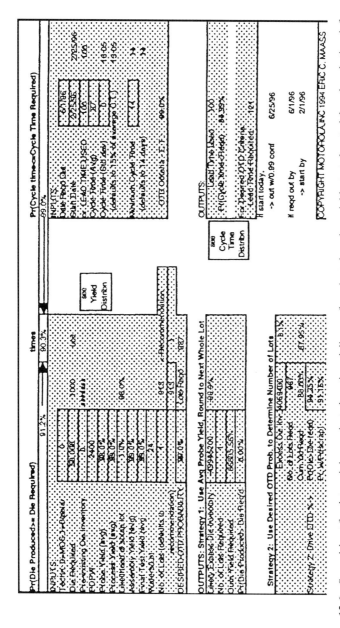

Fig. 12.6 Spreadsheet for determining projected on-time delivery of products based on yield, started material, and cycle time information.

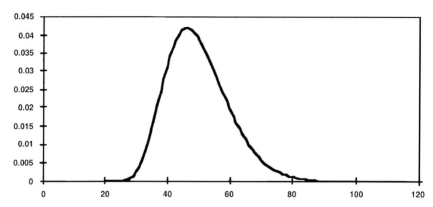

Fig. 12.7 Gamma model for cycle time, in which the average cycle time is 50 min, the standard deviation is 10 min, and the minimum theoretical cycle time is 20 min.

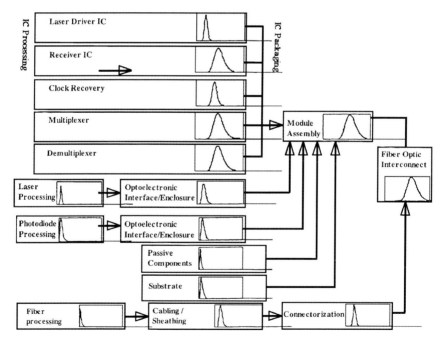

Fig. 12.8 Cycle time distributions associated with the manufacturing of an optoelectronic link.

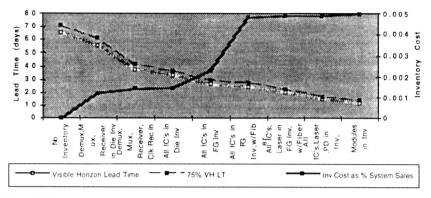

Fig. 12.9 Inventory cost and lead time trade-offs associated with manufacturing an optoelectronic link.

The manufacturing company can reduce this visible horizon lead time by storing inventory (perhaps called "safety stock") for critical, longer cycle time processes. In Fig. 12.8, the progression in adding inventory is in the order of maximum favorable impact on lead time; that is, the cycle time in the longest step of the critical path (or "rate determining step," for those with a more chemical bent) is reduced using inventory. The lead time is reduced, and the inventory cost increases. Because the jumps are rather sharp in places, the manufacturing company that constructs such a diagram can easily see reasonable trade-offs: In this example, the manufacturing company may be willing to store all integrated circuits in finished goods inventory but use just-in-time techniques for the remainder of the items required for manufacturing the optoelectronic link.

12.2.1.4. Meeting Customer Requirements: Price and Cost Trade-Offs

Optoelectronic data communication systems can be expected to be continuously under strong price pressure. Data communication systems will not only experience the learning curve expectations associated with computer systems, in general, and the price pressure from competitors but also experience price pressures from competing solutions, such as cable or other "copper wire" electronic solutions.

The learning curve expectations associated with computer systems in general, which can be expected to be applied to the associated data communication systems, have historically been what is called a "70% learning curve." This means that, for every doubling of cumulative volume shipped to customers, the price is expected to drop by 30% (Fig. 12.10); alternatively,

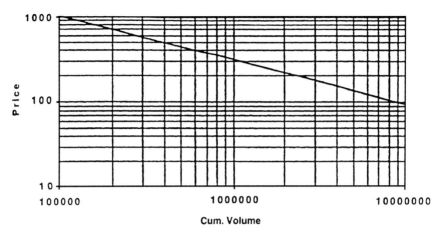

Fig. 12.10 Learning curve, with 70% historical expectations: reduction of price expected for increasing cumulative volume.

the price expected at any cumulative volume is

New price expectation = (previous price) × .7 ^ (Ln (new cum. volume/previous cum. volume)/Ln(2)) (12.3)

The competitive pressures that will influence pricing of optoelectronic systems come not only from competing optoelectronic manufacturers and solutions but also from nonoptoelectric technologies. Cable and other copper wire-based solutions can be expected to put competitive pressure on optoelectronic solutions. Fortunately for optoelectronic solutions, cable and copper wire solutions can be expected to encounter barriers that hinder high-speed operation, such as capacitative and inductive coupling that impacts noise immunity, as well as legal requirements such as electromagnetic interference. However, optoelectronic solutions will compete with cable and copper wire solutions at lower speeds. Also, it may be wise to consider the possibility that innovations in cable and wire solutions will surmount the high-speed barriers.

The competitive pressures and customer expectations influencing pricing will need to be addressed by minimizing the manufacturing costs associated with optoelectronic data communication systems. These include materials, direct and indirect labor, equipment and other fixed asset costs or depreciation, utilities, chemicals, and other consumables.

The materials, chemicals, and other consumables are considered variable costs: They increase directly with volume. Equipment, building, and

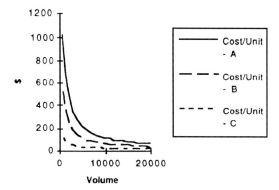

Fig. 12.11 Manufacturing cost per unit.

similar asset costs (or depreciation) are considered fixed costs; They will not vary with the volume of product sold. Direct and indirect labor and utilities generally will behave as if they have both a fixed and a variable cost component. This is because running a manufacturing line requires a minimal level of personnel and utilities, such as electrical power, regardless of the volume of product; however, more people and power consumption will be needed for increasing volume.

The cost per unit behaves as shown in Fig. 12.11; the volume impact can be quite substantial, particularly for semiconductor processing portions of the flow such as manufacture of integrated circuits, laser diodes, and photodetectors. In fact, it would be expected that the volume impact is dominant: The greater the volume of optoelectronic systems sold, the lower the cost due to the allocation of fixed costs over a larger number of products sold. It would be very fortunate if this cost reduction with increasing volume met or exceeded the customers' expectations of lower prices, as perhaps represented by the aforementioned 70% learning curve.

The manufacturing costs can be addressed by optimizing the factors associated with the total cost. Optimizing the fixed-cost impact means maximizing the volume — to a point. If the factory is planned and set up in a modular fashion, it may be that the volume impact is totally benevolent. In many cases, however, there are natural break points in volume: The factory can be effectively run at a certain percentage of maximum available capacity, perhaps approximately 85%. Increasing beyond this percentage utilization may involve cycle time impacts that cause customer dissatisfaction that will impact future sales. Increasing beyond this point may involve purchase of additional fixed assets. In some cases, the purchase of additional equipment may more than offset the cost reduction expected from increased volume of sales. This would be especially true if the capacity increase requires the construction of an additional factory. Hence, the up-front planning of the factory may be critical in allowing the increase in capacity in a controlled, perhaps modular, manner.

Optimization of direct and indirect labor costs involves trade-offs of labor rates versus location and automation. For example, labor costs in various countries may be lower, but the impact on cycle time may offset this: The most desirable cycle time would likely be achieved if all the factories are essentially co-located. Also, building manufacturing sites in countries with lower labor rates can involve other risks, such as political instability that can impact production or even lead to the loss of the factory, and the potential that the presence of the factory in the country will result in dissemination of intellectual knowledge that will ultimately lead to the creation of a competitor.

Reduction in material costs often involves negotiation with suppliers or process optimization that reduces the amount of wasted material. Reduction in material costs by using less material than required for quality and reliability is very unwise. Another means of reducing material costs can be by using newer technologies to miniaturize components — for example, using a new integrated circuit process with smaller geometries to increase the potential die per wafer, minimizing the cost per die (where a die is an unpackaged integrated circuit). In this case, the miniaturization generally improves performance because the device performance/speed of operation generally increases an integrated circuit geometries are reduced.

Reduction in utilities, chemicals, and other consumable costs is also largely a matter of negotiation with suppliers and process optimization that minimizes waste.

In summary, a cost reduction plan should be developed in order to meet the customers' expectations of price reductions with increased volumes. This cost reduction plan should include such items as maximizing utilization of the factories, process optimization, negotiation with suppliers, minimization of waste, and miniaturization of components. The projected cost savings should be calculated, with the results (if believable) prioritized in order of maximum favorable impact, allowing for negative side effects as risk factors. The cost savings plan should be highly visible, and people should be assigned direct responsibility to see that the cost reduction plan is successfully executed.

12.2.2. CHALLENGES

12.2.2.1. "Chicken and Egg" Syndrome

One key issue, which is especially relevant in emerging technologies, is the timing and risk of the investment in the technology. This issue can be called the chicken and the egg syndrome.

Customers do not want to commit to using an emerging technology without some assurance that it will be fully supported in a manufacturing sense. In the case of optoelectronic communications equipment, the customer would be extremely uncomfortable to have its product success be dependent on the production of something for which there is no factory.

On the other hand, the supplier does not want to commit to putting the emerging technology into production without having a customer that is committed to using the technology.

What often follows is a seemingly endless cycle that delays the implementation of the new technology — the supplier trying to get customers to commit to the new product/technology without a factory, and the customers showing interest but backing off gently when pressed for a commitment.

There appear to be three ways to resolve this chicken and egg syndrome:

1. A brave customer can commit, despite the risk
2. A brave supplier can commit and build the factory without a committed customer
3. The customer and supplier can share the risk: build the factory under a financial agreement that spreads the risk between the customer and supplier
4. An external agent can assume the risk for both parties — perhaps a venture capitalist, a government agency, or a potential partner of the supplier.

12.2.2.2. Diversity of Technologies

The technologies involved in optical fiber for communications are very diverse, as illustrated by Table 12.3.

The diversity of technologies generally results in using different manufacturing facilities for the different processing and assembly manufacturing involved in developing a complete fiber optics system. The integrated circuits may be manufactured in semiconductor "wafer fab" lines in the

Table 12.3 **Technologies Involved in Optoelectronics for Data Communications**

Technology	Example in optoelectronics	Manufacturing step
Mechanical engineering	Fiber/laser alignment	Last step
Polymer chemistry	Laser package/waveguide	Laser assembly
Digital electronics	Multiplexer IC	IC manufacturing
Analog electronics	Receiver IC	IC manufacturing
Materials science	Laser diode	Laser manufacturing
Optics	Photodiode, waveguide	Photodiode manufacturing Waveguide manufacturing
Thermodynamics	Laser diode packaging	Laser assembly

United States, the lasers may be manufactured in Japan, the assembly of the components may be accomplished in Asia, and the fiber itself may be drawn, sheathed, and connectorized in three different locations around the globe.

The implications of the diversity of manufacturing locations include effects on cycle time, quality, and cost. Although some of the trade-offs are obvious, the issue of colocation of the key steps of manufacturing, or even the entire manufacturing process, needs to be explored up-front.

12.2.2.3. Colocation?

The most capital-intensive parts of the overall manufacturing process are the IC processing, followed by the laser diode processing, the photodiode processing, and the fiber processing. Complete colocation would therefore require building the new manufacturing near an existing wafer fabrication facility or else building an entirely new set of manufacturing facilities, including wafer fabrication, in the selected colocation site. Due to the very high capital costs of building a new wafer fabrication area (or several areas, if the processes for laser drivers, receivers, and multiplexers are not compatible), the former alternative seems more reasonable if colocation is desired.

The benefits of colocation can include measurable improvements, such as total cycle time, and less tangible improvements in areas such as communication between the parties involved and speed in resolving problems.

The benefits of colocation on cycle time can be estimated from the example used earlier. Without colocation, the cycle time includes the time to prepare the wafers, die, components, fibers, or similar item for shipping, the time required for transportation, and the time to unpack the components. For manufacturing sites within the same continent, this cycle time would be on the order of a few days — within the process flow for an optoelectronic data communication system illustrated in Figs. 12.8 and 12.9, the total cycle impact would be approximately 1 to 3 weeks, or approximately a 10–30% reduction.

The cycle time impact on the customer would be minimized by setting up storage facilities at the outgoing or the incoming manufacturing sites or both. The components must sit in storage until needed — this has minimal effect on cycle time, but affects costs through the costs of space and people to control this inventory.

Colocation also improves communication, which translates into more rapid detection and resolution of problems and improved yield and reliability. It also improves the development time of new generations of optoelectronic products and minimizes the unused, unsold inventory of components that may be rendered obsolete by the next generation of product.

Effectively, the colocation decision is a cost decision: If the labor costs in a geographically separate facility are significantly lower than the costs of maintaining and expediting inventories of components, or if the capital costs for building a manufacturing facility for some components are prohibitive so that the components should be purchased from an outside supplier in a distant location, then colocation may not be feasible.

Generally, colocation may make sense at some level in the overall manufacturing flow but not for the complete manufacture of the system. For example, semiconductor manufacturing facilities are extremely capital intensive, whereas assembly facilities are much less capital intensive.

Consequently, it may make sense to obtain semiconductor devices such as integrated circuits from a vendor — and similarly purchase the laser and photodetector devices and the fiber — setting up storage facilities for these components, but colocating the facilities for assembling the module as well as the optoelectronic interfaces and enclosures.

12.2.2.4. Integration

Integration seems to be the natural direction of electronics. The historical trend of electronic components has been from discrete devices (e.g., diodes and transistors) to small-scale integration (SSI; e.g., simple gates), to medium-scale integration (MSI; e.g., arithmetic logic unit), to large-scale integration, (e.g., microprocessor), to very large-scale integration (e.g., microcomputer chip).

However, the path toward increasing scales of integration and complexity has always been aligned with the directions of increased speed of operation and decreased cost. In fact, these latter two directions seem to be the driving force behind integration of circuitry.

In digital systems, the alignment of cost, speed, and complexity derives from shrinking geometries due to improved manufacturing processes, especially photolithography and etch processing.

Smaller geometry MOS transistors perform at higher speed owing to the shorter distance from source to drain that must be transversed by the charge carriers as well as reduced parasitic capacitances with the smaller transistors.

Lower cost is achieved by the smaller devices taking up less area on the wafer (commonly referred to as "real estate") so that more integrated circuits can be produced on the same-size wafer. This lower cost is complicated by the costs for equipment capable of resolving the small geometries, resulting in higher wafer costs, the trend to larger wafers, and the minimum die size set by the requirements of assembly (pad sizes and number of pads that must be wire bonded).

Smaller geometry devices also increase the level of complexity or integration that can be achieved. More devices can be integrated in the same area, allowing more functions to be integrated and eliminating the requirement to package some of the integrated circuits that were previously separate. The reduction in package count can reduce the total cost to the customer for populating the boards.

Although the directions of reduced cost, increased speed, and increased complexity are aligned very well in the fully digital applications, the directions are not so clearly aligned in the mixed-signal world associated with data communication.

Mixed-signal applications, in which there are both analog and digital functions, impose additional requirements on the semiconductor technology. On the receive side of a data communication system, the transimpedance and postamplifier functions are analog functions, the clock recovery function generally involves a phase-locked loop, a feedback circuit that in turn involves the analog function of a voltage-controlled oscillator (VCO), the digital function of a frequency divider (generally based on digital flip flops), and phase detector, and the analog function of an operational amplifier or charge pump to convert the digital signal from the phase detector into a voltage to control the VCO frequency.

An analog function such as a transimpedance amplifier requires high noise immunity and high transconductance. The necessity of handling a range of signals, dependent on the distance the light has travelled through the fiber or other medium, imposes requirements on noise/stray signal immunity that conflict with the need to handle an increasing number of signals on the same chip as the integration is increased.

The MOS transistors that have been so effective for digital applications experience difficulties in achieving these noise immunity and transconductance requirements. Integration of these functions may require a bipolar or BiCMOS process with excellent isolation — a more complex and expensive process compared to standard CMOS processes.

An alternative integration path is to integrate the transimpedance amplifier with the photodetector. This has the advantage of minimizing the parasitic inductances and capacitance experienced by the signal as it is conducted from the photodetector to the transimpedance amplifier, allowing higher frequency performance.

The analog functions, including the transimpedance amplifier, the post-amplifier, the VCO, and the charge pump or operational amplifier, also require passive elements such as capacitors, varactors, and inductors. These passive elements need high impedance per unit area and generally high Q values (where Q can be thought of as the ratio of how much the passive inductor behaves as an ideal inductor to how much it act as a capacitor and resistor). The optimization of these passive elements adds to the process complexity and, therefore, to the cost.

However, there may be a performance advantage of integrating passive elements — the requirement to go off-chip to connect to external capacitors and inductors adds the parasitic capacitances and inductances of bond wires, for example, into the total capacitance and inductance of the passive element. These parasitic capacitances and inductances generally degrade the effective Q of the passive element.

The added complexity required to integrate mixed-signal functions will generally increase the wafer cost, and the integration of multiple functions handling multiple signals may degrade the performance. As a consequence, the trend toward higher levels of integration is not as straightforward as it is in the digital arena.

The impact of integration on cycle time is also less favorable than might be thought. The integrated circuits are produced in parallel; therefore, the effective cycle time is approximately the same as the cycle time of the longest integrated circuit process. Integrating the functions onto one complex integrated circuit process would either keep the cycle time the same or possibly increase the cycle time, if the complex, mixed-signal integrated circuit process requires significantly more cycle time.

12.2.2.5. Assembly of Components

Assembly processes are involved in at least two stages of the overall manufacturing flow for optoelectronic systems. The individual components, such as integrated circuits, are assembled into packages, such as small outline integrated circuit or quad flat pack, and then the packaged components are assembled into a module such as a printed circuit board.

The purposes of the integrated circuit packages are to protect the integrated circuits and to facilitate the handling and attachment of the integrated circuits to the printed circuit boards.

For optoelectronic systems, however, the packaging of integrated circuits actually has some detrimental side effects on cycle time, performance, and cost. The cycle time required to assemble integrated circuits into packages is in series with the integrated circuit process time, adding to the total cycle time. This cycle time impact is generally on the order of a few days; however, because the package assembly sites are often in different countries (in order to minimize the associated labor costs), the cycle time impact may be on the order of a few weeks, allowing for transportation and customs inspections requirements.

The integrated circuit packages add parasitic capacitances, resistances, and inductances that impact high-frequency performance. Although these parasitic properties can be modeled, the Q of the capacitances and inductances may be such that the performance requirement of low jitter may be impacted by the phase noise of the PLL due to the parasitics, and the signal detection and low bit error rate may be impacted by the noise generated by the parasitic, low Q capacitances and inductances. The signal may be further degraded at high frequencies if, for example, the bond wires act as quarter-wavelength antennas.

Naturally, the assembly of integrated circuits into packages involves costs — at the SSI and even the MSI levels, these costs can dominate the costs of the complete integrated circuit, despite the high capital costs associated with wafer processing.

A reasonable manufacturing trend, therefore, might be to eliminate the packaging of the integrated circuits from the total manufacturing flow. The cost, cycle time, and performance advantages of this direction are obvious from the previous discussion; however, there are three issues raised by this elimination of unnecessary steps: the impact on handling and attachment during module assembly, the protection of the integrated circuits (now in die or wafer form rather than packaged form), and the thoroughness of testing of the integrated circuits.

The elimination of packaging of integrated circuits changes the module assembly from the simple process of attaching packaged parts onto a printed circuit board to a process that includes die attach and wire bonding or solder bumping of integrated circuits in die form. This migration is effectively a change from printed circuit boards to MCMs. The MCM process can be considered a hybrid of the printed circuit board process and the integrated circuit packaging process: Die are attached to the MCM substrate rather than to the leadframe of an integrated circuit package, and the bond pads of the integrated circuits are wire bonded or soldered (solder bump) to the MCM connections rather than wire bonded to the leads of the integrated circuit package.

The die are protected in the MCM assembly operation much as they would be in the integrated circuit package assembly operation. The testing requirements in wafer form, however, are more intense: Whereas packaged integrated circuits can be tested in wafer form and packaged form, with the more difficult tests perhaps being accomplished on packaged integrated circuits, the MCM approach requires that all testing be in wafer form, prior to MCM assembly, in order to minimize the costs. Yield loss after MCM assembly increases costs to include the costs of all other components assembled onto the MCM as well as the MCM assembly itself. Thorough testing in wafer form of DC and functional requirements, and some or all AC performance, is referred to as the known good die (KGD) approach. Testing AC performance at high frequency in wafer form may involve some challenges — contacting the pads in wafer form sufficiently well for RF and AC measurements, without damaging the pads, is very difficult. The KGD approach may also involve efforts to avoid reliability problems by testing the integrated circuits in wafer form at high voltages, currents, and/or temperatures.

If, as expected, the MCM costs follow a respectable learning curve, the advantages of improved performance, reduced cycle time, and the improved costs by the elimination of process steps (the packaging of components such as integrated circuits) make the MCM approach very attractive and perhaps the preferred path.

12.2.2.6. Flexibility

There is often a trade-off between flexibility and cost in manufacturing. Flexibility refers to both flexibility in the production volume and flexibility in the product manufactured.

Generally, low cost can best be achieved by constant, high-volume manufacturing of a standard product. The manufacturing process itself can be optimized for this standard process, and equipment can be purchased and optimized for the standard process.

Deviations from this low-cost ideal include variations in the volume and variations in this standard process. Variations in the volume affect total cost and unit cost as shown in Fig. 12.12. In Fig. 12.12 (right), the total manufacturing cost is depicted. The fixed costs, which include depreciation expenses for the equipment and the manufacturing facility, correspond to the y-intercepts for cases A–C. The variable costs, which include materials used in the manufacturing process, correspond to the slopes. In Fig. 12.12

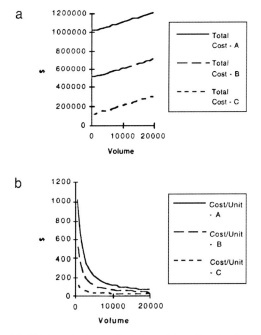

Fig. 12.12 (a), Total cost of manufacturing; (b), unit costs of manufacturing.

(left), the manufacturing cost per unit, which affects the price that the customer can be quoted while maintaining profitability (a worthy goal!), is depicted. Note that in Fig. 12.12 (left), the cost per unit rapidly increases at low volumes in case A, corresponding to the case of high fixed costs.

From the analysis of the cost impact of volume, the flexibility in production volume is constrained by upper and lower limits. The lower limit is set by the price acceptable to the customers, the profitability acceptable to the manufacturer, and the cost determined by a manufacturing cost per unit curve derived in Fig. 12.12 (left). The key determinant of the flatness of the cost per unit curve is the fixed costs — minimizing these fixed costs, therefore, is critical to flexibility. The upper limit is set by some natural break points in volume, as mentioned earlier: The factory can be effectively run at a certain percentage of maximum available capacity, perhaps approximately 85%. Increasing beyond this percentage utilization may involve cycle time impacts or the purchase of additional fixed assets.

Flexibility in terms of flexibility in the product manufactured involves the versatility of the manufacturing line both in running several different product types simultaneously and in terms of introducing new products, or product evolutions, into manufacturing. This flexibility also involves cost, primarily in terms of equipment and movement of material through the process but also in terms of planning and controlling the manufacturing process to minimize errors associated with the additional complexity introduced by maintaining multiple process flows.

12.2.2.6.1. "Job Shop" vs Flow Assembly Line

A manufacturing area that essentially runs only one product, efficiently, is often referred to as a "flow shop," whereas a manufacturing area that deals with a variety of products, each treated as a custom product, is often referred to as a job shop. These two extremes will generally have different manufacturing approaches, reflected most obviously in the factory layout.

A manufacturing area can be considered to consist of a material flow and an information flow. For any individual product type, the material flow can be drawn as a flowchart indicating the movement of material from workstation to workstation (a workstation generally being associated with each piece of equipment used in the process), with queues available between workstations.

In a flow shop, the layout of the manufacturing area will generally reflect the material flowchart in a cellular or modular layout. Equipment will be laid out to ease the movement of material between workstations in the sequence indicated by the flowchart. A simple example for an optoelectronic system is illustrated in Fig. 12.13. A modular layout that reflects the flowchart shortens the cycle time by minimizing the delay in moving material from workstation to workstation and by allowing operators at each workstation to readily see and respond to cycle time problems such as excessively large queues or equipment downtime at the next process. The flow layout also minimizes errors due to skipped processes and miscommunication between operators at adjacent work stations. The advantages of modular layout include

- Reduced material handling
- Reduced setup time
- Reduced in-process inventory
- Reduced need for expediting
- Improved operator expertise
- Improved communication, therefore human interactions

The disadvantages include

- Reduced shop flexibility
- Possible reduced machine utilization
- Possible increased cycle times

In a job shop, the layout of the manufacturing area will often be a functional layout in which similar equipment or types of manufacturing processes arc grouped together. Because the sequence of manufacturing operations varies from product to product type, the factory layout may concentrate on minimizing equipment downtime and maximizing equipment utilization and operator expertise in a particular type of equipment rather than reflecting the varying flowcharts. As simple example of this for an optoelectronic system is illustrated in Fig. 12.14.

For either job shops or flow shops, the movement of material between workstations can be accomplished by several means, including manual or automatic transport systems.

12.2.2.6.2. *Allowance for Product Evolution*

The modular layout is optimal for one particular product type and the layout is derived from the material flowchart for one particular product. If all the products and anticipated new products will have this same flow, then the modular layout is clearly the preferred choice.

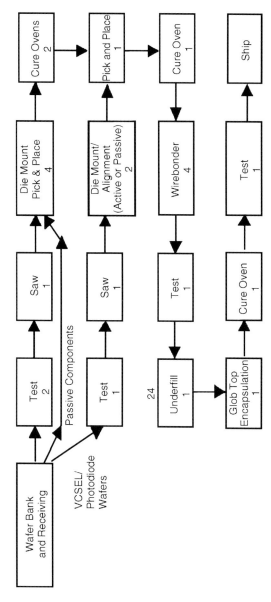

Fig. 12.13 Cellular or modular layout for optoelectronics assembly.

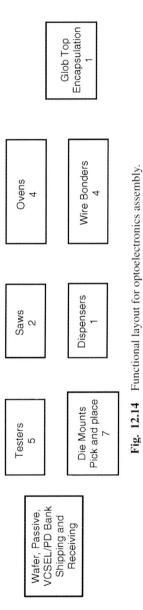

Fig. 12.14 Functional layout for optoelectronics assembly.

If several products and newer products are expected to differ from the material flowchart, the modular layout may experience problems. It is important to note that products that have material flowcharts that are nearly the same may actually involve worse problems than products with much different flowcharts because the very slight differences are likely to be missed by operators used to performing routine tasks the same way virtually every time.

The functional layout is not tied to any particular material flow — all flows are simply variations in which the material is transported between the groupings of similar equipment after each step in the process flow is completed.

Although the functional layout may appear to be the most flexible manufacturing layout, the cycle time and communication issues make this less than optimal, even for meeting the flexible needs associated with process evolution.

Instead, the optimal layout for a flexible optoelectronic manufacturing line is most likely a variation of the hybrid layout, in which some equipment is arranged in a modular layout (Fig. 12.13) and some equipment in a functional layout (Fig. 12.14). The intention is to anticipate the product evolution through development of a product road map, study the product road map to determine the portions of the material flow that will be common to all product types, and determine the portions of the material flow that will involve variations in the flowchart. The modular layout will be determined by the commonalities of all the flows, whereas the functional layout will reflect the steps in the material flow that will vary.

12.3. Manufacturing Alternatives

12.3.1. AUTOMATION/ROBOTICS

Automation can be applied to the manufacture of optoelectronic systems in several areas: automating the process operations at individual workstations such as attachment of the die to the package or MCM module, automating the transport of material between workstations, and automating the information flow (planning, control, and information gathering at each of the workstations).

Automation can provide advantages in terms of minimized cycle time (in terms of the expected cycle time or the variability of cycle time), minimized variability of processing, and perhaps reduced cost in complex labor-intensive operations. For example, the alignment of photonic devices such as light-emitting diodes (LEDs), or photodetectors to the fiber (either directly or indirectly through a lens or waveguide) can be a complex operation for an operator, whether the alignment is passive or active.

In the ideal situation, the entire manufacturing process would be automated. The information flow would plan and control the transport of the optoelectronic system from workstation to workstation, initiate the downloading of information to the workstation on the processing needed, provide for error prevention through the application of Poka Yoke (mistake proofing) approaches, initiate the processing of the material, provide feedback through the process consistent with error prevention, and initiate the uploading of information from the workstation regarding the processing or information on the results at a test operation. Finally, the information flow would initiate the transport of the material to the next workstation. After the last process is completed, the complete optoelectronic system would be transported to a staging area for delivery to the customer, with the information on the performance and processing readily available to the customer.

12.3.2. DESIGN FOR MANUFACTURABILITY

In designing an optoelectronic system for manufacturability, the first goal is to ensure that the processing variability does not impact meeting the customers' specifications. The second goal is to make the processing as simple as possible while still achieving the first goal.

Ensuring that the processing variability does not impact meeting the customers' requirements involves a combination of modeling and statistics: determining which processing variables affect which customer specifications, monitoring and obtaining measures of the variance of the relevant processing variables, obtaining an equation or simulation for the relationship between the processing variables and the specifications, and using this relationship combined with the variance of the processing variables to predict the impact of processing variables on the customers' requirements.

12.3.3. DESIGN FOR ASSEMBLY

To achieve the second goal, making the processing as simple as possible, the approach is to keep the first goal, achieving specifications allowing for process variability, and additionally determine the simplest assembly flow with the least amount of unique parts of processes. Achieving a simple assembly process for an optoelectronic system would involve the following:

1. Minimize the number of parts.
    This could be achieved by integrating the ICs where appropriate; for example, the parallel-to-serial conversion, clock generation, and laser or LED driver could be integrated into one integrated circuit on an appropriate high-frequency semiconductor process (small geometry CMOS or BiCMOS, for example). On the receive side, the photodiode and transimpedance amplifier could be integrated into one part (perhaps in GaAs) and the post- or limiting amplifier, serial-to-parallel conversion, and clock recovery and decision circuitry on another integrated circuit (again in small geometry CMOS or BiCMOS).
    Integrating the passive components onto the ICs where possible, or otherwise minimizing the number of separately packaged passive or discrete components through system design or use of multiple passive elements packaged together could also help.
2. Maximize automatic or self-alignment and similar part handling.
    If possible, the alignment of the optical elements, lasers and photodiodes, should be a passive alignment. The optical elements can have alignment pins, notches, or other features molded in so that they can be automatically aligned with the complementary alignment structures on the module.
3. Minimize the number of separate attachment/fastening/connection processes.
    The attachment of integrated circuits, optical elements, and passive and discrete components could be performed by robotic pick-and-place operations followed by a combined solder attachment heat cycle.
    Alternatively, in some MCM schemes, the integrated circuits can be processed to have solder bumps so that the pick-and-place operation is performed on the die upside down. A combined heat treatment performs the solder attachment, which is a combined die attachment and electrical connection to each of the pads. The MCM schemes effectively eliminate one set of assembly processes — the wire bonding operations associated with integrated circuit packaging.

12.3.4. FLEXIBLE MANUFACTURING APPROACHES

The "spine" approach shows promise for flexible manufacturing. The spine approach organizes the material flow and the information flow in a parallel fashion, analogous in some ways to the human spine, or to some topologies in information networking.

The spine is a linear axis, the main route of material movement, with branches for individual operations. The storage for materials and finished goods is also placed in the spine. Material handling equipment, such as conveyors or automated guided vehicle systems, can be incorporated into the spine.

The material handling equipment can be interfaced with automatic identification equipment, such as bar code reader systems, that can ensure appropriate movement of the materials and prevent errors. The automatic identification equipment can also be interfaced with a network communication system along the spine, which controls the movement of materials and obtains information on the procesesing of the equipment.

The modular nature of the spine flow is consistent with the cellular or modular flow discussed earlier and organizes the associated information flow of planning, control, and data acquisition along the spine. The modular nature allows for flexibility in the system and simplifies addition of modules to the spine for process evolution. Figure 12.15 shows an example of a spine approach adapted for optoelectronic systems.

12.3.5. DESIGN FOR TOTAL PRODUCT COST

Obviously there is a good deal of overlap between the design of a product and its manufacturing cost. This often-neglected area can be a source of significant cost reductions in a well-controlled manufacturing environment. The concept that strategic product design and manufacturing is the key to increased hardware profitability (or reduced cost as a percentage of revenue) is sometimes known as Design for Total Product Cost (DTPC). There are seven major elements that can be influenced in a product design to impact final cost, as follows:

- Development Expense: this represents the total development investment in people, test equipment, prototypes, etc.
- Manufacturing Cost: this represents the cost to manage procurement of parts, inventory, system assembly and test, and shipping costs.

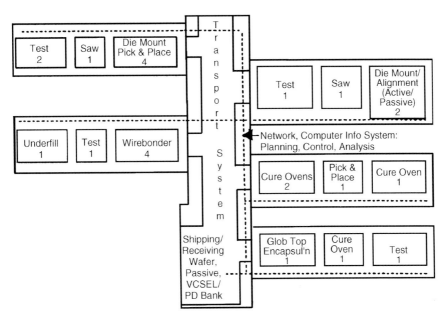

Fig. 12.15 Example of a spine approach adapted for optoelectronic systems.

- Parts Cost: the sum of all bill of material components used in a design.
- Ongoing inventory management cost: the cost to maintain and manage an inventory, as compared with procuring parts for inventory stocking.
- Obsolescence and scrap cost: this can be written off as a tax expense and lowers inventory, but also lowers gross profit margin.
- Supply chain costs: this cost is variable over time. It costs less if parts can be moved quickly through the supply chain; thus a more complex supply chain that is slower will cost more. This includes shipping and distribution expenses for raw material or assemblies from suppliers and delivery of some final product to the end customer.
- Serviceability and Warranty: this represents the cost associates with installing products (parts and labor) and maintaining them during the specified warranty period.

These elements are often managed as part of the Gross Profit equation, namely, Revenue minus Cost of Goods Sold (COGS) equals Gross Profit (GP). Subtracting costs for Sales, Distribution, Development Expense (this includes basic research), Interest, and other miscellaneous expenses yields

Pre-Tax Income. Further subtracting taxes yields the Net Profit. In this expression, the COGS is composed of Parts Cost, Manufacturing Cost, Serviceability and Warranty Cost, Obsolescence and Scrap, and a portion of the Supply Chain Cost (usually called the Product Cost Apportionment or PCA). Lower total cost enables manufacturers to price their products more competitively, which should drive additional market share and growth. Inventory is not an explicit part of the gross profit equation, but it follows that faster inventory turnover will lower the COGS. For example, Parts Cost is tracked as the price of parts when they are purchased, not what they cost the day they are assembled into a product; since the cost of most parts trends downward over time, it is advantageous to turn parts inventory quickly to take advantage of lower costs. Furthermore, lower inventory tends to accompany lower manufacturing, service, and obsolescence costs.

Note that if we attempt to minimize parts cost only without regard for other factors, design changes will likely increase cost in other areas; the result can be a net decrease in the gross profit margins. Also note that design for manufacturability implies that the product design will incorporate as many common building blocks as possible early in the manufacturing process, with customized design steps being configurable late in the manufacturing process. This is also called "design for postponement" — product designs that give manufacturing the ability to postpone assembly of specialized parts until the final step in the manufacturing process. To reduce both parts and manufacturing cost, it is desirable to move from a build-to-order model into a more standardized process in which sub-assemblies are purchased from suppliers rather than components, and final authorized assembly is done by strategically placed fulfillment centers. This will also reduce system assembly time, although it implies a fundamentally different approach to manufacturing; suppliers now become your manufacturers. This often implies using industry standard preferred parts where possible (when it does not compromise performance). Common parts also reduces the product variability, meaning it's possible to reduce safety stock (components or assemblies held as inventory to cover all possible order configurations). Once again, suppliers can play a role if they agree to hold inventory; their cost will be reflected as increased parts cost. In this way, lower inventory requirements also influence parts costs. Because inventory tends to follow the 80/20 rule (80% of inventory is tied up in orderable features, 20% in base functional parts), reducing the number of features to those that are absolutely necessary or provide obvious product differentiation can significantly lower inventory. This is also related to supply chain cost; in general, supply chain cost can be reduced by reducing unique part numbers, implementing more commonly stockable parts or subassemblies, using more industry standard parts (which in principle are easier to obtain with shorter lead times, and easier to dispose of if excess inventory is purchased), and by design for postponement (this helps the supply chain to remain flexible and responsive to a customer's changing needs at lower cost). Taken together, these inter-related parts of the total cost picture can be managed to improve hardware design profit margins in any business, including manufacturing of fiber optic systems.

Appendix A: Measurement Conversion Tables

English-to-Metric Conversion Table

English unit	Multiplied by	Equals metric unit
Inches (in.)	2.54	Centimeters (cm)
Inches (in.)	25.4	Millimeters (mm)
Feet (ft)	0.305	Meters (m)
Miles (mi)	1.61	Kilometers (km)
Fahrenheit (F)	(°F − 32) × 0.556	Celsius (C)
Pounds (lb)	4.45	Newtons (N)

Metric-to-English Conversion Table

Metric unit	Multiplied by	Equals English unit
Centimeters (cm)	0.39	Inches (in.)
Millimeters (mm)	0.039	Inches (in.)
Meters (m)	3.28	Feet (ft)
Kilometers (km)	0.621	Miles (mi)
Celsius (C)	(°C × 1.8) + 32	Fahrenheit (F)
Newtons (N)	0.225	Pounds (lb)

Absolute Temperature Conversion

Kelvin (K) = Celsius + 273.15

Celsius = Kelvin − 273.15

Area Conversion

1 square meter = 10.76 square feet = 1550 square centimeters

1 square kilometer = 0.3861 square miles

Metric Prefixes

Yotta = 10^{24}

Zetta = 10^{21}

Exa = 10^{18}

Peta = 10^{15}

Tera = 10^{12}

Giga = 10^{9}

Mega = 10^{6}

Kilo = 10^{3}

Hecto = 10^{2}

Deca = 10^{1}

Deci = 10^{-1}

Centi = 10^{-2}

Milli = 10^{-3}

Micro = 10^{-6}

Nano = 10^{-9}

Pico = 10^{-12}

Femto = 10^{-15}

Atto = 10^{-18}

Zepto = 10^{-21}

Yotto = 10^{-24}

Appendix B: Physical Constants

Speed of light = c = 2.99 792 458 × 10^8 m/s
Boltzmann constant = k = 1.3801 × 10^{-23} J/K = 8.620 × 10^{-5} eV/K
Planck's constant = h = 6.6262 × 10^{-34} J/S
Stephan–Boltzmann constant = σ = 5.6697 × 10^{-8} W/m²/K⁴
Charge of an electron = 1.6 × 10^{-19} C
Permittivity of free space = 8.849 × 10^{-12} F/m
Permeability of free space = 1.257 × 10^{-6} H/m
Impedance of free space = 120 π ohms = 377 ohms
Electron volt = 1.602 × 10^{-19} J

Appendix C: Index of Professional Organizations

AIP, American Institute of Physics
ANSI, American National Standards Institute
APS, American Physical Society
ASTM, American Society for Test and Measurement
CCITT, International Telecommunications Standards Body (see ITU)
CDRH, U.S. Center for Devices and Radiological Health
DARPA, Defense Advanced Research Projects Association (also referred to as ARPA)
EIA, Electronics Industry Association
FCA, Fibre Channel Association
FDA, U.S. Food and Drug Administration
HSPN, High Speed Plastic Network Consortium
IEC, International Electrotechnical Commission
IEEE, Institute of Electrical and Electronics Engineers
IETF, Internet Engineering Task Force
IrDA, Infrared Datacom Association
ISO, International Standards Organization
ITU, International Telecommunications Union (formerly CCITT)
NBS, National Bureau of Standards
NFPA, National Fire Protection Association
NIST, National Institute of Standards and Technology
OIDA, Optoelectronics Industry Development Association

OSA, Optical Society of America
OSHA, U.S. Occupational Health and Safety Administration
PTT, Postal, Telephone, and Telegraph Authority
SI, System International (International System of Units)
SIA, Semiconductor Industry Association
SPIE, Society of Photooptic Instrumentation Engineers
TIA, Telecommunications Industry Association
UL, Underwriters Laboratories

Appendix D: OSI Model

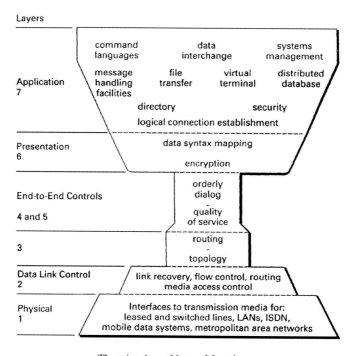

The wineglass of layered functions.

Appendix E: Network Standards and Documents

The IEEE defines a common path control (802.1) and data link layer (802.2) for all the following LAN standards; although FDDI is handled by ANSI, it is also intended to fall under the logical link control of IEEE 802.2 (the relevant ISO standard is 8802-2).

IEEE LAN Standards

802.3-code sense multiple access/collision detection (CSMA/CD) also known as Ethernet. A variant of this is Fast Ethernet (100BaseX). 802.3z is the emerging gigabit Ethernet standard. IEEE 802.3ae is the proposed 10 Gbit/s Ethernet standard.

802.4-Token bus (TB)

802.5-Token ring (TR)

802.6-Metropolitan area network (MAN) [also sometimes called switched multimegabit data service (SMDS) to which it is related]

802.9-Integrated services digital network (ISDN) to LAN interconnect

802.11-Wireless services up to 5 Mb/s

802.12-100VG AnyLAN standard

802.14-100BaseX (version of Fast Ethernet)

ANSI Standards
   Fast Ethernet: ANSI X3.166
   FDDI: ANSI X3.263
   Fiber distributed data interface (FDDI): ANSI X3T9.5 (the relevant ISO standards are IS 9314/1 2 and DIS 9314/3)
      Physical layer (PHY)
      Physical media dependent (PMD)
      Media access control (MAC)
      Station management (SMT)
   Because the FDDI specification is defined at the physical and data link layers, additional specifications have been approved by ANSI subcommittees to allow for FDDI over single-mode fiber (SMF-PMD), FDDI over copper wire or CDDI, and FDDI over low-cost optics (LC FDDI). A time-division multiplexing approach known as FDDI-II has also been considered.
   Serial byte command code set architecture (SBCON): ANSI standard X3T11/95-469 (rev.2.2., 1996); follows IBM's enterprise systems connectivity (ESCON) standard as defined in IBM documents SA23-0394 and SA22-7202 (IBM Corporation, Mechanicsburg, PA)
   Fibre channel standard (FCS)
      ANSI X3.230-1994 rev. 4.3, physical and signaling protocol (FC-PH)
      ANSI X3.272-199x rev. 4.5 (June 1995) Fibre Channel arbitrated loop (FC-AL)
   High-performance parallel interface (HIPPI)
      ANSI X3.183-mechanical, electrical, and signaling protocol (PH)
      ANSI X3.210-framing protocol (FP)
      ANSI X3.218-encapsulation of ISO 8802-2 (IEEE 802.2) logical link protocol (LE)
      ANSI X3.222-physical switch control (SC)
      Serial HIPPI has not been sanctioned as a standard, although various products are available.
      HIPPI 6400 is currently under development.
   Synchronous optical network (SONET): originally proposed by

Bellcore and later standardized by ANSI and ITU (formerly CCITT) as ITU-T recommendations G.707, G.708, and G.709.

Other standards

Asynchronous transfer mode (ATM): controlled by the ATM Forum

Appendix F: Data Network Rates

The "fundamental rate" of 64 kb/s derives from taking a 4-kHz voice signal (telecom), sampling into 8-bit wide bytes (32 kb/s), and doubling to allow for a full-duplex channel (64 kb/s). In other words, this is the minimum data rate required to reproduce a two-way voice conversation over a telephone line. All of the subsequent data rates are standardized as multiples of this basic rate.

DS0	64 kb/s	
T1 = DS1	1.544 Mb/s	24 × DS0
DS1C	3.152 Mb/s	48 × DS0
T2 = DS2	6.312 Mb/s	96 × DS0
T3 = DS3	44.736 Mb/s	672 × DS0
DS4	274.176 Mb/s	4032 × DS0

Note: framing bit overhead accounts for the bit rates not being exact multiples of DS0.

STS/OC is the SONET physical layer ANSI standard. STS refers to the electrical signals and OC (optical carrier) to the optical equivalents. Synchronous digital hierarchy (SDH) is the worldwide standard defined by CCITT (now known as ITU, the International Telecommunications Union); formerly known as synchronous transport mode (STM). Sometimes the notation STS-*XC* is used, where *X* is the number (1, 3, etc.) and *C* denotes the frame is concatinated from smaller frames. For example, three STS-1 frames at 51.84 Mb/s each can be combined to form one STS-3C frame at 155.52 Mb/s. Outside the United States, SDH may be called plesiochronous digital hierarchy (PDH). Note that OC and STS channels are normalized to a data rate of 51.84 Mbit/s, while the equivalent SDH specifications are normalized to 155.52 Mbit/s.

STS-1 and OC-1	51.840 Mb/s	
STS-3 and OC-3	155.52 Mb/s	same as STM-1
STS-9 and OC-9	466.56 Mb/s	
STS-12 and OC-12	622.08 Mb/s	same as STM-4
STS-18 and OC-18	933.12 Mb/s	
STS-24 and OC-24	1244.16 Mb/s	same as STM-8
STS-36 and OC-36	1866.24 Mb/s	
STS-48 and OC-48	2488.32 Mb/s	same as STM-16
STS-192 and OC-192	9953.28 Mb/s	same as STM-64
STS-256 and OC-256	13271.04 Mb/s	same as STM-86
STS-768 and OC-768	39813.12 Mb/s	same as STM-256
STS-3072 and OC-3072	159252.48 Mb/s	same as STM-1024
STS-12288 and OC-12288	639009.92 Mb/s	same as STM-4096

Higher speed services aggregate low-speed channels by time division multiplexing; for example, OC-192 can be implemented as four OC-48 data streams. Note that although STS (synchronous transport signal) is analogous to STM (synchronous transport mode) there are some important differences. The first recognized STM is STM-1, which is eqivalent to STS-3. Similarly, not all STS rates have a corresponding STM rate. The frames for STS and STM are both set up in a matrix (270 columns of 9 bytes each) but in STM frames the regenerator section overhead (RSOH) is located in the first 9 bytes of the top 3 rows. The fourth STM row of 9 bytes is occupied by the administrative unit (AU) pointer, which operates in a manner similar to the H1 and H2 bytes of the SONET line overhead (LOH). The 9 bytes of STM frame in rows 5 through 9 is the multiplex section overhead (MSOH) and is similar to the SONET LOH. In SONET we defined virtual tributaries (VT), while SDH defines virtual containers (VC), but they basically work the same way. A VT or VC holds individual

E1 or other circuit data. VCs are contained in tributary unit groups (TUG) instead of SONET's VT groups (VTG). VCs are defined as follows:

DS-2	VC-2
E3/DS-3	VC-3
DS-1	VC-11
E1	VC-12

Also note that while there are many references to 10 Gbit/s networking in new standards, the exact data rates may vary. As this book goes to press, proposed standards for 10 Gigabit Ethernet and Fibre Channel are not yet finalized. The Ethernet standard has proposed two data rates, approximately 9.953 Gbit/s (compatible with OC-192) and 10.3125 Gbit/s. Fibre Channel has proposed a data rate of approximately 10.7 Gbit/s. There is also ongoing discussion regarding standards compatible with 40 Gbit/s data rates (OC-768).

The approach used in Europe and elsewhere:

E0	64 Kb/s	
E1	2.048 Mb/s	
E2	8.448 Mb/s	4 E1s
E3	34.364 Mb/s	16 E1s
E4	139.264 Mb/s	64 E1s

Thus, we have the following equivalences for SONET and SDH hierarchies:

SONET signal	*SONET capacity*	*SDH signal*	*SDH capacity*
STS-1, OC-1	28 DS1s or 1 DS3	STM-0	21 E1s
STS-3, OC-3	84 DS1s or 3 DS3s	STM-1	63 E1s or 1 E4
STS-12, OC-12	336 DS1s or 12 DS3s	STM-4	252 E1s or 4 E4s
STS-48, OC-48	1344 DS1s or 48 DS3s	STM-16	1008 E1s or 16 E4s
STS-192, OC-192	5376 DS1s or 192 DS3s	STM-64	4032 E1s or 64 E4s
STS-768, OC-768	21504 DS1s or 768 DS3s	STM-256	16128 E1s or 256 E4s

For completeness, the SONET interface classifications for different applications are summarized in the table below. Recently, a new 300-meter Very Short Reach (VSR) interface based on parallel optics for OC-192 data rates has been defined as well (see Optical Internetworking Forum document OIF2000.044.4 or contact the OIF for details). Industry standard network protocols are summarized in the tables below:

	Short reach	Intermediate reach	Long reach			Very long reach			
Distance (km)	<2	15	15	40	60	60	120	160	160
Wavelength (nm)	1,310	1,310	1,550	1,310	1,550	1,550	1,310	1,550	1,550
OC-1	SR	IR-1	IR-2	LR-1	LR-2	LR-3	VR-1	VR-2	VR-3
OC-3	SR	IR-1	IR-2	LR-1	LR-2	LR-3	VR-1	VR-2	VR-3
OC-12	SR	IR-1	IR-2	LR-1	LR-2	LR-3	VR-1	VR-2	VR-3
OC-48	SR	IR-1	IR-2	LR-1	LR-2	LR-3	VR-1	VR-2	VR-3
OC-192	SR	IR-1	IR-2	LR-1	LR-2	LR-3	VR-1	VR-2	VR-3

Common Fiber Optic Attachment Options
(without repeaters or channel extenders)

Channel	Fiber	Connector	Bit rate	Fiber bandwidth	Maximum distance	Link loss
ESCON (SBCON)	SM	SC duplex	200 Mb/s	N/A	20 km	14 dB
	MM 62.5 micron	ESCON duplex or MT-RJ	200 Mb/s	500 MHz-km 800 MHz-km	2 km 3 km	8 dB
	MM 50.0 micron	ESCON duplex or MT-RJ	200 Mb/s	800 MHz-km	2 km	8 dB
Sysplex Timer ETR/CLO	MM 62.5 micron	ESCON duplex or MTRJ	8 Mb/s	500 MHz-km or more	3 km	8 dB
	MM 50.0 micron	ESCON duplex or MTRJ	8 Mb/s	500 MHz-km or more	2 km	8 dB
FICON/ Fibre Channel LX	SM	SC duplex or LC duplex	1.06 Gb/s	N/A	10 km	7 dB
FICON LX	MM w/MCP 62.5 micron	SC duplex or LC duplex	1.06 Gb/s	500 MHz-km	550 meters	5 dB

(continued)

Common Fiber Optic Attachment Options
(without repeaters or channel extenders) (continued)

Channel	Fiber	Connector	Bit rate	Fiber bandwidth	Maximum distance	Link loss
FICON/ Fibre Channel SX	MM w/MCP 50.0 micron	SC duplex or LC duplex	1.06 Gb/s	400 MHz-km	550 meters	5 dB
	MM 50.0 * micron	SC duplex or LC duplex	1.06 Gb/s	500 MHz-km	500 meters	3.85 dB
	MM 62.5 * micron	SC duplex or LC duplex	1.06 Gb/s	160 MHz-km	250 meters	2.76 dB
	MM 62.5 * micron	SC duplex or LC duplex	1.06 Gb/s	200 MHz-km	300 meters	3 dB
	MM 50.0 * micron	SC duplex or LC duplex	2.1 Gb/s	500 MHz-km	300 meters	
	MM 62.5 * micron	SC duplex or LC duplex	2.1 Gb/s	160 MHz-km	120 meters	
	MM 62.5 * micron	SC duplex or LC duplex	2.1 Gb/s	200 MHz-km	150 meters	
Parallel Sysplex coupling links- HiPer- Links	SM	SC duplex or LC duplex	1.06 Gb/s compatibility mode	N/A	10 km (20 km on special request)	7 dB
	SM	LC duplex	2.1 Gb/s peer mode	N/A	10 km (20 km on special request)	7 dB
	MM w/MCP 50.0 micron	SC duplex or LC duplex	1.06 Gb/s	500 MHz-km	550 meters	5 dB
	MM 50.0 * Discontinued May 98	SC duplex	531 Mb/s	500 MHz-km*	1 km	8 dB *
OC-3/ ATM 155	SM	SC duplex	155 Mb/s	N/A	20 km	15 dB
	MM 50 micron	SC duplex	155 Mb/s	500 MHz-km	2 km	

(*continued*)

Common Fiber Optic Attachment Options
(without repeaters or channel extenders) (continued)

Channel	Fiber	Connector	Bit rate	Fiber bandwidth	Maximum distance	Link loss
	MM 62.5 micron	SC duplex	155 Mb/s	500 MHz-km	2 km	11 dB
	MM * 50 micron	SC duplex	155 Mb/s	500 MHz-km	1 km	
	MM * 62.5 micron	SC duplex	155 Mb/s	160 MHz-km	1 km	
OC-12/ ATM 622	MM 50 micron	SC duplex	622 Mb/s	500 MHz-km	500 m	
	MM 62.5 micron	SC duplex	622 Mb/s	500 MHz-km	500 m	
	MM * 50 micron	SC duplex	622 Mb/s	500 MHz-km	300 m	
	MM * 62.5 micron	SC duplex	622 Mb/s	160 MHz-km	300 m	
FDDI	MM 62.5 micron	Media Access Connector (MAC) or SC duplex	125 Mbit/s overhead reduces to 100Mb/s	500 MHz-km	2 km	9 dB
	MM 50 micron	MAC or SC duplex	125 Mbit/s overhead reduces to 100 Mb/s	500 MHz-km	2 km	9 dB
Token Ring *	MM 62.5 micron	SC duplex	16 Mbit/s	160 MHz-km	2 km	
	MM 50 micron	SC duplex	16 Mbit/s	500 MHz-km	1 km	
Ethernet	MM * 50 micron 10Base-F	SC duplex	10 Mbit/s	500 MHz-km	1 km	
	MM * 62.5 micron 10Base-F	SC duplex	10 Mbit/s	160 MHz-km	2 km	
	MM * 50 micron 100Base-SX	SC duplex	100 Mbit/s	500 MHz-km	300 m	
	MM * 62.5 micron 100Base-SX	SC duplex	100 Mbit/s	160 MHz-km	300 m	
Fast Ethernet	MM 50 micron 100Base-F	SC duplex	100 Mbit/s	500 MHz-km	2 km	

(continued)

Common Fiber Optic Attachment Options
(without repeaters or channel extenders) (continued)

Channel	Fiber	Connector	Bit rate	Fiber bandwidth	Maximum distance	Link loss
	MM 62.5 micron 100Base-F	SC duplex	100 Mbit/s	500 MHz-km	2 km	
Gigabit Ethernet IEEE 802.3z	SM 1000BaseLX	SC duplex	1.25 Gb/s	N/A	5 km	4.6 dB
	MM * 62.5 micron 1000BaseSX	SC duplex	1.25 Gb/s	160 MHz-km 200 MHz-km	220 meters 275 meters	2.6 dB *
	MM w/MCP 62.5 micron 1000BaseLX	SC duplex	1.25 Gb/s	500 MHz-km	550 meters	2.4 dB
	MM * 50.0 micron 1000BaseSX	SC duplex	1.25 Gb/s	500 MHz-km *	550 meters	3.6 dB *
	MM w/MCP 50.0 micron 1000BaseLX	SC duplex	1.25 Gb/s	500 MHz-km	550 meters	2.4 dB

Notes:
* Indicates channels that use short wavelength (850 nm) optics; all link budgets and fiber bandwidths should be measured at this wavelength.
* SBCON is the non-IBM trademarked name of the ANSI industry standard for ESCON.
* All industry standard links (ESCON/SBCON, ATM, FDDI, Gigabit Ethernet) follow published industry standards. Minimum fiber bandwidth requirement to achieve the distances listed is applicable for multimode (MM) fiber only. There is no minimum bandwidth requirement for single mode (SM) fiber.
* Bit rates given below may not correspond to effective channel data rate in a given application due to protocol overheads and other factors.
* SC duplex connectors are keyed per the ANSI Fiber Channel Standard specifications.
* MCP denotes mode conditioning patch cable, which is required to operate some links over MM fiber.
* As light signals traverse a fiber optic cable, the signal loses some of its strength (decibels (dB) is the metric used to measure light loss). The significant factors that contribute to light loss are: the length of the fiber, the number of splices, and the number of connections. All links are rated for a maximum light loss budget (i.e., the sum of the applicable light loss budget factors must be less than the maximum light loss budget) and a maximum distance (i.e., exceeding the maximum distance will cause undetectable data integrity exposures). Another factor that limits distance is jitter, but this is typically not a problem at these distances.
* Unless noted, all links are long wavelength (1300 nm) and the link loss budgets and fiber bandwidths should be measured at this wavelength. For planning purposes, the following worse case values can be used to estimate the total fiber link loss. Contact the fiber vendor or use measured values when available for a particular link configuration:
  - Link loss at 1300 nm = 0.50 dB/km
  - Link loss per splice = 0.15 dB/splice (not dependent on wavelength)
  - Link loss per connection = 0.50 dB/connection (not dependent on wavelength)
* HiPerLinks are also known as Coupling Facility (CF) or Inter System Channel (ISC) links.
* All links may be extended using channel extenders, repeaters, or wavelength multiplexers. Wavelength multiplexing links typically measure link loss at 1550 nm wavelength, typical loss is 0.3 dB/km.

Common Ethernet Standards

	Standard	IEEE	Data rate (Mbit/s)	Medium	Max. distance (m) 1/2 duplex	Full duplex
Ethernet	10BaseT	802.3i	10	2 pair 100 ohm Cat 3 UTP copper	100	100
	10BaseFL	802.3j	10	Optical fiber pair	2,000	< 2,000
	10BaseFB	802.3j	10	Optical fiber pair	2,000	
	10BaseFP	802.3j	10	Optical fiber pair	1,000	
Fast Ethernet	100BaseTX	802.3u	100	2 pair 100 ohm Cat 5 UTP copper	100	100
	100BaseFX	802.3u	100	Optical fiber pair	412	2,000
	100BaseT4	802.3u	100	4 pair 100 ohm Cat 3 UTP copper	100	
	100BaseT2	802.3y	100	2 pair 100 ohm Cat 3 UTP copper	100	100
Gigabit Ethernet	1000BaseLX	802.3z	1,000	62.5 micron multimode fiber	316	550
	1000BaseLX	802.3z	1,000	50 micron multimode fiber	316	550
	1000BaseLX	802.3z	1,000	Singlemode fiber	316	5,000
	1000BaseSX	802.3z	1,000	62.5 micron multimode fiber	275	275
	1000BaseLX	802.3z	1,000	50 micron multimode fiber	316	550
	1000BaseCX	802.3z	1,000	Shielded balanced copper jumpers	25	25
	1000BaseT	802.3ab	1,000	4 pair 100 ohn Cat 4 UPT copper	100	100

Proposed 10 Gigabit Ethernet Standard (IEEE 802.3ae) Draft Objectives (final draft target 2002)

Fiber type	distance	Physical layer
SM	2 km	1300 nm serial
SM	10 km	1300 nm serial
SM	40 km	1550 nm serial
MM (all types of fiber, including new fibers such as 50 micron/ 2000 MHz-km)	65 meters (very short reach, VSR)	**850 nm serial ** (preferred)** 1300 nm WWDM 850 nm CWDM
MM installed (62.5 micron, 160/500 MHz-km at 850/1300 nm)	300 meters (very short reach, VSR)	**1300 nm WWDM * (preferred)** 850 nm CWDM 850 nm serial

* Does not require new fiber installation to reach 300 meters distance, and is the preferred, minimal solution for this service class.
** Reaches 65 meters on installed fiber, and up to 300 meters over new MM fiber.
Note: previous guidelines calling for 100 meters over installed MM fiber and 300 meters over MM fiber (including new fiber types) were revised December 2000.

Transmit and Receive Levels of Common Fiber Optic Protocols

Protocol type	I/O Spec.
ESCON/SBCON MM and ETR/CLO MM	TX: -15 to -20.5 RX: -14 to -29
ESCON/SBCON SM	TX: -3 to -8 RX: -3 to -28
FICON LX SM (MM via MCP)	TX: -4 to -8.5 RX: -3 to -22
FICON SX MM	TX: -4 to -9.5 RX: -3 to -17
ATM 155 MM (OC-3)	TX: -14 to -19 RX: -14 to -30
ATM 155 SM (OC-3)	TX: -8 to -15 RX: -8 to -32.5
FDDI MM	TX: -14 to -19 RX: -14 to -31.8

(*continued*)

Transmit and Receive Levels of Common Fiber Optic Protocols (continued)

Protocol type	I/O Spec.
Gigabit Ethernet LX SM (MM via MCP)	TX: -14 to -20
	RX: -17 to -31
Gigabit Ethernet SX MM (850 nm)	TX: -4 to -10
	RX: -17 to -31
HiPerLinks (ISC coupling links, IBM Parallel Sysplex, 1.06 Gbit/s compatibility mode)	TX: -3 to -11
	RX: -3 to -20
HiPerLinks (ISC coupling links, IBM Parallel Sysplex, 2.1 Gbit/s peer mode)	TX: -3 to -9
	RX: -3 to -20

Appendix G: Other Datacom Developments

Because the field of fiber optic data communication is expanding so rapidly, inevitably new technologies and applications will emerge while this book is being prepared for print. In addition, there will be some recent developments that have not been incorporated into the previous chapters. This appendix is an attempt to address these changes by including a partial list of related datacom devices and standards for reference purposes and to include brief descriptions of several emerging datacom technologies that the reader may encounter in the literature (product names and terminology used in this appendix may be copyrighted by the companies that developed them).

Free-Space Optical Links

Recent advances have made free-space optical data links practical in conditions when the weather is not a factor in attenuating the signal. Although bit error rates are typically on the order of 10^{-6} to 10^{-9} at 155 Mb/s, this is adequate for some applications such as voice and video transmission; recent experiments have shown that error rates as low as 10^{-12} can be obtained at 200 Mb/s using ESCON protocols on free-space optical links. Some examples of this technology and its applications include the following:

AstroTerra Corp. currently offers the TerraLink system, a tripod-mounted line-of-sight communication link with an 8-in. telescope aperture capable of 155 Mb/s transmission over 8 km in clear weather.

ThermoTrex Corp. in cooperation with SDL Inc. and funded by the Ballistic Missile Defense Organization, has demonstrated 1.2 Gb/s free-space communication over 150 km. This technology may form the basis of an aircraft-to-aircraft communications system and is scheduled to fly aboard the Space Technology Research Vehicle in 1998 to demonstrate communications among satellites.

Portable free-space laser communications are being developed by companies such as Leica Technologies, which has demonstrated a prototype binocular communications system for military applications at 100 kb/s over 3–5 km.

Optical Connectors

Older connector types, including the optical SMA connector, DIN, and LightRay MPX (a 2- to 12-element multifiber connector incompatible with the MTP/MPO) still resist being replaced and may be found in some installations. Many other legacy connectors continue to be used in large quantities, including the SC and ST connectors (recall that the ST was developed and trademarked by AT&T over 15 years ago). As new connector options emerge, many types of parallel or multifiber connectors have been proposed, most of which are mutually incompatible. For example, the MTP connector and its variations (MPO or MPX) do not mate with the SMC connector used on the Paroli parallel optical transceivers, yet both are being deployed as part of HIPPI 6400 installations. Recently, there has been increased interest in HIPPI 6400 for technical computing applications. Formerly known as SuperHIPPI and now known officially as Gigabyte System Network (GSN, a trademark of the High Performance Networking Forum), the physical layer of this protocol is now available as a draft standard (ANSI NC ITS T11.1 PH draft 2.6, dated December 2000, ISO/IEC reference number 11518-10) or online at www.hippi.org. This link provides a two-way, 12-channel-wide parallel interface running at 6400 Mbit/s (an increase from the standard HIPPI link rate of 800 Mbit/s). The link layer uses a fixed size 32-byte packet, 4B/5B encoding, and credit-based flow control, while the physical layer options include parallel copper (to 40 meters) or parallel optics (several hundred meters to 1 km).

Optical backplane connectors are likewise not yet standardized, and may include variants such as the MAC-II (an improved version of the original 12-fiber AT&T MAC connector) or the mini-MAC (a Bellcore approved version based on the MT ferrule). There have been various proposals for stacking multiple MT ferrules in a single connector; for example, a stack of 6 MTs forming a 72-fiber connector has been developed by LoDan West Corporation, and multichannel optical transceivers capable of utilizing this capacity in a full duplex 320 Gbit/s link were demonstrated at a recent Optical Fiber Conference (OFC). Research efforts continue to push the limits of stacked one-dimensional and two-dimensional connectors; an example is the 144-fiber "super-MT" developed by the POINT consortium between Amp, Allied Signal, Honeywell, and University of Columbia. Other examples include the 4×12 element Diamond MF multiple fiber connector.

Plastic Optical Fiber

At the Optical Fiber Conference 1997 in Dallas, Texas, plastic fiber applications were demonstrated for 155-Mb ATM and 100 Mb/s Ethernet LANs. This technology was developed by the High Speed Plastic Network Consortium (HSPN), a DARPA-funded venture consisting of Boeing, CEL, Boston Optical Fiber, Honeywell, Packard Hughes, and Lucent. Graded-index plastic fiber made from materials such as PMMA now offers bandwidth >3 Gb/s to 100 m and may be an alternative media for premises wiring. HSPN has submitted a proposal to the Optoelectronics Micro Networks OMNET program to continue development of plastic fibers.

Although some manufacturers have presented data indicating that plastic fiber will be unable to meet industry standard TIA/EIA-568A at 100 m, a number of companies are already developing plastic fiber applications. Some companies currently marketing plastic fiber or components including the following:

Amp Inc. offers plastic fiber connectors and components for airplanes and automobiles.

CEL (California Eastern Laboratories), a distributor for NEC Japan, offers the HiSpot transceiver with a visible red LED over step-index plastic fiber at 155 Mb/s using an F07 connector; applications include Fast Ethernet, ATM, and IEEE 1394 (Firewire).

Spec Trans Specialty Optics (distributor for Toray Industries of Japan) offers 200-μm step-index fiber; they have proposed to the ATM Forum a standard for 200-μm hard clad silica at 155 Mb/s, 100 m and plastic fiber for 50 Mb/s, 50 m.

Transceivers

Although discussion over the next generation of Fibre Channel continues, some companies are now offering an alternative, the gigabit interface connector (GBIC). This is a gigabit module that offers the convenience of being able to unplug a copper interface and replace it with an optical interface at a later time for increased distance and bandwidth. It features a parallel rather than a serial interface; the copper and optical modules plug into the same electrical connection point, which includes a pair of rails to guide the modules during plugging. Found mainly in workstation applications, the GBIC is available from several sources including IBM Rochester, Minnesota.

Low-cost bidirectional transceivers for fiber in the loop and similar applications (fiber to the curb or home) are also being developed. For example, a recent device reported by Amp in collaboration with Lasertron, Digital Optics Corp., GTE Labs, BroadBand Technologies, and the University of Colorado offers a connectorized, single-mode bidirectional transceiver with a source, modulator, and detector integrated into a single package. Each module includes an InP diode laser, silicon CMOS modulator/demodulator electronics, a PIN photodiode, an integrated beam splitter, and a built-in SC for ferrule-based fiber attachment. The beam splitter is used for bidirectional transmission at 1.55-μm wavelength and reception at 1.3-μm wavelength. Data rates up to 1.2 Gb/s is at 1.3 μm and a burst mode, 50 Mb/s receiver at 1.55 μm, are available in a package measuring $85 \times 17 \times 10$ mm.

There are a number of trends in the future development of datacom transceiver technology. Among these is the migration to lower power supply voltages, following the trend of digital logic circuits. Most digital logic has migrated from 5 to 3.3 V and is on a path toward 2.5-V operation; optical transceivers are emerging that will be able to operate at 1 Gb/s speeds using a single 3.3-V power supply, although many sources will remain at 5 V for some time, especially VCSELs.

Although the telecommunications market continues to have distinct requirements from the datacommunications market, there is certainly some merging of requirements between these two areas as well. The telecom market is currently driving development of analog optical links for cable television applications, including bidirectional wavelength multiplexed links that transmit in one direction at 1300 nm and in the other at 1550 nm. This is achieved by incorporating some form of optical beam splitter in the transceiver package. Many of these devices incorporate optical fiber pigtails rather than connectorized transceiver assemblies. Packaging is another evolving area; existing datacom industry standards, such as the 1×9 pin serial transceiver or 2×9 pin transceiver with integrated clock recovery, may give way to surface-mount technology in the near future. In the telecom industry, the standard optical transceiver package has been the dual in-line pin (DIP) package (14 pins and two rows); recent developments in the so-called mini-DIP (8 pins and two rows) may develop into a new packaging standard for the industry. Finally, integrated optoelectronic integrated circuits are advancing to the point where they may replace the standard T0 can and ball lens assembly for optical device packaging and alignment in the next 5 years.

VCSELs hold a great deal of promise as low-cost optical sources for data communications, especially for optical array interconnections. Research continues on VCSELs that can operate at long wavelengths (1300 nm); although some devices have been demonstrated under laboratory conditions, the technology is probably several years away from commercial applications.

Parallel optical transceivers will face new challenges in the area of laser safety. Although many parallel transceivers will be able to meet U.S. class 1 safety standards, it is more difficult to meet the international class 1 limits specified by IEC 825. One possible alternative is international certification as a class 3A product, which would require a warning label on the transceiver; it is unclear whether the IEC will allow short-wavelength lasers to hold this classification or whether the market will accept such products instead of class 1 sources.

Acronyms

2R	retime/reshape
3R	retime/reshape/regenerate
AAL	ATM adaptation layer
AC	alternating current
ACK	acknowledgment
ACTS	advanced communication technologies and services
A/D	analog to digital
AFNOR	Assoc. Française de NORmalisation (ISO member)
AIP	American Institute of Physics
ANS	American National Standards
ANSI	American National Standards Institute
AOI	add or invert
AON	all-optical network
APC	auto power control
APD	avalanche photodiode
API	application programmer interface
APS	American Physical Society
AR	antireflective
ARP	address resolution protocol
ARPA	Advanced Research Project Agency
ASCI	Advanced Strategic Computing Initiative
ASCII	American Standard Code for Information Interchange (developed by ANSI)
ASE	amplified spontaneous emission
ASIC	application specific integrated circuit
ASTM	American Society for Test and Measurement
ATM	asynchronous transfer mode
ATMARP	ATM address resolution protocol
AUA	another useless acronym (from Preface to the Second Edition)
AWG	array waveguide grating

BD	solid bore inner diameter
BER	bit error rate
BGA	ball grid array
BGP	border gateway protocol
BH	buried heterostructure
BiCMOS	bi-junction transistor/complementary metal oxide semiconductor
BIP	bit interleaved parity
BIS	Bureau of Indian Standards
B-ISDN	broadband integrated services digital network
bit/s	bits per second
BLIP	background limited in performance
BLM	ball-limiting metallurgy
BLSR	bidirectional line switched ring
BNR	Bell Northern Research
BOL	beginning of lifetime
BPS	bytes per second
BSI	British Standards Institute
BSY	control code for busy, as defined by the ANSI Fibre Channel Standard
BTW	behind the wall
BUS	broadcast and unknown server
BWPSR	bi-directional path-switched rings
C4	controlled collapse chip connection
CAD	computer-aided design
CAM	computer-aided manufacturing
CATV	community antenna television (cable TV)
C-band	conventional wavelength band used in wavelength multiplexing (1520–1570 nm)
CBD	connector body dimension
CBGA	ceramic ball grid array
CBR	constant bit rate
CCD	charge-coupled devices
CCGA	ceramic column grid array
CCITT	Consultative Committee for International Telephone and Telegraph (International Telecommunications Standard Body) — now ITU
CCW	channel control words

CD	compact disc
CDL	converged data link
CDMA	code-division multiple access
CDR	clock and data recovery
CDRH	U.S. Center for Devices and Radiological Health
CE mark	Communaute Europeene mark
CECC	CENELEC Electronic Components Committee
CEL	California Eastern Laboratories
CENELEC	Comitéé Européen de Normalisation ELECtrotechnique or 'European committee for electrotechnical standardization'
CFR	Center for Devices and Radiological Health
CICSI	Building Premises Wiring Standard Body
CIP	carrier-induced phase modulation
CLEC	competitive local exchange carrier (local telephone service)
CLEI	common language equipment identification (Bellcore)
CLI	common line interface
CLO	control link oscillator
CLP	cell loss priority bit
CML	current-mode logic
CMOS	complementary metal oxide semiconductor
CMU	Carnegie Mellon University
CNS	Chinese National Standards
COB	chip on board
COGS	cost of goods sold
COSE	Committee on Optical Science & Engineering
CPR	coupled power range
CPU	central processing unit
CRC	clock recovery circuit
CRC	cyclic redundancy check
CS	convergence sublayer
CSMA/CD	carrier sense multiple access/with collision detection
CSP	channeled substrate planar
CTE	coefficient of thermal expansion
CVD	chemical vapor deposition
CW	continuous wave
CWDM	coarse wavelength-division multiplexing (aka. wide-WDM)
CYTOP	a transparent fluorpolymer

Acronyms

DA	destination address
DARPA	Defense Advanced Research Projects Association (see also ARPA)
DAS	dual-attach station
DASD	direct access storage device
DBR	diffraction Bragg reflector
DBR	distributed Bragg reflector
DC	direct current
DCD	duty cycle distortion, also data carrier detect (modems)
DCF	dispersion compressing fiber
DCN	data communications network
DDM	directional division multiplexing
DDOA	dysprosium-doped optical amplifier
DE	diethyl (as in diethylzinc-DEZn)
DEC	Digital Equipment Corporation
DEMUX	demultiplexer
DFB	distributed feedback
DH	double heterojunction/double heterostructure
DHHS	Department of Health Services in the IEC
DIN	Deutsches Institut für Normung, (the German Institute for Standardization) or Data Input
DIP	dual in-line package
DJ	deterministic jitter
DLL	delay locked loop
DM	dimethyl (as in dimethylzinc-DMZn)
DMD	differential mode delay
DMM	digital multimeter
DP	data processing
DRAM	dynamic random access memory
DS	Dansk Standardiseringsrad
DSF	dispersion-shifted fiber
DSL	digital subscriber line
DSO	digital sampling oscilloscope
DSP	digital signal processing
DTPC	design for total product cost
DUT	device under test
DVD	digital video disk
DWDM	dense wavelength division multiplexers

EBP	end of bad packet (delimiter for Infiniband data packets)
EC	European Community
ECF	echo frame
ECH	frames using the ECHO protocol
ECHO	European Commission Host Organization
ECL	emitter-coupled logic
EDFA	erbium-doped fiber amplifier
EELD	edge-emitting semiconductor laser diode (also ELED)
EFA	end face angle
EGP	end of good packet (delimiter for Infiniband data packets)
EH	hybrid mode where electric field is largest in transverse direction
EIA	environmental impact assessments
EIA standard	Electronics Industry Association
EIA	Electronics Industry Association
EIA/TIA	Electronics Industry Association/Telecommunications Industry Association
ELAN	emulate local area network
ELED	edge-emitting semiconductor laser diode (also EELD)
ELO	epitaxial liftoff
EMBH	etched-mesa buried heterostructure
EMC	electromagnetic compatibility
EMI	electromagnetic interference
ENIAC	Electronic Numeric Integrator and Computer
EOF	end of frame
EOL	end of lifetime
EOS	everything over SONET
ESCON	Enterprise system (IBM) connection
ESD	electrostatic discharge
ESPRIT	European Strategic Program for Research in Information Technology
ETDM	electronic time division multiplexing
ETR	external timing reference
ETSI	European Telecommunication Standards Institute
FBG	fiber Bragg grating
FBS	fiber beam spot
FC	fibre channel
FC	frame control

FC	connector-threaded fasteners
FC	flip chip, a manufacturing chip bonding procedure
FCA	Fibre Channel Association
FC-AL	Fibre Channel arbitrated loop
FCC	Federal Communications Commission
FCE	ferrule/core eccentricity
FCS	Fibre Channel Standard
FCSI	Fibre Channel Standard Initiative
FCV	Fibre Channel Converted/FICON conversion vehicle
FD	ferrule diameter
FDA	Food and Drug Administration
FDDI	fiber-distributed data interface
FDMA	frequency division multiple access
FET	field effect transistor
FF	ferrule float
FICON	Fibre Channel Connection or Fiber Connection
FIFO	first in, first out
FIG	fiber image guides
FIT	failures in time
FJ	Fiber-Jack
FLP	fast link pulse
flops	floating point operation performed per second
FO	fiber optic
FOCIS	fiber optic connector intermatability standard
FOTP	fiber optic test procedures
F_Port	fabric port, as defined by the ANSI Fibre Channel Standard
FP	Fabry-Perot
FPA	Fabry Perot amplifier
FPGA	field-programmable gate array
FQC	Fiber Quick Connect (feature in the IBM Glabal Services structured cable offering)
FRPE	flame-retardant polyethylene
FSAN	full service access network
FSR	free spectral range
FTC	fiber quick connect
FTS	fiber transport services
FTTC	fiber-to-the-curve
FTTH	fiber-to-the-home
FTTO	fiber-to-the-office

FWHM	full width at half maximum
FWM	four wave mixing
GBE or GbE	Gigabit Ethernet
GBIC	gigabit interface connector
GDPS	geographically dispersed Parallel Sysplex
GFC	generic flow control
GHz	gigahertz
GI	graded index
GLC	Gigalink Card
GOSS	State Committee of the Russian Federation for Standardization Metrology and Certification
GP	gross profit
GRIN	graded index
GRP	glass reinforced plastic
GSN	gigabit serial network (also known as HIPPI 6400)
GVD	group velocity dispersion
HAN	home area network
HC	horizontal cross-connect
HE	hybrid mode where magnetic field is largest in transverse direction
HEC	header error correction
HFET	heterostructure field effect transistor
HIBITS	high bitrate ATM termination and switching (a part of the EC RACE program)
HIPPI	high-performance parallel interface
HMC	hardware management console
HOCF	hard polymer clad (glass) fiber
HOEIC	hybrid optoelectronic integrated circuit
HPC	high-performance computing
HPCF	hard poloymer clad fiber
HR	highly reflective
HSPN	High Speed Plastic Network Consortium
IB	InfiniBand
IBM	International Business Machines
IC	integrated circuit
IC	intermediate cross-connect

ICCC	International Conference on Computer Communications
ID	identification
IEC	International Electrotechnical Commission
IEEE	Institute of Electrical and Electronics Engineers
IETF	Internet Engineering Task Force
IFB	imaging fiber bundles
IHS	Information Handling Services
ILB	inner lead bond
ILEC	incumbent local exchange carrier (local telephone service)
ILMI	interim local management interface
InATMARP	inverse ATM address resolution protocol
I/O	input/output
IODC	Institute for Optical Data Communication
IP	internet protocol
IPI	intelligent physical protocol
IPT	integrated photonic transport (inband)
IR	infrared reflow
IrDA	Infrared Datacom Association
IRED	infrared emitting diode
ISC	intersystem channel gigabit links
ISI	intersymbol interference
ISO	International Organization for Standardization
ISP	internet service provider
IT	information technology
ITO	indium tin oxide
ITS	Institute for Telecommunication Sciences
ITU	International Telecommunications Union
IVD	inside vapor deposition
IXC	interexchange carrier (long distance telephone service)
JEDEC	Japanese-based telecom/datacom standards
JIS	Japanese Industrial Standards
JSA	Japanese Standards Association
KGD	known good die
LAN	local area network
LANE	local area network emulation
L-band	long wavelength band used in wavelength multiplexing (1560–1610 nm)
LC	late counter

LC	"Lucent Connector"
LCOS	liquid crystal on silicon
LCT	link confidence test
LD	laser diode
LEAF	large effective area fiber
LEC	LAN emulation client
LECID	LEC (LAN emulation client) identifier
LED	light-emitting diode
LES	LAN emulation server
LIA	Laser Institute of America
LIS	logical IP subnet
LLC	logic link control
LOH	line overhead
LOS	loss of signal
LP	linear polarized mode
LPE	liquid phase epitaxy
LSZH	low smoke zero halogen cable
LUNI	LANE user-network interface
LVD	low-voltage directive
LVDS	low-voltage differential signal
LX	long-wavelength (1300 nm) transmitter
MAC	media access control
MAN	metropolitan area network
MBE	molecular beam epitaxy
MBGA	metal ball grid array
MC	main cross-connect
MCM	multichip module
MCP	mode conditioning patch cables
MCVD	modified chemical vapor deposition
MDF	main distribution facility
MDI	medium-dependent interfaces
MEM	microelectromechanical
MESFET	metal semiconductor field effect transistor
METON	metropolitan optical network
MFD	modal field diameter
MIB	management information base
MIC	media-interface connector
MII	media-independent interface
MIMD	multiple-instruction multiple data stream

520   Acronyms

MJS	methodology for jitter specification
MM	multi-mode
MMF	multi-mode fiber
MOCVD	metal organic chemical vapor deposition
MONET	multiple wavelength optical network
MOS	metal oxide semiconductor
MOSIS	MOS implementation system
MPE	maximum permissible exposure
MPLS	multi-protocol label switching
MPLmS	multi-protocol lambda switching
MPO	trade name for multifiber optical connector using a 12 fiber interface; also known as MPX
MPX	trade name for multifiber optical connector using a 12 fiber interface; also known as MPO
MQW	multiquantum well
MS	multi-standard
MSA	multi-source agreements
MSI	medium-scale integration
MSM	metal-semiconductor-metal
MT	multi-fiber termination
MTP	multi-fiber terminated push-on connector
MT-RJ	multi-termination RJ-45 latch (a type of optical connector)
MT-RT	multi-fiber RJ-45 latched connector
MU	multi-termination unibody
MUX	multiplexcr
MZ	Mach-Zehnder
NA	numerical aperture
NAND	not and (logic gate)
NAS	network attached storage
NBS	National Bureau of Standards (now NIST)
NC	narrowcast
NC&M	network control and management
NEBS	network equipment building system
NEC	National Electric Code
NEP	noise equivalent power
NEXT	near end cross-talk
NFPA	National Fire Protection Agency
NGI	next-generation Internet
NGIO	next-generation input/output

NIC	network interface card
NIF	neighbor information frame
NIST	National Institute of Standards & Technology
NIU	network interface units
NLOG	non-linear optical gate
NMOS	negative-channel metal oxide semiconductor
NMS	network management system
NOR	not or (logic gate)
N_Port	nodal port, as defined by the ANSI Fibre Channel Standard
NPN	negative-positive-negative
NRC	National Research Council
NRZ	nonreturn to zero
NSPE	National Society of Professional Engineers
NTIA	National Telecommunication and Information Administration
NTT	Nippon Telegraph and Telephone Corporation
NUMA	non-uniform memory architecture
NZDSF	non-zero dispersion shifted fiber
OA	optical amplifiers
OADM	optical add drop multiplexer
OAMP	operations, administration, maintenance, and provisioning (sometimes OAM&P)
OCI	optical channel interface
OCI	optical chip interconnect
OCLD	optical channel laser detector
OCM	optical channel manager
ODC	optical data center
ODSI	Optical Domain Service Interconnect coalition
OE	optoelectronic
O/E/O	optical/electrical/optical
OFB	ordered fiber bundles
OFC	optical fiber control system
OFN	optical fiber nonplenum/nonriser/nonconductive
OFNP	optical fiber nonconductive plenum listing
OFNR	optical fiber nonconductive riser listing
OFSTP	optical fiber system test procedure
OIDA	Optoelectron Industry Development Association
OIF	Optical Internetworking Forum
OIIC	optically interconnected integrated circuits

OLIVES	optical interconnections for VLSI and electronic systems (a part of the EC ESPRIT program)
OLS	optical label switching
OLTS	optical loss test set
OMA	optical modulation amplitude
OMB	office of management and budget
OMC	operational management committee
OMNET	optical micro-network
OMX	optical multiplexing modules
ON	Osterreisches Normungistitut (Austrian Standards Institute)
ONTC	Optical Network Technology Consortium
OPFET	optical field effect transistor
OSA	open system adapter
OSA	optical subassembly, also called the coupling unit
OSA	Optical Society of America
OSHA	Occupational Safety & Health Administration
OSI	open systems interconnection
OSIG	optical signal
OSNR	optical signal-to-noise ratio
OSPF	open shortest path first
OTDM	optical time division multiplexer
OTDR	optical time domain reflectometer
OTF	optical transfer function
OTN	optical transport network
OUT	output
OUTB	average output
OVD	outside vapor deposition
OXBS	optical crossbar switches
OXC	optical cross connects
PA	preamble
PA	pointing angle
PANDA	polarization maintaining and absorption reducing fiber
PAROLI	parallel optical link
PBGA	plastic ball grid array
PBX	private branch exchange
PC	personal computer
PC	polycarbonate-a thermoplastic compound
PCA	product cost apportionment

PCB	printed circuit board
PCI	peripheral component interconnect
PCM	physical connection management
PCVD	plasma-assisted chemical vapor deposition
PD	photodiode
PDF	probability density function
PDG	polarization dependent gain
PDL	polarization dependent loss
PECL	postamplifier emitter-coupled logic unit
PECVD	plasma-enhanced chemical vapor deposition
PF	perfluorinated
PGA	pin-grid arrays
PHY	physical layer of OSI model
PIN	positive-intrinsic-negative
PLCC	plastic leaded chip carrier
PLL	phase-lock loop
PLOU	physical layer overhead unit
PLS	primary link station
PMD	physical medium-dependent sublayer
PMD	polarization mode dispersion
PMF	parameter management frame
PMF	polarization maintaining fiber
PMI	physical media-independent sublayer
PMMA	polymethylmethacrylate—a thermoplastic compound
PMT	photomultiplier tube
POF	plastic optical fiber
POI	parallel optical interconnects
POINT	polymer optical interconnect technology
PON	passive optical network
POP	post office protocol
POTS	plain old telephone system
PPL	phase locked loop
PPRC	peer-to-peer remote copy
PR	plug repeatability
PRBS	pseudo-random binary sequence
PSTN	packet-switched telephone network
PT	payload type
PTT	local telecom authority
PVC	polyvinyl chloride (a plastic)
PVL	parallel vixel link

QCSE	quantum confined Stark effect
QE	quantum efficiency
QFP	quad flat pack
QOS	quality of service
RACE	Research in Advanced Communications for Europe
RAID	redundant array of inexpensive disks
RARP	reverse address resolution protocol
RC	resistor capacitor
RCDD	registered communication distribution designers
RCE	resonant-cavity enhanced
RCLED	resonant-cavity light-emitting diode
RECAP	resonant-cavity enhanced photodetector
RF	radio frequency
RFI	radio frequency interference
RHEED	reflection high-energy electron diffraction
RIE	reactive ion etching
RIN	reflection-induced intensity noise
RISC	reduced instruction set computing
RJ	random jitter
RJT	control code for reject, as defined by the ANSI Fibre Channel Standard
RMAC	repeater media access control
RMS	root mean square
ROSA	receiver optical subassembly
RSOH	regenerator section overhead
Rx	receiver
SA	source address
SA	Standards Association of Australia
SAGCM	separate absorption, grading, charge sheet and multiplication structure
SAGM	separate absorption grading and multiplication
SAM	separate absorption and multiplication layers
SAM	sub-assembly misalignment
SAN	storage area network
SAR	segmentation and reassembly sublayer
SAS	single attach station
SASO	Saudi Arabian Standards Organization

S-band	short wavelength band used in wavelength multiplexing (1450–1510 nm)
SBCCS	single byte command code set
SBCON	single byte command code sets connection architecture
SBS	source beam spot
SBS	stimulated Brillouin scattering
SC connector	subscriber connector (spring latch fasteners)
SC-DC	subscriber connector, dual-connect ferrule
SCI	scalable coherent interface
SC-QC	subscriber connector, quad-connect ferrule
SCSI	small computer system interface
SD	shroud dimension
SD	start delimiter
SDH	synchronous digital hierarchy
SDM	space division multiplexing
SDO	standards developing organizations
SEAL	simple and efficient adaptation layer
SEED	self electrooptic effect devices
SETI	search for extraterrestrial intelligence
SFF	small form factor
SFP	small form factor pluggable
SFS	Suomen Standardisoimislitto Informaatiopalvelu (Finland Standards Information)
SI	System International (International System of Units)
SIA	Semiconductor Industry Alliance
SIC	Standard Industrial Classification
SIF	status information frame
SIMD	single instruction-stream, multiple data-stream
SiOB	silicon optical bench (non-hermetic)
SIPAC	Siemens Packaging System
SIS	Standardiseringkommisiionen I Sverige (Swedish Standards Commission)
SJ	sinusoidal jitter
SL	superlattice
SLA	secure level agreement
SL-APDs	superlattice avalanche photodiodes
SM	single mode
SMC	surface mount component
SMD	surface-mount device

SMDS	switched megabit data services
SMF	single-mode fiber
SMP	shared memory processor
SMT	station management frames
SMT	surface-mount technique
SMU	Sanwa multi-termination unibody
SNMP	simple network management protocol
SNR	signal-to-noise ratio
SNZ	Standards Association of New Zealand
SOF	start of frame
SOHO	small office/home office
SOJ	small outline packages with J leads
SONET	Synchronous Optical NETwork
SP	shelf processor
SPC	statistical process control
SPE	synchronous payload envelope
SPIBOC	standardised packaging and interconnect for inter- and intra-board optical connections (a part of the EC ESPRIT program)
SPIE	Society of Photooptic Instrumentation Engineers
SQNR	signal-to-quantum-noise ratio
SQW	single quantum well
SRAM	static random access memory
SRM	sub-rate multiplexing
SRS	stimulated Raman scattering
SSA	serial storage architecture
SSA	storage system architecture
S-SEED	symmetric self electrooptic effect devices
SSI	small-scale integration
SSP	storage service provider
ST connector	subscriber termination
STP	shielded twisted pair
STS	synchronous transport signal
SWDM	sparse WDM
SWF	surface wave filter
SX	short-wavelength (850 nm) transmitter
TAB	tape access bonding
TAB	tape automated bonding

TAXI	transparent asynchronous transceiver/receiver interface
TBGA	Tape ball grid array
TC	transmission convergence sublayer
TCM	time compression multiplexing
TCP/IP	transmission control protocol/internet protocol
TDFAs	thulium doped fiber amplifiers (to be used in the wavelength range of 1450–1510 nm)
TDMA	time-division multiple access
TE mode	transverse electric (electric field normal to direction of propogation)
TE	triethyl (as in triethylgallium-TEGa)
TERKS	trade name of a data base used by Telcordia (formerly Bell Labs) to track inventory and depreciation on telecommunications hardware
THC	through-hole component
THT	token holding timer
THz	terahertz
TIA	Telecommunications Industry Association
TIA	trans-impedance amplifier (also TZA)
TM mode	transverse magnetic (magnetic field normal to direction of propogation)
TM	trimethyl (as in trymethylgallium-TMGa)
TO	transistor outline
TOSA	transmitter optical subassembly
TP	twisted pair
TRT	token rotation timer
TRX	transceiver
TTI	time to installation
TTL	transistor-transistor logic
TTRT	target token rotation time
TUG	tributary unit groups
TWA	traveling wave amplifier
Tx	transmitter
TZA	trans-impedance amplifier (also TIA)
UDP	user datagram protocol
UHV	ultra-high vacuum
UI	unit interval
UL	Underwriters Laboratories

ULP	upper-level protocol
UNI	user-network interface
USNC	United States National Committee
UTP	unshielded twisted pair
UV	ultra-violet
VAD	vapor axial deposition
VBR	variable bit rate
VC	virtual circuit
VCC	virtual channel connection
VCI	virtual channel identifier
VCO	voltage-controlled oscillator
VCSELs	vertical-cavity surface-emitting lasers
VDE	Verband Deutscher Electrotechniker (Association of German Electrical Engineers)
VDI	Vereins Deutscher Ingenierure (German Standards)
VF-45	trade name of a small form factor optical connector developed by 3M Corporation with an RJ-45 latch
VLIW	very long instruction word
VLSI	very large-scale integration
VPI	virtual path identifier
VPN	virtual private network
VPR	vapor phase reflow
VT	virtual tributary
VTG	virtual tributary group
WAN	wide area network
WDM	wavelength-division multiplexing
WWSM	wide spectrum WDM
XDF	extended distance feature
XGP	Multi-Gigabit Ethernet pluggable
XPR	cross plug range
XRC	extended remote copy
ZBLAN	fluorozirconate

Some useful acronym searches on the World Wide Web
The One Look Dictionary www.onelook.com
Stammtisch Beau Fleuve Acronyms http://www.plexoft.com/SBF/(Best accessed through The One Look Dictionary)
Acronym Finder http://www.acronymfinder.com/

Glossary

1M	a laser safety classification
absorption	the loss of light when passing through every material, due to conversion to other energy forms, such as heat.
acoustooptic tunable filter	a light filter tuned by acoustic (sound) waves. This is accomplished using polarizers and an acoustic diffraction grating, which creates a resonant structure that rotates the polarization.
active region	the area in a semiconductor that either absorbs or emits radiation.
aggregate capacity	a measure of the total information-handling capability of a smart pixel array. It combines the individual channel data rate with the total number of channels in the array to produce an aggregate information carrying capacity.
alignment	the connection of optical components to maximize signal transmitted.
alignment sleeve	part of an ESCON connector into which the ferrules are inserted, assuring accurate alignment.
annealing	the process of heating and slowly cooling a material. This makes glass and metal stabilize their optical, thermal, and electric properties, and can reverse lattice damage from doping to semiconductors.
architecture	overall structure of a computer system, including the relationship between internal and external components.
asynchronous	a form of data transmission where the time that each character, or block of characters, starts is arbitrary. Asynchronous data has a start bit and a stop bit, since there is no regular time interval between transmissions, and no common clock reference across the system.
attenuation	the decrease in signal strength caused by absorption and scattering. The power or amplitude loss is often measured in dB.

Auger nonradiative recombination
: when an electron and hole recombine, and then pass excess energy and momentum into another electron or hole. This process does not generate any additional radiation.

avalanche photodiodes
: photodiodes where the primary carriers, electron/holes pairs created by the incident photon, are accelerated through a voltage, and then collide with neutral atoms, creating more secondary carriers. These secondary carriers may accelerate and create new carriers. This process, known as photomultiplication, can multiply the signal by up to two orders of magnitude.

backbone network
: a primary conduit for traffic that is often both coming from, and going to, other networks.

backplane
: circuit board with sockets to connect other cards, especially communication channels.

back-reflections
: reflections back to the laser from devices in its path.

band gap
: the energy difference between the top of the valence band and the bottom of the conduction band of a solid. The size of the band gap will determine whether a photon will eject a valence electron from a semiconductor.

bandwidth
: the range of frequencies over which a fiber optic medium or device can transmit data. This range, expressed in hertz, is the difference between the highest and the lowest frequencies for optical filter elements. For MM fibers, the range is expressed as a product of the bandwidth and distance, in MHz-km.

baud
: the signaling speed, as measured by the maximum number of times per second that the state of the signal can change. Often a "signal event" is simply the transmission of a bit. Baud rate is measured in $\sec^{-1}$.

beacon process
: a token ring process that signals all remaining stations that a significant problem has occurred, and provides restorative support. Like a string of signal fires, the beacon signal is passed from neighbor to neighbor on the ring.

Glossary

beating	when superimposing waves of different frequencies, beating occurs when the maximum amplitude of these waves match up. Coherent beating occurs when the data signal beats with itself out of phase, and can cause cross-talk.
bias voltage	the voltage applied when biasing. (*See* biasing)
biasing	applying a voltage or a current across a junction detector. This changes the mode of the detector, which can affect properties such as the noise and speed. (*See* photoconductive and photovoltaic mode)
bit	either 0 or 1, the smallest unit of digital communications.
bit error rate	the probability of a transmitted bit error. It is calculated by the ratio of incorrectly transmitted bits to total transmitted bits.
bit synchronization	when a receiver is delivering retimed serial data at the required BER.
Bragg reflector	(*See* distributed Bragg reflector)
breakdown voltage	the bias voltage in an avalanche photodiode where the current gain begins to approach infinity. In other words, the bias voltage when an avalanche photodiode is no longer responding to incident photons.
Brillouin scattering	the scattering of photons (light) by acoustic phonons (sound waves). A special case of Raman scattering.
broadcast-and-select network	a network that has several nodes connected in a star topology. (*See* topology)
bulkhead splice, splice bushing	a unit that allows two cables with unlike connectors to mate.
burn-in	the powering of a product before field operation, to test it and stabilize its characteristics.
burst mode switching networks	an approach that minimizes latency by transmitting large amounts of data in a short time over the network.
bus	(*See* optical bus)
bus topology	(*See* topology)
butt coupling	a mechanical splice between a fiber and a device. This technique is done with the signal on, and the fiber secured into position when measured signal is maximized. (*See* splice)

butterfly package	(*See* dual-in-line pin package)
byte	eight bits, numbered 0 to 7. (*See* octet)
cable jacket	the outer material that surrounds and protects the optical fibers in a cable.
cable plant	passive communications elements, such as fiber, connectors and splices, located between the transmitter and receiver.
calibration	the comparison to a standard or a specification. Also to set a device to match a specification.
channel	a single communications path or the signal sent over that path.
chatter	any transient response greater than 0.5 volts, ingoing between the signal detect negative level to assert its level, and of any duration that can be sensed by the channel logic.
"chicken and egg" syndrome	when it is hard to find someone to take the first risk on a new technology, which delays its implementation. For example, when customers don't want to commit to using an emerging technology without some assurance that it will be fully supported by manufacturing, and suppliers don't want to commit to putting emerging technology into production without having a customer who is committed to using it.
chirp	a pulsed signal whose frequency lowers during the pulse.
cholesteric liquid crystals	(*See* liquid crystals)
chopping	when an optical signal is varied through use of a mechanical or electrical "chopper" to block the signal at a certain times. (*See* lock-in amplifier)
chromatic dispersion	(*See* dispersion)
cladding	the material of low refractive index used to cover an optical fiber, which reflects escaping light back into the core, as well as strengthens the fiber.
cleaving	to split with a sharp instrument along the natural division in the crystal lattice.

Glossary

client	a computer or program that can download data or request a service over the network from the server. (*See* server)
clock generation/ multiplication	synchronizing a terminal's internal clock to the received bit stream.
clock recovery	reconstructing the order of operations after serial/parallel conversion.
coherent beating	(*See* beating)
correlators	a device that detects signal from noise by computing correlation functions, which is similar to transforms.
costs of goods sold	the sum of parts cost, manufacturing cost, serviceability and warranty cost, obsolescence and scrap and a portion of the supply chain cost. (*See* parts cost, manufacturing cost, supply chain cost)
coupled power range	the allowable difference between the minimum and maximum allowed power.
coupling	connecting two fibers or the connector between two fibers.
cross-plug range	the difference between the measured lowest power and highest power for matings between multiple connectors and the same transceiver.
cross-talk	the leaking of or interference between signal in two nearby pixels, wires, or fibers.
cutoff wavelength	the wavelength for which the normalized power becomes linear to within 0.1 dB.
daisy chain	a bus wiring scheme where devices are connected to each other in sequence (device A is wired to device B is wired to device C, etc.), like a chain of daisies. (*See* topology: bus)
dark current	the current measured from a detector when no signal is present.
data dependent jitter	(*See* jitter)
data rate	short for data transfer rate, the speed devices transmit digital information. Units include bits per second, but are more likely to be in the range of megabits

534 Glossary

	per second (Mbit/s), or even gigabits per second (Gbit/s) in the field of fiber optics.
de-adjustment	the variation of the coupled power after several matings with the same connector.
decoding	(*See* encoding/decoding)
detectivity	the reciprocal of the noise equivalent power (NEP). This can be a more intuitive figure of merit, because it is larger for more sensitive detectors.
deterministic jitter (DJ)	(*See* jitter)
development expense	the total development investment in people, test equipment, prototypes, etc.
device	any machine or component that attaches to a computer.
dielectric mirror	a multilayer mirror that is an alternative to a distributed Bragg reflector (DBR) in semiconductor laser manufacture.
differential mode delay	delay variation caused by differences in group velocity among the different propagation modes of an optical fiber.
diffraction	when a wave passes through an edge or an opening, secondary wave patterns are formed that interfere with the primary wave. This can be used to create diffraction gratings, which work similarly to prisms.
directional	division multiplexing. (*See* multiplexer)
dispersion	the distortion of a pulse due to different propogation speeds. Can be chromatic, which is caused by the wavelength dependence of the index of refraction, or intermodal, caused by the different paths traveled by the different modes.
dispersion-flattened fiber	specialty optical fiber whose cross-section is designed to minimize dispersion over a broad range of input wavelengths.
dispersive self-phase modulation	nonlinear property of an optical fiber, in which the light passing through the fiber induces a change in the refractive index of the core, and chromatic dispersion causes either an up-chirp or down-chirp in the optical frequency.
distortion	the change in a signal's waveform shape.

Glossary

distributed Bragg reflector (DBR)
: a mirror used in the manufacture of semiconductor lasers and photodetectors. It is made from multiple layers of semiconductors that have a band gap in the wavelength of interest.

distribution panel
: a central panel from which a signal is routed to points of use. (*See* patch panel)

divergence
: the bending of light rays away from each other, for example the spreading of a laser beam with increased distance.

doping
: adding an impurity to a semiconductor.

double-clad optical amplifier
: an optical fiber amplifier in which the pump light is delivered via an outer coating on the same fiber that performs the amplification.

down-chirp
: a linear decrease in frequency over time.

dual homing architecture
: a type of network with two hubs that fail over to each other in the event of a disaster at either location. (*See* topology)

dual hubbed rings
: variation on a hubbed ring architecture in which all traffic flows through not one but two hub sites in a network. (*See* topology)

dual-in line pin package, butterfly package
: a cavity package for a semiconductor laser that is wire bonded and then the plastic is molded around the body and leads of the package.

duplex
: a communications line that lets you send and receive data at the same time.

duty cycle
: the pulse duration times the pulse repetition frequency. In other words, the percentage of time an intermittent signal is on.

duty cycle distortion (DCD)
: the ratio of the average pulse width of a bit to the mean of twice the unit interval. (*See* unit interval)

dynamic range
: the ratio of the largest detectable signal to the smallest detectable signal, such as the receiver saturation charge to the detection limit (also known as linear dynamic range).

effective sensing area
: (*See* active area)

electrooptic transducer
: a device that converts an electric signal to an optical signal and vice versa. An example of this is the photocell and laser in a transceiver. (*See* transducer)

emulation	the use of program to simulate another program or a piece of hardware.
encoding/ decoding	encoding is the process of putting information into a digital format that can be transmitted using communications channels. Decoding is reversing the process at the end of transmission.
end node	a node that does not provide routing, only end user applications.
epitaxy	method of growing crystal layer on a substrate with an identical lattice, which maintains the continuous crystal structure. Styles include molecular beam epitaxy, vapor phase epitaxy, etc.
etalon	an optical device with two reflective mirrors facing each other to form a cavity.
etched mesa	a flat raised area on an electronic device, created during the photolithography process.
extinction ratio	the ratio of the power level of the logic '1' signal to the power of the logic '0' signal. It indicates how well available laser power is being converted to modulation or signal power.
eye diagram	an overlay of many transmitted responses on an oscilloscope, to determine the overall quality of the transmitter or receiver. The relative separation between the two logic levels is seen by the opening of the eye. Rise and fall times can be measured off the diagram. Jitter can be determined by constructing a histogram of the crossing point.
fabric, switched network	a FCS network in which all of the station management functions are controlled at the switching point, rather than by each node. This approach removes the need for complex switching algorithms at each node. The telephone system, where the dialer supplies the phone number, can be used as an analogy.
Fabry-Perot resonance	resonance that occurs when waves constructively interfere. Fabry-Perot resonance occurs in a cavity surrounded by mirrors. Semiconductor lasers, such as VCSELs, take advantage of Fabry-Perot resonance to stimulate emission in their active region.

Glossary 537

facet passivation	growing a thin oxide layer over the semiconductor (for datacom the EELD) facet, to limit environmental exposure and natural oxidation.
Fast Ethernet	A LAN running at speeds up to 100 MBd. (*See* LAN)
Fermi level	the maximum energy of the electrons in a solid, which determines the availability of free electrons. If the Fermi level is in the conduction (top) band, the material is a conductor (metal). If the Fermi level is in the valence (lower) band, the material is an insulator. If the Fermi level is between the conduction and the valence band, the material is a semiconductor.
ferrule	a cylindrical tube containing the fiber end that fits within high tolerance into the flange of the transceiver port.
fiber optic link	the transmitter, receiver, and fiber optic cable used to transmit data.
fiber ribbon	multifiber cables and connectors.
Fibre Channel Connection (FICON)	an I/O interface standard that mainframe computers use to connect to storage devices. This standard is eight times faster than the previous fiber optics standard, ESCON, due to a combination of new architecture and faster link rates. (*See* architecture)
field installable connectors	optical connectors that can be installed on-site at a customer location, as opposed to factory installed connectors that can only be attached at an authorized manufacturing center.
flicker noise	noise with a 1/f spectrum, which occurs when materials are inhomogeneous.
flip chip mounting	a semiconductor substrate in which all of the terminals are grown on one side of the substrate. It is then flipped over for bonding onto a matching substrate.
flip flop	a device that has two output states, and is switched by means of an external signal.
flow shop	a manufacturing area that makes only one product.
footprint	the amount of desk or floor space used by a component.
frame	1. In the SONET transmission format, a 125 microsecond frame contains 6480 bit periods, or 810 octets (bytes), that contain layers of information. These layers include the overhead, or instructions

538   Glossary

	between computers, and the voice or audio channel. SONET can be used to support telephony.
	2. A technique used by web pages to divide the screen into multiple windows.
frame errors	errors from missing or corrupted frames, as defined by the ANSI Fibre Channel Standard.
frequency agile	capable of being easily adjusted over a range of operating frequencies.
frequency chirping	inducing either an up-chirp or down-chirp in an optical signal. (*See* up-chirp and down-chirp)
frozen process	a process where, if there are major changes of equipment, process parameters, agents or even parts, a partial or total requalification of the process must be performed.
fusion splice	(*See* splice)
gain	amplification. In a photodetector, the number of electron–hole pairs generated per incident photon.
GGP fiber	a specialty optical fiber, proprietary to 3M Corporation, which is more resistant to mechanical fractures when bent; used in optical connectors such as the VF-45.
Gigabit Ethernet (GBE)	a standard for high-speed Ethernet, that can be used in backbone environments to interconnect multiple lower speed internets. (*See* backbone network)
Gigalink card	a laser-based transceiver card that runs at approximately 1 Gb/s and works as a transponder.
graded index fiber	a fiber with a refractive index that varies with radial distance from the center.
gross profit	revenue minus cost of goods sold (COGS).
group velocity	the transmission velocity of a wave packet, which is made of many photons with different frequencies and phase velocities.
hermetic seal	a seal that air and fluids cannot pass through.
heterostructure	a semiconductor layered structure, with lattice matched crystals grown over each other.
Hill gratings	the first types of in-fiber Bragg diffraction gratings, named after the researcher who discovered them.

Glossary 539

homologation	confirming that a product follows the rules of each country in which it is used.
hubbed ring	optical network architecture in which all data traffic flows through a single common location or hub. (*See* topology)
hybrid integration	a technique of developing smart pixels where optical devices are grown separately from the silicon electronic circuitry, and then are bonded together. (*See* smart pixels)
image halftoning	an image compression technique whereby a continuous-tone, gray-scale image is printed or displayed using only binary-valued pixels.
Infiniband	an architecture standard for a high-speed link between servers and network devices. It will initially run at 0.3 GB/sec, but eventually scale as high as 6.0 GB/sec. It is expected to replace peripheral component interconnect (PCI). (*See* architecture)
intelligent optical network	optical network that also controls higher level switching and routing functions above the physical layer.
intermodal dispersion	(*See* dispersion)
intersymbol interference	the distortion by a limited bandwidth medium on a sequence of symbols which causes adjacent symbols to interfere with each other.
intrinsic	without impurities.
jitter	The error in ideal timing of a threshold crossing event. The CCITT defines jitter as short-term variations of the significant instants (rising or falling edges) of a digital signal from their ideal position in time. Jitter can be both deterministic and random. Low-frequency jitter can be tracked by the clock recovery circuit, and does not directly affect the timing allocations within a bit cell. Other jitter will affect the timing. Data-dependent jitter, a type of deterministic jitter (DJ), includes intersymbol interference (ISI). DJ may also include duty cycle distortion (DCD),

sinusoidal jitter, and other non-Gaussian jitter. Random jitter (RJ) can be defined as the peak-to-peak value of the bit error rate (BER) of 10^{-12}, or approximately 14 times the standard deviation of the Gaussian jitter distribution. (*See* intersymbol interference, duty cycle distortion, bit error rate)

job shop — a manufacturing area that deals with a variety of products, each treated as a custom product.

jumper cable — an optical cable that provides a physical attachment between two devices or between a device and a distribution panel. In this way it is different from trunk cables.

junction capacitance — the capacitance formed at the pn junction of a photodiode.

lambertian — scattering that obeys Lambert's cosine law — the flux per unit solid angle leaving a surface in any direction is proportional to the cosine of the angle between that direction and the normal to the surface. Matt (not shiny) surfaces tend to be lambertian scatterers.

latency — the delay in time between sending a signal from one end of connection to the receipt of it at the other end.

lattice-matched — when growing a crystal layer on a substrate, the junction is lattice-matched if new material's crystal structure fits with the substrate's structure.

launch reference cable — a known good test cable used in loss testing.

layer — a program that interacts only with the programs around it. When a communications program is designed in layers, such as OSI, each layer takes care of a specific function, all of which have to occur in a certain order for communication to work. (*See* layered architecture)

layered architecture — a modular way of designing computer hardware or software to allow changes in one layer not to affect the others.

legacy product — a product an organization has already invested in, or has currently installed. New technology must be compatible with legacy products.

line state	a continuous stream of a certain symbol(s) sent by the transmitter that, upon receipt by another station, uniquely identifies the state of the communication line. For example, Q for quiet or H for halt.
linearity range	the range of incident radiant flux over which the signal output is a linear function of the input.
link	the fiber optic connection between two stations, including the transmitter, receiver, and cable, as well as any other items in the system, such as repeaters.
link budget	range of acceptable link losses.
link-level errors	errors detected at lower level of granularity than frames, as defined by the ANSI Fibre Channel Standard.
link loss analysis	a calculation of all the losses (attenuation) on a link.
liquid crystals, nematic, smectic, and cholesteric	a material that has some crystalline properties and some liquid properties. In optics these materials usually have elongated molecules that are rod shaped. If they are oriented randomly they have different optical properties than when they are aligned. Nematic liquid crystals tend to have the rods oriented parallel but their positions are random. Cholesteric liquid crystals are a subcategory of nematic, in which the molecular orientation undergoes a helical rotation about the central axis. Smectic liquid crystals have a permanent dipole moment that can be switched by an externally applied electric field. This gives them many photonics applications, such as smart pixel arrays.
local area network (LAN)	a network limited to about 1 km radius.
lock-in amplifier	a device used to limit noise by encoding input data with a known modulation, made by chopping the signal. The amplifier "knows" from this reference signal when the signal to be detected is on and when it is off. This allows the detector to low-pass filter the data, which narrows the bandwidth of the detector, making it more precise. (*See* chopper)
long wavelength	approximately 1300 nm (1270–1320) or approximately 1550 nm.

Glossary

loopback testing	looping a signal back across a section of the network to see if it works properly. If a transceiver passes a unit loopback test, but fails a network loopback test, the problem is in the cables, not the transceiver.
loss	attenuation of optical signal.
machine capability	the statistical safety for processes performed by a tool or a machine.
macro-bending losses	losses due to nanometer size deviations in the fiber.
manufacturing cost	the cost to manage procurement of parts, inventory, system assembly and test, and shipping costs.
margin	the amount of loss, beyond the link budget amount, that can be tolerated in a link.
master/slave	an architecture where one device (the master) controls other devices (the slaves).
mechanical splice	(*See* splice)
meshed rings	network topology in which any node may be connected to any other node. (*See* topology)
metrology	the science of measurement.
metropolitan area network (MAN)	interconnected LANs with a radius of less than 80 km (50 miles).
micro-bending losses	losses due to visible bends in the fiber.
mini-zip	a type of zipcord fiber cable with a smaller outer diameter than standard zipcord.
mode mixing	the changing of the modal power distribution following a splice.
mode partition noise	within a laser diode, the power distribution between different longitudinal modes will vary between pulses. Each mode is delayed by a different amount due to the chromatic dispersion and group velocity dispersion in the fiber, which causes pulse distortion. (*See* distortion, dispersion, and group velocity)
mode scrambled launch	a type of optical coupling into a multimode fiber or waveguide that attempts to uniformly excite all modes in the target waveguide or fiber.

modulate/ demodulate	modulation is when one wave (the carrier) is changed by another wave (the signal). Demodulation is restoring the initial wave.
monolithic arrays	a technique of simultaneous fabrication of electronic and optical circuits on the same substrate, to produce high-speed smart pixels. (*See* smart pixels)
Monte Carlo simulation	any type of statistical simulation that accounts for the probability of various events. This technique is useful when dealing with non-Gaussian probability distributions. An example is the probability of absorption and scattering of photons traveling through a medium.
multimode fiber	a type of fiber in which light can travel in several independent paths.
multiplexer	a device that combines several signals over the same line. Wavelength division multiplexing (WDM) sends several signals at different wavelengths. Directional division multiplexing (DDM) combines the laser diode and photodiode using a coupler. Time compression multiplexing is similar to DDM, when a diode is in the transmission mode and the other in the receiving mode ("ping-pong transmission"). Space division multiplexing (SDM) requires two fibers — one for upstream transmission, and one for downstream transmission. Time-division multiplexing (TDM) transmits more than one signal at the same time by varying the pulse duration, pulse amplitude, pulse position, and pulse code, to create a composite pulse train.
narrowcast (NC)	to direct a program to a specific, well-defined audience.
nematic liquid crystals	(*See* liquid crystals)
neural networks	a system of programs and data structures that simulates the brain. They use a large number of simple processors in parallel, each with local memory. The neural network is "trained" by feeding it data and rules about relationships between the data.

noise equivalent power (NEP)	the flux in watts necessary to give an output signal equal to the root mean square noise output of a detector.
noise floor	the amount of noise self-generated by a device.
numerical aperture (NA)	The NA defines the light-gathering ability of a fiber, or an optical system. The numerical aperture is equal to the sine of the maximum acceptance angle of a fiber.
Nyquist frequency	the highest frequency that can be reproduced when a signal is digitized at a given sample rate. In theory, the Nyquist frequency is half of the sampling rate.
octet	eight bits in a row. Also known as a byte.
ongoing inventory management cost	the cost to maintain and manage an inventory, as compared with procuring parts for inventory stocking.
opened rings	network topology in which a ring has been opened at one point and turned into a linear network with add/drop of channels at any point. (*See* topology)
optical bus	a facility for transferring data between several fiber-connected devices located between two end points, when only one device can transmit at a time.
optical bypass	an optical switch that diverts traffic around a given location.
optical interface	where the optical fiber meets the optical transceiver.
optical internetworking	allows IP switching layer to operate at the same line rate as a DWDM network, typically using an OC-48c connection.
optical power	the time rate of flow of radiant energy of a signal, expressed in watts.
optical seam	part of a network that does not allow passthrough of traffic meant for other nodes.
overshoot	waveform excursions above the normal level.
packet switching	protocols where data is encoded into packets, which travel independently to a destination where they are decoded.
paradigm	a model, example or pattern.
parallel optical links	links that transform parallel electrical bit streams directly into parallel optical bit streams. Possible

mechanisms include transceivers made from VCSELs and array detectors sending data through fiber ribbon. (*See* VCSEL and fiber ribbon)

parts cost — the sum of all bill of material components used in a design.

passivation layer — to coat a semiconductor with an oxide layer, to reduce contamination by making the surface less reactive.

patch panel — a hardware unit that is used as a switchboard, to connect within a LAN, and to outside for connection to the internet or a WAN.

phase-locked loop — a circuit containing an oscillator whose output phase locks onto and tracks the phase of a reference signal. The circuit detects any phase difference between the two signals and generates a correction voltage that is applied to the oscillator to adjust its phase. This circuit can be used to generate and multiply the clock signal. (*See* clock generation/multiplication)

phase velocity — the speed of a wave, as determined by a surface of constant phase.

phasor formalism — a polar method of displaying complex (real and imaginary) quantities.

photoconductive — the mode of a reverse-biased detector. This reduces the capacitance of the detector, and thus increases the speed of response of the diode. It is the preferred mode for pulsed signals.

photoconductor — a non-junction type semiconductor detector where incident photons produce free charge carriers, which change the electrical conductivity of the material. Lead sulfide and lead selenide are examples of this type of detector, as well as most MSMs.

photolithography — a process for imprinting a circuit on a semiconductor by photographing the image onto a photosensitive substrate, and etching away the background.

photomicrograph — a photograph of an object that is magnified more than 10 times its size.

photonics — the study of photon devices and systems.

photoresist implant mask — used in photolithography to cover parts of a photosensitive medium when imprinting a circuit on a semiconductor.

photovoltaic	the mode of a junction unbiased detector. Since 1/f noise increases with bias, this type of operation has better NEP at low frequencies.
pigtail	a short fiber permanently fixed on a component, used for connecting the component to the fiber optic system.
PIN (or p-i-n) photodiode	a diode made by sandwiching p-type (doped with impurities so the majority of carriers as holes), i-intrinsic (undoped), and n-type (doped with impurities so the majority of carriers as electrons) semiconductor layers. Photons absorbed in the intrinsic region create electron-hole pairs that are then separated by an electric field, thus generating an electric current in a load circuit.
plenum-rated cable	cable that can be used in a duct work system (plenum) which has smoke-retardant properties.
plug repeatability	the variation in coupled power for multiple connections between the same components.
point-to-point transmission	carrying a signal between two endpoints without branching to other points.
polarized connector	a connector that can only plug in one position, so that it is aligned properly.
polarized light	light whose waves vibrate along a single plane, rather than randomly. There are several polarization modes such as TE, where the electric field is in the direction of propagation, and TM, where the magnetic field is in the direction of propagation.
preamplifier	a low-noise amplifier designed to be located very close to the source of weak signals. Often the first stage of amplification.
private mode	when a port or repeater receives only packets addressed to the attached node. Also known as normal mode.
process window	a defined variation of process parameters that characterize equipment and production used in series production.
processor	the part of a computer that interprets and executes instructions. Is sometimes used to mean microprocessor or central processing unit (CPU), depending on context.

product cost apportionment (PCA) or supply chain costs	shipping and distribution expenses for raw material and assemblies from suppliers and delivery of some final product to the end customer.
promiscuous mode	when a port or repeater forwards all packets, not only those addressed to the attached node.
protocol	the procedure used to control the orderly exchange of information between stations on a data link, network, or system. There are several standards, in code set; such as ASCII, transmission mode; asynchronous or synchronous, and non-data exchanges; such as contract, control, failure detection, etc.
pump laser diode	a short wavelength (typically 900 nm range) laser used to provide a pump input or gain in an optical fiber amplifier.
quantum confined Stark effect	a mechanism for changing the optical absorption of a quantum well by applying an electric field. Because of this effect quantum wells are used in optical modulators.
quantum efficiency (QE)	the ratio of the number of basic signal elements produced by detector (usually photoelectrons) to the number of incident photons.
quantum well	a heterostructure with sufficiently thin layers that quantum effects begin to affect the movement of electrons. This can increase the strength of electrooptical interactions by confining the carriers to small regions.
raised floor	a floor made of panels that can be removed for easy access to the wiring and plumbing below.
Raman scattering	scattering off phonons (quanta of vibration). A special case is Brillouin scattering, which is when an acoustic phonon (sound wave) is involved.
random jitter (RJ)	(*See* jitter)
refractive index	the ratio of the speed of light in a vacuum to the speed of light in a material at a given wavelength.
regeneration	insures that there is sufficient optical power for a signal to reach its destination.

repeater	a device placed in a data link to amplify and reshape the signal in mid-transmission, which increases the distance it can travel.
reshaping	removes pulse distortion caused by dispersion.
resonant cavity photodetectors	a photodetector made by placing a photodiode into a Fabry-Perot (FP) cavity to enhance the signal magnitude.
response time	the time it takes a detector's output to rise when subjected to a constant signal.
responsivity	the ratio of the detector output to the radiation input. It is usually expressed as a function of wavelength.
retiming	restores a timing reference to a signal to remove jitter and improve clock/data recovery
ring topology	(*See* topology)
ringing	waveform overshoot and oscillations. (*See* overshoot)
roadmap	a long-range projection for the future of a type of product, for example, "The National Technology Roadmap for Semiconductors."
run length	the number of consecutive identical bits, such as the number of 1's or 0's in a row, in the transmitted signal. The pattern 010111100 has a run length of four.
running disparity	the difference between the number of 1's and 0's in a character, which is often tracked as a special parameter in network management software.
saturation	when the detector begins to form less signal output for the same increase of input flux.
scalability	the ability to add power and capability to an existing system without significant expense or overhead.
Schottky-barrier photodiodes	a variation on the PIN photodiode where the top layer of semiconductor material has been eliminated in favor of a reverse biased, metal-semiconductor-metal (MSM) contact. This results in faster operation, but lower signal. The advantage of this approach is improved quantum efficiency, because there is no recombination of carriers in the surface layer before they can diffuse to either the ohmic contacts or the depletion region. (*See* PIN photodiode)

scintillation	rapid changes in the irradiance of a laser beam.
serializer/ deserializer	a serializer is a device that converts parallel digital information into serial. A deserializer converts it back.
server	a central computer where data is deposited, and can be accessed over the network by other computers, known as clients.
serviceability	the ease in which a product can be serviced and inspected.
short wavelength	780–850 nm.
Shot noise	noise made by the random variations in the number and speed of the electrons from an emitter.
shunt resistance	the resistance of a silicon photodiode when not biased.
signal-to-noise ratio	the ratio of the detector signal to the background noise.
simplex	a one-way communications line, which cannot both send and receive data.
single-mode fiber	fiber where the light can only propagate through one path.
skew	the tendency for parallel signals to reach an interface at different times. The skew for a copper cable is 2-nanosecond bit periods over 20 m, while for fiber ribbons it is under 10 picoseconds/m.
slope efficiency	the differential quantum efficiency of the laser combined with the losses of the optical coupling.
smart pixel array	an array of optical devices (either detectors, modulators, or transmitters) which are directly connected to logic circuits. By integrating both electronic processing and individual optical devices on a common chip, one may take advantage of the complexity of the electronic processing circuits and the speed of the optical devices.
smectic liquid crystals	(*See* liquid crystals)
space division multiplexing	(*See* multiplexer)
splice bushing	(*See* bulkhead splice)
splice loss	the loss in optical power due to splicing cables.

550 Glossary

splices	to join together two pieces at their ends to form a single one. When applying to optical cable, to form a permanent joint between two cables, or a cable andbreak a port. There are two basic types of splice, mechanical splices and fusion splices. Mechanical splices place the two fiber ends in a receptacle that holds them close together, usually with epoxy. Fusion splices align the fibers and then heat them sufficiently to fuse the two ends together.
spoofing	sending acknowledgments of data transfer before data is actually received at its destination; usually done to artificially reduce latency in a network.
spun fiber	a manufacturing process in which the fiber is rotated during the drawing process in an effort to remove polarization dependence.
star topology	(*See* topology)
Stark effect	the splitting of spectral lines due to an incident electric field. The quantum confined Stark effect is a special case of that. It uses an electric field to wavelength modulate sensitivity of quantum well detectors. (*See* quantum confined Stark effect)
statistical process control	the use of statistical techniques to analyze, monitor, and control a process. Quality is often monitored in this way.
storage area network (SAN)	a high-speed network, or section of an enterprise network, of storage devices, for access by local area networks (LAN) and wide area networks (WAN).
strain relief	a design feature that relieves the pressure on a connector, which could otherwise cause it to crack or unplug.
striping	simultaneously allowing data transfer through multiple ports.
Strowger switch	developed in the late 1800s, this early telecommunications switch replaced human switchboard operators by automatically making phone connections using electrical signals on the phone line.
substrate	the base layer of support material on which crystals are grown. Products grown on substrates include semiconductor detectors and integrated circuits. A silicon wafer is an example of a substrate.

Glossary

supply chain costs — (*See* product cost apportionment [PCA])

surface wave filter — a surface acoustic wave (SAW) phase filter often used to generate and multiply the clock signal. (*See* clock generation/multiplication)

switched base mode — operating mode of the IBM 2029 Fiber Saver DWDM device, in which unprotected channels are passed through a dual fiber optical switch that protects availability in the case of a fiber cut only, not in the case of equipment failure.

switched network — (*See* fabric).

synchronous — a system is synchronous if it can send and receive data at the same time, using a common timing signal. Since there is regular time interval between transmissions, there is no need for a start or stop bit on the message.

thermal noise — noise caused by randomness in carriers generation and recombination due to thermal excitation in a conductor; it results in fluctuations in the detector's internal resistance, or in any resistance in series with the detector. Also known as Johnson or Nyquist noise.

time compression multiplexing — (*See* multiplexer)

time division multiplexing — (*See* multiplexer)

timing reference, trigger — signals used by the oscilloscope to start the waveform sweep.

TO can — metal packaging in the form of transistor outline for semiconductor lasers.

token — a frame with control information, which grants a network device the right to transmit.

topology — the physical layout of a network. The three most common topologies are bus, star, and ring. In a bus topology all devices are connected to a central cable. (*See* optical bus and backbone) In a ring topology in which the terminals are connected serially point-to-point in an unbroken circle. In a star topology

	all devices are connected to a central hub, which copies the signal to all the devices.
transceiver	a package combining a transmitter and a receiver.
transducer	a device that converts energy from one form to another. (*See* electrooptic transducer)
transimpedance amplifier	an operation amplifier and variable feedback resistance connected between the input and output of the amplifier. They are used to amplify low photodiode signals, and provide high dynamic range, good sensitivity and bandwidth.
transponder	a receiver/transmitter that can reply to an incoming signal. Passive transponders allow devices to identify objects, such as credit card magnetic strips. Active transponders can change their output signal, such as radio transmitter receivers. Satellite systems use transponders to uplink signal from the earth, amplify it, convert it to a different frequency, and return it to the earth.
trigger	(*See* timing reference)
troubleshooting	a systematic method to find the reason for a problem.
trunk fiber	a fiber between two switching centers or distribution points.
two beam holographic exposure	a technique used to fabricate gratings using an expanded laser beam that is divided by a beamsplitter and then recombined, creating an interference pattern. This pattern is transferred photolithographically onto the surface of a semiconductor substrate.
ultrasonic	sound waves at a frequency too high for humans to hear. These waves can be used to excite metals used in wire bonding, or to clean items before placing them in vacuum.
undershoot	waveform excursions below the normal level.
unit interval (UI)	the shortest nominal time between signal transition. The reciprocal of baud, it has units of seconds.
up-chirp	a linear increase in frequency over time.
vertical-cavity surface-emitting lasers (VCSEL)	a semiconductor laser made from a bottom distributed Bragg reflector (DBR), an active region, and a top DBR.

Glossary

virtual tributary	a SONET format with lower bandwidth requirements, to allow services like a DS1 or T1 signal to be carried on a SONET path without remultiplexing the voice channels.
wafer	a flat round piece of silicon that is used as a substrate on which to manufacture integrated circuits.
wavelength-division multiplexing	(*See* multiplexer)
wavelength-routed networks	a network with several routers which choose the light path of the signal in the network by its wavelength. This architecture is being developed for all-optical networking, and has WDM applications.
wide area network (WAN)	a network that is physically larger than a LAN, with more users.
wire bond	connecting wires to devices.
word	four contiguous bytes.
zipcord	a type of optical fiber consisting of two unibody cables connected in the middle by a thin, flexible outer coating.

Some useful dictionaries on the World Wide Web
The One Look Dictionary www.onelook.com
The Photonics Dictionary www.photonicsdictionary.com
The Academic Press Dictionary of Science and Technology www.harcourt.com/dictionary/browse/

Index

2-D error diffusion neural network, 406
2R regenerators, 158
2R repeaters, 156
3D-OESP (3D OptoElectronic Stacked Processors), 249
3M, 64
3R regeneration, 157
3R repeaters, 156

Acoustooptic tunable filter, 432, 529
Acronyms, 489–490, 511–528
Active emitters, smart pixel technology, 367–368
Advanced Communications Technologies and Services (ACTS), 426
Advanced Free-Light fiber, 114
Advanced Research Initiative in Microelectronics (MEL-ARIOPTO), 253
Aggregate capacity, 373–374, 529
Agile lasers, 152
AlGaAs, 355
All Optical Network Consortium, 170, 426
All-optical networks (AONs), 136, 424, 425
All-optical regenerators, 157
Allied topology, 273
Allwave fiber, 118
American Optical Corp., 10, 13, 14
Amplified spontaneous emission (ASE), 157, 434
Animation, 24
ANSI standards, 493–494
AONs. *See* All-optical networks
Area array technology, 312–313
ARPAnet, 22
Array waveguide grating (AWG), 145
Arthur, J.R., 13
Artificial neural networks, 391, 406
Asahi Chemical Industry Co., Ltd., 227
Asahi Glass Co., 123, 227, 228

Asahi Kasei Co., 227
ASCI White, 297
ASE. *See* Amplified spontaneous emission
AstroTerra Corp., 505
Asymmetric Fabry-Perot modulator, 358
Atipa ATServer 6000, 277
ATM/SONET protocols, 136–137, 423
AT&T, 8, 19, 117, 353, 354
AT&T Bell Laboratories, 11, 12, 13, 19, 388
AT&T Submarine Cable Systems, 17
Attenuated cables, 102–103
Automation, in manufacture, 479–480
Avalon processor, 277
Average Optical Power, 331
AWG. *See* Array waveguide grating

Babbage, Charles, 7
Backplanes, 230–241, 386–388, 530
Baird, John L., 11
Ball bonding, 308, 309
Ball grid array (BGA) packages, 313
Bandwidth, 22, 530
Bardeen, John, 12
BBN butterfly, 273, 276
"Behind the wall" connectors. *See* BTW connectors
Bell Atlantic, 171
Bell Laboratories, 7, 15
Berkeley Millennium, 27
BGA packages. *See* Ball grid array (BGA) packages
Bi-directional wavelength path-switched ring (BWPSR), 162
Binary-valued pixels, 391
Bit-to-bit skew, InfiniBand, 333–334
Blaze Photonics, 128
"Blue Gene," 298
Blue Tiger cables, 118
Board interconnects, 217, 218, 241–246
Bohr, Niels, 12

555

556 Index

Boltzmann constant, 488
Brattain, Walter, 12
Broadcast-and-select networks, 424, 531
BTW connectors, 74, 75
Bulk infrared optical fibers, 94
Business planning, 37–41
Butterfly topology, 273
BWPSR. *See* Bi-directional wavelength path-switched ring

Cable. *See* Fiber optic cable
Cage, 80
California Eastern Laboratories, 507
Carrier-induced phase modulation (CIP), 165
Cascaded Mach-Zehnder tunable filter, 431
Cascaded multiple Fabry-Perot filters, 431
CA Unicenter, 163
CBGA package. *See* Ceramic ball grid array (CBGA) package
CDC-7600 processor, 275
CDMA. *See* Code division multiple access
CEL, 507
Celsius-to-Kelvin conversion, 487
Ceramic ball grid array (CBGA) package, 314, 315
Ceramic column grid array package, 314
Ceramic ferrule, 83, 84, 120
Cerf, Vinton, 22
ChEEtah/OMNET, 235
"Chicken and egg" syndrome, 467, 532
Chip interconnections, 246–255
Chirping, 165, 532
Cho, A.Y., 13
Ciena Corp., 170
CIP. *See* Carrier-induced phase modulation
Cladding, 97–98, 532
Cladding mode coupling device, 97
CLO. *See* Control Link Oscillator
CM-2 Connection Machine, 276
C.mmp processor, 272, 275
CMOS-SEED technology, smart pixel technology, 367, 393–402
CMU Warp processor, 273, 283
Coarse WDM (CWDM), 140, 222, 225
Coaxial cables, 7
Code division multiple access (CDMA), 209
COGS. *See* Cost of Goods Sold
Coherent bundle, single-mode fiber, 95
Collector network, 206
Colocation, 468–469

Columbia QCDSP, 277
Commercial Cabling Standard TIA-568-B, 64
Communication, history, 3–31
Communication networks, 13, 15. *See also* Data communication
Communication storage, 17
Compaq AlphaServers, 277
Computer networks, 21
  HANs, 229
  LANs, 492, 539
  MANs, 136, 137
  SANs, 138, 203, 546
  WANs, 196–200
Computers
  "Blue Gene," 298
  computing tasks, 274
  Deep Blue, 297
  deep computing, 296
  ENIAC, 7, 297
  history, 7, 18–19
  parallel processor design, 270–299
  supercomputers, 296–298
Connection Machine CM-1, 276
Connectors
  BTW connectors, 75, 506–507
  "elite MT" connectors, 94–95
  Fiber-Jack (FJ) connectors, 77–78
  InfiniBand, 345–347
  LC connectors, 64, 74–77, 81, 85–87
  "Mini-GBJC" connector, 80
  mini-MT connectors, 77
  "mini-SC" connectors, 79
  MMC connectors, 233
  MPO connector, 506
  MPX connector, 506
  MTP connectors, 99, 100, 506
  MT-RJ connectors, 64–68, 81–87
  MU connectors, 78
  optical backplane connectors, 506
  Opti-Jack connectors, 77–78
  SC-DC connectors, 64, 68–70, 80–82, 86, 87
  SC-DC UniCam connectors, 70
  SC-QC connectors, 68–70, 82
  small form factor connectors, 63–87
  SMU connectors, 78
  standards, 64
  VF-45 connectors, 64, 71–74, 81–83, 86, 87
Control charting, 451
Control Link Oscillator (CLO), 288
Conversion tables, units of measure, 486–487

Index   557

Copper wire transmission, 8, 9
Corning, 92, 93, 104, 117, 118
Corning Glass Works, 11
Cost of Goods Sold (COGS), 483, 484, 532
Couplers, 97
Coupling links, 286
CP-PACS/2048, 281, 285
Cray processors, 276–278
Cross-connect switches, 160
Cross-gain modulation, 442
Cross-phase modulation, 442
Cross-system coupling facility (XCF) communication, 292
Cross-talk, 434, 439, 533
Crystal Fibre Co., 128
CSPI MultiComputer 2741, 278
Curtiss, Lawrence, 11
CWDM. *See* Coarse WDM
Cycle time, manufacturing, 460–463, 479
CYTOP fiber, 227

DARPA, 235
Data communication, 134–140
  all-optical networks (AONs), 136, 424, 425
  broadcast-and-select networks, 424
  costs, 463–466
  history, 3–31
  inventions, 5–9
  market analysis, 48–50
  optical amplifier, 433–436
  pricing, 463
  trends, 53–55
  tunable receiver, 429–433
  tunable transmitter, 426–429
  wavelength converter, 440–443
  wavelength division multiplexing. *See* Wavelength division multiplexing
  wavelength multiplexer/demultiplexer, 436–437
  wavelength-routed networks, 425
  wavelength router, 437–440
Data rates, 19, 495–504, 533
DBR. *See* Distributed Bragg reflector
DCF. *See* Dispersion compensated fiber
DDM. *See* Directional-division multiplexing
"Deep Blue," 297
Deep computing, 296
Demultiplexers, 436–437
Denneau, Monty, 298

Dense wavelength division multiplexing (DWDM), 17, 136, 138, 139–140, 141, 207
  design, 147
  ITU grid standard wavelengths, 140–144
  latency, 158–160
  network management, 162–163
  protection and restoration, 160–162
Design
  DWDM design, 147
  manufacturing and, 456–457, 480–485
  parallel processor design. *See* Parallel processor design
  smart pixel technology, 374–381, 382–383
  WDM, 147, 149–152
Design for postponement, 484
Design for Total Product Cost (DTPC), 482–483
Desurvire, Emmanuel, 13
Deterministic jitter, 333, 533
Differential mode delay (DMD), 224
Digital image halftoning, smart pixel technology, 391–402
Digital signal processors (DSPs), 19
Digital television, 23
Digital wrappers, 194–196
DIP. *See* Dual Inline Package
Direct epitaxy, smart pixel technology, 363–365
Direct interboard interconnection, 238
Directional-division multiplexing (DDM), 55
Disaster recovery, 134
Dispersion compensated fiber (DCF), 117
Dispersion controlling fiber, 114–119
Dispersion-flattened fibers, 118
Dispersion shifted fiber (DSF), 115
Dissimilar fibers, 98
Distributed Bragg reflector (DBR), 427, 534
Distributed computing, 134
Distributed photonic A/D conversion, smart pixels, 406–409
DMD. *See* Differential mode delay
Doped-fiber amplifier, 434–435
"Double crucible" method, fiber optic cable, 93
Down-chirp, 115
DSF. *See* Dispersion shifted fiber
DSPs. *See* Digital signal processors
DTPC. *See* Design for Total Product Cost
Dual contact connectors. *See* SC-DC connectors
"Dual homing" architecture, 153
Dual hubbed ring WDM topology, 155
Dual Inline Package (DIP), 309–310
DuPont, 235

558 Index

Duty cycle distortion, 333
DWDM. *See* Dense wavelength division multiplexing

E-business, 23
Echelle grating, 145–146
EDFAs. *See* Erbium-doped fiber amplifiers
EH mode, 111
Einstein, Albert, 12
Electroabsorption modulators, 357–358
Electromechanical calculator, 7
Electronic Time Division Multiplexing. *See* ETDM
"Elite MT" connectors, 94–95
ELO. *See* Epitaxial lift-off
Emitters, smart pixel technology, 361–363, 367–368
Encapsulated fiber, 103
English-to-metric conversion, 486
ENIAC, 7, 297
Enterprise 10000, 280
Entertainment video, 23
EOS. *See* "Everything over SONET"
Epitaxial lift-off (ELO), smart pixel technology, 368–370, 382
ER. *See* Extinction Ratio
Erbium-doped fiber, 13
Erbium-doped fiber amplifiers (EDFAs), 114, 167
ESCON protocol, 99, 151, 152, 158, 208–209, 498, 503
ESPRIT program, 170, 234
Etched-grating demultiplexer, 436–437
ETDM, 423
Ethernet, 500
Ethernet standards, 497, 502–503
ETR. *See* External Timer Reference
"Everything over SONET" (EOS), 194
Exotic topology, 273
EXtended Remote Copy (XRC) protocols, 294
External cavity tunable lasers, 427
External Timer Reference (ETR), 287
Extinction Ratio (ER), 331

Fabry-Perot amplifiers (FPA), 433–434
Fabry-Perot filters, 431
Fast Ethernet, 500–501, 502
FBGs. *See* Fiber Bragg gratings
FC. *See* Flip chip bonding
FCA. *See* Fibre Channel Association

FCS. *See* Fibre Channel Standard
FCSI. *See* Fibre Channel Systems Initiative
FDDI, 33, 500, 503
FDMA. *See* Frequency division multiple access
Ferroelectric liquid crystals, 371
Ferrules
  ceramic ferrule, 83, 84, 120
  defined, 535
  Fiber-Jack connector, 7–78
  glass ceramic ferrule, 83
  MT ferrules, 507
  MT-RJ connector, 64, 65
  multimode ferrules, 120
  plastic/polymer ferrules, 120
  SC-DC/SC-QC connectors, 69
FGPAs. *See* Field-programmable gate arrays
Fiber Bragg gratings (FBGs), 119, 169–170
Fiber bundle imaging, 11
Fiber cabled backplane, 231, 233
Fiber image guides (FIGs), 250, 251
Fiber-Jack (FJ) connectors, 77–78
Fiber links, 13
Fiber optic cable, 15, 89–130. *See also entries beginning with* Optical
  attenuated cables, 102–103
  dispersion controlling fiber, 114–119
  fabrication, 89–98
  Fiber Transport Services, 98–111
  history, 7–8
  InfiniBand, 348–350
  jacket types, 94
  mode conditioning patch cables (MCP), 106–111
  optical mode conditioners, 105–111
  photosensitive fibers, 119–120
  plastic optical fiber, 120–123
  polarization controlling fibers, 111–114
  types, 95, 96
  WDM and cable TV, 102–103
Fiber optic communication, emerging technologies, 22–31, 422–424
Fiber optic components. *See also entries beginning with* Optical
  market forces, 26–29
  North American consumption, 44
Fiber Optic Connector Intermatability Standard (FOCIS), 64
Fiber optic connectors. *See* Connectors
Fiber optic links, 135

Index   559

Fiber optic networks, 9, 15–17
  for datacom, 153
  future of, 22–31
  topologies, 153, 154–156
  trends, 53–55
Fiber optic power splitter, 97
Fiber optics industry. *See also entries beginning with* Optical
  drawbacks, 51
  history, 9–31
    industry description and outlook, 42–45, 46
    professional organizations, 489–490
    world statistics, 47–48
Fiber optic transceivers, 508
Fiber Quick Connect (FQC), 82, 98
Fiber-to-the-curve configuration. *See* FTTC configuration
Fiber-to-the-home configuration. *See* FTTH configuration
Fiber-to-the-office configuration. *See* FTTO configuration
Fiber Transport Services (FTS), 98–111
Fiber trunk switch, 161
Fibre Channel, 497, 536
Fibre Channel Arbitrated Loop, 153
Fibre Channel Association (FCA), 60
Fibre Channel Connection. *See* FICON
Fibre Channel MFS, 333
Fibre Channel Standard (FCS), 52, 53
Fibre Channel Systems Initiative (FCSI), 60
FICON, 158, 209, 294, 498–499, 503, 536
Field-programmable gate arrays (FGPAs), 252–253
FIGs. *See* Fiber image guides
First-generation CMOS-SEED, 394–397
First-generation DWDM systems, 140, 175
Fixed attenuators, 102
FJ connectors. *See* Fiber-Jack connectors
Flash lamps, 11, 12
"Flat panel" video screens, 24
Flip-chip bonding, 308, 366, 382, 536
"Flow shop," 475, 536
Fluorozirconate glass, 124
FOCIS. *See* Fiber Optic Connector Intermatability Standard
Four photon mixing, 116
Fourth-generation WDM systems, 141
Fourth-order Bessel-Thompson filter, 329–330
Four wave mixing (FWM), 116, 164, 443
FPA. *See* Fabry-Perot amplifiers

FQC. *See* Fiber Quick Connect
Frame-to-frame interconnections, 217–230
Free-space backplane, 236–238
Free-space communications, 506
Free-space OCIs, 248
Free-space optical links, 128, 505–506
Free spectral range, 430
Frequency chirping, 165
Frequency division multiple access (FDMA), 209
FTS. *See* Fiber Transport Services
FTTC configuration, 55
FTTH configuration, 55
FTTO configuration, 55
Fujitsu VPP700, 278
Fujitsu VPP5000, 278
Full width at half maximum (FWHM), 430
Functional layout, 478, 479
FWM. *See* Four wave mixing

Gain equalization, 435–436
Gallium arsenide, 12–13
Garage, 80
GBIC. *See* Gigabit interface converter
GDPS, 139, 286
General Electric Co., 8, 15
Geographically Dispersed Parallel Sysplex. *See* GDPS
GGP fiber, 72
Gigabit Ethernet, 219, 222, 501–504, 536
Gigabit interface converter (GBIC), 152, 508
Gigabit links, 286
Gigabyte System Network (GSN), 506
Glass, history, 14
Glass ceramic ferrule, 83
Glass-clad fiber bundle imaging, 11
Glass fibers, history, 14
Glass-glass-polymer, 72
"Glob-top" encapsulation, 314
Gordon, James, 12
GR-326-CORE, 64
Group velocity dispersion (GVD), 115, 157, 165–166
GSN. *See* Gigabyte System Network
Guided-wave system demonstrator, 252–255
GVD. *See* Group velocity dispersion

Hall, Robert N., 15
HANs. *See* Home area networks
Hansell, Clarence W., 11, 14

Hard Polymer Clad fiber. *See* HPCF
Harper, Marion, Jr., 34
Hayashi, Izuo, 13
HE mode, 111
Heavy-metal fluoride fibers, 94
Hecht, Jeff, 14
HIBITS project, 234
Hicks, Wilbur, 10, 11, 14
Hierarchical networks, 204–206
High Bitrate ATM Termination and Switching, 234
High Performance Networking Forum, 506
High Speed Plastic Network Consortium (HSPN), 507
High-speed switching, smart pixel technology, 388–391
Hill gratings, 169
HiPerLinks, 286–290, 504
HIPPI, 506
Hitachi SR2001, 278
Hitachi SR8000, 278
Hockham, George, 10, 14
HOIEC packaging. *See* Hybrid optoelectronic integrated circuit (HOEIC) packaging
Home area networks (HANs), 229
Hopkins, Harold H., 11
HP 9000 Superdome, 279
HP Openview, 163
HPC-1, 280, 284–285
HPCF, 123
HSPN. *See* High Speed Plastic Network Consortium
Huang, A., 353
Hubbed ring WDM topology, 154, 183
Hughes Research Laboratories, 15
Hybrid integration, 365–373, 382, 537
Hybrid modes, 111
Hybrid optoelectronic integrated circuit (HOEIC) packaging, 20
Hypercube topology, 273
Hyperplane, 385–387

IBM, 139, 296, 298
IBM 2029 Fiber Saver, 176–193
IBM 2938 Array Processor, 275
IBM 3838 Array Processor, 275
IBM 9729 Optical Wavelength Division Multiplexer, 175–176
IBM Parallel Sysplex, 139, 279, 286–296, 499
IBM RS/6000 S80, 279
IBM RS/6000 SP, 279
IBM System/390 Enterprise Servers, 135, 286
IB Signal Conditioner, 325
ICL DAP, 275
IEEE LAN standards, 492
IFBs. *See* Imaging fiber bundles
Imaging fiber bundles (IFBs), 250
Incoherent bundles, 95
Incom, 10
Infineon, 20, 233, 244
InfiniBand, 321–350, 537
  bit-to-bit skew, 333–334
  connectors, 345–347
  electrical interface, 324–326
  fiber optic cable, 348–350
  InfiniBand link layer, 323–326
  optical jitter specification, 331–333
  optical signals, 325–331
  optical specifications, 334–344
  packet format, 323–324
"Inside process," fiber optic cable fabrication, 90
Inside vapor deposition (IVD), fiber optic cable, 92
Integrated circuit, history, 20, 21
Integrated waveguides, 234–236
Intel iWarp, 276, 283
Intelligent optical internetworking, 193–207
Internet
  growth rate, 135–136
  history, 22
  IP over WDM, 194–196, 198
  next-generation Internet (NIGH), 159
  "optical Internet," 137
Intersymbol interference (ISI), 333
InterSystem Channel (ISC), 139, 286
ISC. *See* InterSystem Channel
ISI. *See* Intersymbol interference
ITU grid standard wavelengths, for DWDM, 140–144
IVD. *See* Inside vapor deposition
iWarp, 276, 283

Jenkins, B.K., 353
Jitter, 331–333, 537
"Job shops," 475–476, 538

Kahn, Robert, 22
Kao, Charles K., 10, 11, 14
Kapany, Narinder, 11
Karbowiak, Antoni E., 10

KDD (Japan), 17
Keck, Donald, 11, 15
Kelvin-to-Celsius conversion, 487
Kentucky Linux Athlon Testbed 2. *See* KLAT2
Kerr effect signal distortion, 157
KGD. *See* Known good die (KGD) approach
Kilby, Jack St. Clair, 20
KL10 TOPS-10 processor, 276
KLAT2 processor, 277
Known good die (KGD) approach, 473
Kotler's rule, 35
Kroemer, Herbert, 12

Lambadanet, 424
Lamm, Heinrich, 11
LANs, 492, 539
Large effective area fiber. *See* LEAF
Laser Diode Labs, 15
Laser diodes, history, 15
Lasers, 11–12
Latency, 158–160, 296–298, 538
LBOS. *See* Liquid crystal on silicon
LC connectors, 64, 74–77, 81, 85–87
LEAF, 117–118
LEDs, smart pixel technology, 383, 404, 405
Leica Technologies, 506
LIGA fabrication technology, 234
Lightby 40 channel AWG mux-demux, 145
LightRay MPX, 506
Liquid crystal on silicon (LBOS), smart pixel technology, 370–373, 379, 382
"Low birefringence" fiber, 113
Low-loss optical fiber, history, 10–11, 15
LP mode, 111
Lucent Corp., 64, 103, 104, 114, 118, 145, 170, 171
Lucina fiber, 123

MAC-III, 506
MacChesney, John, 11
Mach-Zehnder interferometer, 431
Maiman, Theodore, 11–12, 15
MANs, 136, 137, 539
Manufacturing. *See also* Packaging
  assembly processes, 472–474
  automation/robotics, 479–480
  "chicken and egg" syndrome, 467
  colocation, 468–469
  costs, 463–466, 474–475, 482, 539
  cycle time, 460–463, 479

delivery, 448
design, 456–457, 480–485
diversity of technologies, 467–468
flexibility, 474–479
integration, 469–472
"job shop" vs "flow shop," 475–476
modular layout, 476–477, 479
on-time delivery, 457
performance, 447–448, 450–466
pricing, 463–466
reliability, 448
service expectations, 449
"visible horizon lead time," 460
Market analysis, 32–62
  fiber optic communication, 26–29
  market survey, 34–37, 55–58
  need for product, 32–33
  sample, 42–62
  technology infrastructure, 33–34
  transmitter optical subassembly, 42–62
Marketing strategy, defining, 36–37
Market survey, 34–37, 55–58
Masers, 12
Massachusetts Institute of Technology, 129
Maurer, Robert, 11, 15
MBE. *See* Molecular-beam epitaxy
MBGA package. *See* Metal ball grid array (MBGA) package
MCM process, 473
MCP cable. *See* Mode conditioning patch cables
MCVD. *See* Modified chemical vapor deposition
ME-ARIOPTO. *See* Advanced Research Initiative in Microelectronics
Mears, P.J., 13
Measurement units, conversion tables, 486–487
Mechanical computation machines, 7
MESFETs, monolithic integration, 362
Meshed ring WDM topology, 155
Metal ball grid array (MBGA) package, 315
Metric prefixes, 487
Metric-to-English conversion, 486
MetroCor fiber, 118
Metropolitan Optical Network (METON), 426
Microstructured fibers, 125
Microwave relay communication, 8
Microwave research, 12
MIMD, 274–275
"Mini-GBJC" connector, 80
Mini-MAC, 506
Mini-MT connectors, 77

562   Index

"Mini-SC" connectors, 79
Mitel Corp., 145
Mitsubishi Rayon Co., Ltd., 227
MMC connectors, 233
MOCVD. *See* Modified chemical vapor deposition
Mode conditioning patch cables (MCP), 106–111
Modified chemical vapor deposition (MCVD), fiber optic cable, 90, 92
Modular layout, 475–477, 479
Modulators, smart pixel technology, 360, 366–367
Module assembly, optoelectronic packaging, 317–318
Molecular-beam epitaxy (MBE), 13
MONET project, 170–172, 426
Monolithic grating spectrometer, 432
Monolithic integration, smart pixel technology, 359–363, 382
Montgomery, Jeff D., 15
Mosaic Fabrications, 10, 14
Mosaics, 96–97
MPlambdaS, 199
MPLS, 199, 423
MPO connector, 506
MPX connector, 506
MQW devices. *See* Multiple quantum well devices
MT ferrules, 507
MTP connectors, 99, 100
MT-RJ connectors, 64–68, 81–87
MU connectors, 78
Multifiber Termination Push-on connectors. *See* MTP connectors
Multi MC connectors. *See* MMC connectors
Multimode ferrules, 120
Multimode fiber, 104–106, 540
Multiple quantum well (MQW) devices, 354–359, 383
Multiple Wavelength Optical Network (MONET), 170–172, 426
Multiplexing, 140, 540. *See also* Wavelength division multiplexing
Multi-Protocol Label Switching. *See* MPLS
Multiwavelength transport network (MWTN), 426

NAS. *See* Network attached storage
NCube, 276

Near end crosstalk (NEXT), 116
NEC Corp., 129
NEC SX-5, 279–280
Negative frequency chirp, 115
Neodymium-doped chalcogenide fibers, 124, 167
Netview, 163
Network attached storage (NAS), 138
Networked computers, 21
Network interface units (NIUs), 159
Network topologies, WDM, 153, 154–160
Network topology, parallel processor design, 272–273
Neural networks, 391, 406, 541
NEXT. *See* Near end crosstalk
Next-generation Internet (NIGH), 159
Next-generation multimode fiber, 104–106
NIGH. *See* Next-generation Internet
NIUs. *See* Network interface units
NLOG. *See* Nonlinear optical gate
Non-entertainment video, 25
Nonlinear effects, wavelength-division multiplexing (WDM), 163–170
Nonlinear optical gate (NLOG), 157
Non-Uniform Memory Access (NUMA) architecture, 284
Non-zero dispersion shifted fiber (NZDSF), 116
Nortel Networks, IBM 2029 Fiber Saver and, 176–193

OADM. *See* Optical add/drop multiplexing
OC-3/ATM 155, 499–500, 503
OC-12/ATM 622, 500
Occam, 283
OCI card. *See* Optical Channel Interface (OCI) card
OCIs. *See* Optical chip interconnects
OCLD card. *See* Optical Channel Laser and Detector (OCLD) card
OCM card. *See* Optical Channel Manager (OCM) card
ODC, 138
OFBs. *See* Ordered fiber bundles
OFC protocol, 139
OIIC. *See* Optically Interconnected Integrated Circuits
OLIVES, 234, 426
OLS. *See* Optical label switching
OMA. *See* Optical Modulation Amplitude
Omnidirectional fibers, 129

Index

"1+1" SONET-type protection switching, 160
One-dimensional VCSEL array, 219–222
ONTC. *See* Optical Network Technology Consortium
Opened rings, 153
Open Fiber Control protocol. *See* OFC protocol
Openview, 163
OptiAir, 170
Optical add/drop multiplexing (OADM), 153, 156, 162
Optical amplifiers, 10–11, 13, 16–17, 123–125, 167–168, 433–436
Optical attenuators, 102
Optical backplane connectors, 506
Optical backplanes, 230–241
Optical Channel Interface (OCI) card, 180
Optical Channel Laser and Detector (OCLD) card, 180
Optical Channel Manager (OCM) card, 180–181
Optical chip interconnects (OCIs), 246–255
Optical computer, 353
Optical connectors, 506
Optical crossbar switches (OXBS), 159
Optical Data Center. *See* ODC
Optical fiber. *See also* Fiber optic cable; Fiber optic networks
  dispersion controlling fiber, 114–119
  encapsulated fiber and flex circuits, 103
  fabrication, 94
  microstructured fibers, 125
  next-generation multimode fiber, 104–106
  omnidirectional fibers, 125
  photonic crystal fiber, 125–128
  photosensitive fibers, 119–120
  plastic optical fibers (POFs), 120–123, 226–230, 507
  polarization controlling fibers, 111–114
  rare earth-doped fiber, 124–125
Optical gating wavelength conversion, 442
Optical Interconnections for VLSI and Electronic Systems. *See* OLIVES
Optical interconnects, 216–257
  board interconnects, 217, 218, 241–246
  chip interconnections, 246–255
  direct interboard interconnection, 238
  frame-to-frame interconnections, 217–230
  optical backplanes, 230–241
  smart pixel technology, 384–391
"Optical Internet," 137

Optical internetworking, 193–207
  digital wrappers, 194–196
  hierarchical networks, 204–206
  IP over WDM, 194–196
  standards, 200–203
  WAN traffic engineering, 196–200
Optical jitter specification, 331–333
Optical label switching (OLS), 199
Optically Interconnected Integrated Circuits (OIIC), 253
Optically interconnected parallel supercomputers, 296–298
Optical Micro-Networks program. *See* ChEEtah/OMNET
Optical mode conditioners, 105–111
Optical Modulation Amplitude (OMA), 331
Optical Network Technology Consortium (ONTC), 170, 426
Optical packet switches, 27, 29
Optical regenerators, 157
Optical signal (OSIG) mode, 171
Optical 3R regenerator, 157
Optical Time Division Multiplexing. *See* OTDM
Optical transfer function (OTF), 152
Optical transport networks (OTNs), 136
Optic splitters, 97
OptiFlex2, 103
Opti-Jack connectors, 77–78
Optoelectronic conversion, 440
Optoelectronic packaging, 315–320
Optoelectronics. *See also* Fiber optic cable; Fiber optic links
  InfiniBand, 321–350
  manufacturing, 447–485
  packaging, 303–320, 359–373
  professional organizations, 489–490
  smart pixel technology, 236, 352–410
Ordered fiber bundles (OFBs), 250
OSI model, 491
OTDM, 423
OTF. *See* Optical transfer function
OTNs. *See* Optical transport networks
"Outside process," fiber optic cable fabrication, 92
Outside vapor deposition (OVD), fiber optic cable, 92
OVD. *See* Outside vapor deposition
OXBS. *See* Optical crossbar switches
OXCs, 162

564   Index

Packaging, 303–320, 472–473
  first level packages, 305, 311–315
  hierarchy, 305
  optoelectronic packaging, 315–320
  printed circuit board (PCB), 305
  second level assembly techniques, 305–311
  smart pixel technology, 359–373
  surface-mount technology (SMT), 304, 311–313
  Through Hole Technology (THT), 304–305
Packet switches, 27, 29, 541
PANDA fiber, 113
Panduit Corp., 77
Panish, Morton, 13
Parallel coupled processor architecture, 139
Parallel links, 20, 541
Parallel optical transceivers, 509
Parallel processor design, 270–299
  CMU Warp processor, 273, 283
  GDPS, 139, 286
  network topology, 272–273
  Parallel Sysplex, 139, 279, 286–296, 499
  processors, 274–282
  SKY HPC-1, 280, 284–285
  supercomputers, 296–298
  transputer, 273, 276, 283–284
  Tsukuba CP-PACS/2048, 281, 285
Parallel processors, history, 275–282
Parallel Sysplex, 139, 279, 286–296, 499
PAROLI transceivers, 244
PAROLI transmitter, 233
Payne, David, 13
PBGA. See Plastic ball grid array (PBGA) packages
PCA. See Product Cost Apportionment
PCB. See Printed circuit board
PCVD. See Plasma-assisted chemical vapor deposition
PDG. See Polarization dependent gain
PDL. See Polarization dependent loss
Peer-to-Peer Remote Copy (PPRC) protocols, 293
"Perfect mirror" technology, 129
Perfluorinated plastic optical fibers, 228
Photonic A/D conversion, smart pixels, 406–409
Photonic crystal fiber, 125–128
Photonic networks, 156–157
Photonics, 353, 542
Photosensitive fibers, 119–120

Photosensitivity, 169
Physical constants, 488
Pipeline topology, 272–273
Pirelli, 114, 170
Planck, Max, 12
Planck's constant, 488
Plasma-assisted chemical vapor deposition (PCVD), fiber optic cable, 90, 92
Plastic ball grid array (PBGA) packages, 313–315
Plastic optical fibers (POFs), 120–123, 226–230, 507
Plastic/polymer ferrules, 120
Pluggable transceivers, 80
Pluribus processor, 275
PMD. See Polarization mode dispersion
PMF. See Polarization maintaining fiber
PMMA, 121, 227, 228, 235
POFs. See Plastic optical fibers
POINT project, 235
Point-to-point WDM topology, 154, 156
Polarization controlling fibers, 111–114
Polarization dependent gain (PDG), 114
Polarization dependent loss (PDL), 114
Polarization Maintaining and Absorption Reducing Fiber. See PANDA fiber
Polarization maintaining fiber (PMF), 112
Polarization mode dispersion (PMD), 157
Polarization preserving fiber, 112
Polyguide, 235
Polymer ferrules, 120
Polymer optical fibers. See Plastic optical fibers
Polymer Optical Interconnect Technology project. See POINT project
Polymer waveguides, 242, 245
Poole, S.B., 13
Positive frequency chirp, 115
PPRC protocols. See Peer-to-Peer Remote Copy (PPRC) protocols
Praseodymium doped amplifiers, 124, 167
Prefixes, metric, 487
Printed circuit board (PCB), 305
Printing press, history, 5, 17–18
Product Cost Apportionment (PCA), 484, 543
Product development
  market analysis, 32–62
  market survey, 34–37, 55–58
  marketing strategy, 36–37
  need for product, 32–33
  supporting technology, 33–34

Professional organizations, 489–490
Protocols
    ATM/SONET protocols, 136–137, 423
    defined, 543
    ESCON, 99, 151, 152, 158, 208–209, 498
    OFC protocol, 139
    PPRC protocol, 293
    for WDM, 150–151
    XRC protocol, 294
Psaltis, D., 353
Pump diode, 16

QCSE. *See* Quantum confined Stark effect
Quad Flat Package (QFP), 310
Quantum confined Stark effect (QCSE), 355, 543
Quattro contact connectors. *See* SC-QC connectors

RACE project, 426
Rainbow, 424
Rare earth-doped fiber, 124–125, 167
RCA, 8
RCA Sarnoff Laboratories, 15
RCE LED. *See* Resonant cavity-enhanced (RCE) LED
Reality-based programs, 25
Regeneration, 156
Remote DMA (Remote Direct Memory Access), 285
Reshaping, 156
Resonant cavity-enhanced (RCE) LED, 383, 404, 405
Retiming, 156
Robotics, in manufacture, 479–480
Rocket vehicle technology, 8
Russel, Phillip, 125

"Safety stock," 463
SANs, 138, 203, 546
Sanwa Corp., 78
Satellite microwave communication, 9
Saturable absorber, 442
SBCON, 498
SBS. *See* Stimulated Brillouin scattering
SC-DC connectors, 64, 68–70, 80–82, 86, 87
SC-DC UniCam connectors, 70
Schockley, William, 12
Schultz, Peter, 11, 15
SC-QC connectors, 68–70, 82

SDH. *See* Synchronous digital hierarchy
SDM. *See* Space-division multiplexing
Second-generation CMOS-SEED, 397–402
Second-generation DWDM systems, 141
SEED. *See* Self-electrooptic effect device
Segmenting, 35
Self-electrooptic effect device (SEED), 357
Self-healing, 162
Semiconductor amplifier, 433–434
Semiconductor devices, history, 12
Serial-to-parallel converter, 210
SETI, 296
SFF connectors. *See* Small Form Factor connectors
SFP transceivers, 80
SGI 2000, 280
SGI 3800, 280
Shelf Processor card (SP card), 180
SIC manual. *See* Standard Industrial Classification manual
Siecor, 68
Siemens, 12
Signal amplification, history, 19
Signal processing, smart pixel technology, 391–409
SIMD, 274–275
Single-mode fiber,11, 95
Sinusoidal jitter, 333
SIPAC system, 232
SKYchannel, 284
SKY HPC-1, 280, 284–285
Small Form Factor connectors (SFF connectors), 63–87
    BTW connectors, 74, 75
    comparing, 80–87
    Fiber-Jack (FJ) connectors, 77–78
    LC connectors, 64, 74–77, 81, 85–87
    "Mini-GBJC" connector, 80
    mini-MT connectors, 77
    "mini-SC" connectors, 79
    MT-RJ connectors, 64–68, 81–87
    MU connectors, 78
    multi-termination unibody, 78
    Opti-Jack connectors, 77–78
    SC-DC connectors, 64, 68–70, 80–82, 86, 87
    SC-DC UniCam connectors, 70
    SC-QC connectors, 68–70, 82
    SMU connectors, 78
    transceivers, 79–80, 84
    VF-45 connectors, 64, 71–74, 81–83, 86, 87

566 Index

Smart pixel technology, 236, 352–410, 545
  aggregate capacity, 373–374
  analog-to-digital conversion, 402–409
  applications, 381, 384–409
  array size, 379–380
  backplane interconnections, 386–388
  complexity, 374–375
  design, 375–381, 382–383
  device speed, 378–379
  digital image halftoning, 391–402
  direct epitaxy, 363–365
  epitaxial lift-off (ELO), 368–370, 382
  flip-chip bonding, 366, 382
  future trends, 409–410
  high-speed switching, 388–391
  history, 353–354
  hybrid integration, 365–373, 382
  liquid crystal on silicon (LBOS), 370–373, 379, 382
  monolithic integration, 359–363, 382
  multiple quantum well devices, 354–359
  optical interconnects, 384–391
  performance metrics, 373–376
  signal processing, 391–409
  system integration, 380
SMF-LS fibers, 17
SMT. *See* Surface-mount technology
SMU connectors, 78
Snitzer, Elias, 10, 13, 14, 16
Solder bumps, 308, 310–311
SONET, 52, 53, 186, 497, 498
Source Synchronous interfaces, 322
Space-division multiplexing (SDM), 55
Sparse WDM (SWDM), 140
SP card. *See* Shelf Processor card
Spec Trans Specialty Optics, 507
Speed of light, 488
SPIBOC project, 234
"Spoofing" the channel, 160
Spun fiber, 113
SR2001, 278
SR8000, 278
SRS. *See* Stimulated Raman scattering
S-SEED. *See* Symmetric SEED
Stacked planar optics, 236
Standard Industrial Classification (SIC) manual, 35
Standardised Packaging and Interconnect for Inter- and Intra-Board Optical Interconnections. *See* SPIBOC project

Standards
  ANSI standards, 493–494
  connectors, 64
  Ethernet standards, 497, 502–503
  IEEE LAN standards, 492
  optical internetworking, 200–203
Standard Telecommunications Laboratories (STL), 10, 14, 15
STARAN processor, 275
Stephan-Boltzmann constant, 488
Stimulated Brillouin scattering (SBS), 166
Stimulated Raman scattering (SRS), 166, 167
STM. *See* Synchronous transport mode
Storage area networks. *See* SANs
Strowger switch, 7, 21
STS/OC, 495
STS-*XC*, 496
Sun Enterprise 10000, 280
Sun Ultra HPC 4500, 280
Sun XDBus, 284
Supercomputers, 296–298
SuperHIPPI, 506
Superluminal waveguides, 129–130
Super-MT, 507
Supply Chain Cost, 484
Surface-mount technology (SMT), 304, 311–313
SWDM. *See* Sparse WDM
Switchable grating, 432
Switch-based private networks, 203
SX-5, 279–280
Symmetric SEED (S-SEED), 358–359
Synchronous digital hierarchy (SDH), 52, 53, 495, 497
Synchronous optical networks. *See* SONET
Synchronous transport mode (STM), 495–496
Sysplex Timer, 287, 291, 498
$System^6$, 354
$System^7$, 354
Systimax LazrSPEED fiber, 104

TAB. *See* Tape-automated bonding
Tape-automated bonding (TAB), 309–310
Tape ball array package, 314
Tape ball grid array (TBGA) package, 314, 315
TBGA package. *See* Tape ball grid array (TBGA) package
TCM. *See* Time-compression multiplexing
TCP/IP, 22
TDM. *See* Time division multiplexing

TDMA. *See* Time division multiple access
Telcordia Corp., 163, 171
Telecommunication technology, history, 7–31
Telephone communications, 134
Telephone signal switching, 21
Television, history, 23–25
Tellium, 171
Temperature conversion table, 487
Terabit-per-fiber transmission, 17, 18
TERKS system, 163
TerraLink system, 505
Terrestrial microwave relay, 8
Texas Instruments, 20
Thermocompression wirebonding, 308
ThermoTrex Corp., 506
Thin film interference filters, 146
Third-generation DWDM systems, 141, 176
Three-wave mixing, 443
Through Hole Technology (THT), 304–305
TIA-568-B, 64
Time compression multiplexing (TCM), 55, 208
Time division multiple access (TDMA), 209, 210
Time division multiplexing (TDM), 147–148
Time-space converter, 210
Timmons, Jeffrey, 37
Tivoli Netview, 163
Token ring topology, 500
Toray Industries, Inc., 227, 507
Torus topology, 273
TOSA. *See* Transmitter optical subassembly
Total jitter, 333
Townes, Charles, 12
Transceivers, 79–80, 84, 334–344, 508, 547
Transistors, 19–20, 26
Transmitter optical subassembly (TOSA), market analysis, 42–62
Transmitters, history, 12–13
Transponder, 150, 547
Transputer, 273, 276, 283–284
Traveling wave amplifiers (TWA), 433, 434
TruePhase family, 114
Tru-Wave fiber, 117, 118
Tsukuba CP-PACS/2048, 281, 285
Tunable laser, 442
Tunable laser diodes, 427
Tunable receiver, 429–433
Tunable transmitter, 426–429
TWA. *See* Traveling wave amplifiers
"Twisted pair" insulated copper wires, 7, 14

Ultra-dense WDM systems, 141
Ultra HPC 4500, 280
Unicenter, 163
Unidirectional switching, 184
Units of measure, conversion tables, 486–487
Up-chirp, 115, 548
U-Tokyo GRAPE-6, 281

Vacuum tubes, 19
VAD. *See* Vapor axial deposition
van Heel, Abraham, 11
Vapor axial deposition (VAD), fiber optic cable, 92, 93
VCs. *See* Virtual containers
VCSELs (vertical-cavity surface-emitting lasers), 60–61, 223, 238, 242, 509
    defined, 548
    monolithic integration, 362
    one-dimensional VCSEL array, 219–222
    smart pixel technology, 362, 383
Very long instruction word (VLIW) microprocessor, 283
VF-45 connectors, 64, 71–74, 81–83, 86, 87
Videoconferencing, 23
Video games, 23–24
Video screen development, 24
Virtual containers (VCs), 496–497
Virtual private network (VPN), 195
"Visible horizon lead time," 460
VLIW microprocessor. *See* Very long instruction word (VLIW) microprocessor
VPN. *See* Virtual private network
VPP700, 278
VPP5000, 278

WANs, traffic engineering, 196–200
Warp processor, 276, 283
Waveguides
    integrated waveguides, 234–236
    polymer waveguides, 242, 245
    superluminal waveguides, 129–130
    two-dimensional waveguide, 250
Wavelength converter, 440–443
Wavelength-division multiplexing (WDM), 55, 102, 140–211
    commercial systems, 170–193
    design, 147, 149–152
    distance and repeaters, 153–158
    first-generation WDM, 175–176
    history, 426

568 Index

Wavelength-division multiplexing (WDM) (cont'd)
   IBM 2029 Fiber Saver, 176–193
   IBM 9729 Optical Wavelength Division Multiplexer, 175–176
   intelligent optical internetworking, 193–207
   latency, 158–160
   network management, 162–163
   network topologies, 153, 154–160
   nonlinear effects, 163–170
   optical amplifiers, 167–168
   protection and restoration, 160–162
   protocols supported by, 150–151
   types, 140–149
   WAN, 172
Wavelength mixing, 443
Wavelength multiplexer/demultiplexer, 436–437
Wavelength-routed networks, 425, 548
Wavelength router, 437–440
WaveMux, 170

WDM. *See* Wavelength-division multiplexing
WDM filters, 154
WDM rings, 153
WDM/WDMA, history, 426
Wedge bonding, 308, 309
Welker, Heinrich, 12
Wide Spectrum WDM (WWDM), 140, 222
Wirebonding, 308, 548
Wolf, Helmut F., 15
WWDM. *See* Wide Spectrum WDM

XDBus, 284
XFC communication. *See* Cross-system coupling facility (XCF) communication
XRC protocols. *See* EXtended Remote Copy (XRC) protocols

ZBLAN glass, 124
Zeiger, Herbert, 12
Zig-zag-type beam propagation, 236
Zimar, Frank, 15

FIBER OPTIC DATA COMMUNICATION:
TECHNOLOGICAL TRENDS AND ADVANCES